General Motors
Chevrolet Malibu
Pontiac G6
Saturn Aura
Automotive Repair Manual

**by Rob Maddox
and John H Haynes**

Member of the Guild of Motoring Writers

Models covered:

Chevrolet Malibu - 2004 through 2012
Pontiac G6 - 2005 through 2010
Saturn Aura - 2007 through 2010

Does not include 2004 and 2005 Chevrolet Classic models
Does not include information specific to hybrid models

(38027-2W9)

J H Haynes & Co. Ltd.
Haynes North America, Inc.
www.haynes.com

Acknowledgements

Technical writers who contributed to this project include Jeff Killingsworth, Joe Hamilton, Mike Stubblefield and John Wegmann. Wiring diagrams originated exclusively for Haynes North America, Inc. by Solution Builders.

A book in the Haynes Automotive Repair Manual Series

ISBN-13: 978-1-62092-282-8
ISBN-10: 1-62092-282-7

Library of Congress Control Number: 2017962456

While every attempt is made to ensure that the information in this manual is correct, no liability can be accepted by the authors or publishers for loss, damage or injury caused by any errors in, or omissions from, the information given.

Contents

Haynes mechanic and photographer with a Pontiac G6

About this manual

Its purpose

The purpose of this manual is to help you get the best value from your vehicle. It can do so in several ways. It can help you decide what work must be done, even if you choose to have it done by a dealer service department or a repair shop; it provides information and procedures for routine maintenance and servicing; and it offers diagnostic and repair procedures to follow when trouble occurs.

We hope you use the manual to tackle the work yourself. For many simpler jobs, doing it yourself may be quicker than arranging an appointment to get the vehicle into a shop and making the trips to leave it and pick it up. More importantly, a lot of money can be saved by avoiding the expense the shop must pass on to you to cover its labor and overhead costs. An added benefit is the sense of satisfaction and accomplishment that you feel after doing the job yourself.

Using the manual

The manual is divided into Chapters. Each Chapter is divided into numbered Sections, which are headed in bold type between horizontal lines. Each Section consists of consecutively numbered paragraphs.

The reference numbers used in illustration captions pinpoint the pertinent Section and the Step within that Section. That is, illustration 3.2 means the illustration refers to Section 3 and Step (or paragraph) 2 within that Section.

Procedures, once described in the text, are not normally repeated. When it's necessary to refer to another Chapter, the reference will be given as Chapter and Section number. Cross references given without use of the word "Chapter" apply to Sections and/or paragraphs in the same Chapter. For example, "see Section 8" means in the same Chapter.

References to the left or right side of the vehicle assume you are sitting in the driver's seat, facing forward.

Even though we have prepared this manual with extreme care, neither the publisher nor the author can accept responsibility for any errors in, or omissions from, the information given.

NOTE

A **Note** provides information necessary to properly complete a procedure or information which will make the procedure easier to understand.

CAUTION

A **Caution** provides a special procedure or special steps which must be taken while completing the procedure where the Caution is found. Not heeding a Caution can result in damage to the assembly being worked on.

WARNING

A **Warning** provides a special procedure or special steps which must be taken while completing the procedure where the Warning is found. Not heeding a Warning can result in personal injury.

Introduction to the Chevrolet Malibu, Pontiac G6 and Saturn Aura

The Malibu models are available in four-door sedan or hatchback body styles. The G6 models are available in four-door sedan or two-door coupe body styles. The Aura models are available in four-door sedan body style. They feature transversely mounted 2.2L and 2.4L DOHC four-cylinder engines, or 3.5L, 3.6L or 3.9L V6 engines.

All models are equipped with an electronically controlled Sequential Fuel Injection (SFI) system.

The engine transmits power to the front wheels through a four-speed automatic transaxle via independent driveaxles.

The front suspension is a MacPherson strut design. The rear suspension utilizes trailing arms, upper and lower suspension arms, toe link arms, shock absorbers and coil springs.

The power-assisted rack-and-pinion steering unit is mounted behind the engine on the front suspension subframe, and is power-assisted on four-cylinder models by an electric speed-proportional power-steering system, or a hydraulic pump-type system on V6 models.

All models are equipped with power assisted front disc brakes, and either rear disc or drum brakes, with an Anti-lock Brake System (ABS) available as an option.

Vehicle identification numbers

1   Modifications are a continuing and unpublicized process in vehicle manufacturing. Since spare parts manuals and lists are compiled on a numerical basis, the individual vehicle numbers are essential to correctly identify the component required.

Vehicle Identification Number (VIN)

2   This very important identification number is stamped on a plate attached to the dashboard inside the windshield on the driver's side of the vehicle (see illustration). It can also be found on the certification label located on the driver's side door post. The VIN also appears on the Vehicle Certificate of Title and Registration. It contains information such as where and when the vehicle was manufactured, the model year and the body style.

VIN engine and model year codes

3   Two particularly important pieces of information found in the VIN are the engine code and the model year code. Counting from the left, the engine code letter designation is the 8th digit and the model year code letter designation is the 10th digit.

On the models covered by this manual the engine codes are:

F ....2.2L four cylinder
5 ....2.4L four cylinder
Z ....2.4L four cylinder
B ....2.4L four cylinder H.O.
O ....2.4L four cylinder E85
U ....2.4L four cylinder E85
              (2011 and later)

1 ....2.4L four cylinder (2011 and later)
8 ....3.5L V6 (2004 to 2006)
N ....3.5L V6 (2007 to 2010)
K ....3.5L V6 (flex fuel)
7 ....3.6L V6
1 ....3.9L V6 (2006 and 2007)

On the models covered by this manual the model year codes are:

4 2004
5 2005
6 2006
7 2007
8 2008
9 2009
A 2010
B 2011
C 2012

Certification label

6   The certification label is attached to the driver's door post (see illustration). The plate contains the name of the manufacturer, the month and year of production, the Gross Vehicle Weight Rating (GVWR), the Gross Axle Weight Rating (GAWR) and the certification statement.

Engine identification numbers

7   The engine serial number can be found attached to the valve cover on four-cylinder engines (see illustration). On 3.5L V6 engines, the engine serial number can be found stamped on the left end of the engine block, near the bellhousing, and on 3.9L V6 engines it can be found on the front and left side of the engine block.

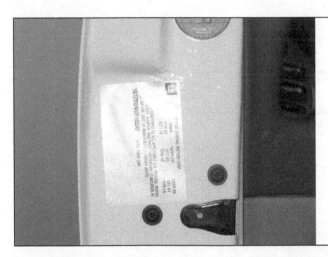

3.2 The vehicle certification label is located on the end of the driver's door

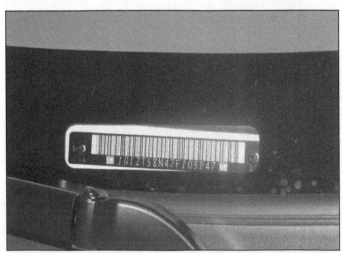

3.6 The Vehicle Identification Number (VIN) is located on a plate on top of the dash (visible through the windshield)

3.7 The engine ID label is attached to the valve cover on four-cylinder engines

Recall information

1   Vehicle recalls are carried out by the manufacturer in the rare event of a possible safety-related defect. The vehicle's registered owner is contacted at the address on file at the Department of Motor Vehicles and given the details of the recall. Remedial work is carried out free of charge at a dealer service department.

2   If you are the new owner of a used vehicle which was subject to a recall and you want to be sure that the work has been carried out, it's best to contact a dealer service department and ask about your individual vehicle - you'll need to furnish them your Vehicle Identification Number (VIN).

3   The table below is based on information provided by the National Highway Traffic Safety Administration (NHTSA), the body which oversees vehicle recalls in the United States. The recall database is updated constantly. For the latest information on vehicle recalls, check the NHTSA website at www.nhtsa.gov, or call the NHTSA hotline at 1-888-327-4236.

Recall date	Recall campaign number	Model(s) affected	Concern
Feb 23, 2004	04V094000	2004 Malibu	Certain passenger vehicles fail to comply with the requirements of Federal Motor Vehicle Safety Standard (FMVSS) No. 208, "Occupant Crash Protection." Some of these vehicles that were not equipped with an Advanced Occupant Restraint System (AORS) have driver and passenger side sun visors with AORS labels. The AORS labels do not include statements and the illustration required by FMVSS No. 208 for vehicles without AORS.
Apr 16, 2004	04V188000	2004 Malibu	On certain passenger vehicles, analysis of a side impact crash test conducted by the NHTSA's New Car Assessment Program (NCAP) indicated that the outboard anchorage of the driver's seat belt could disconnect because of contact between the seat trim and the anchorage connector when the seat was adjusted to its lowest position.
Apr 16, 2004	04V190000	2004 Malibu	Some of these passenger vehicles have an Electronic Control Unit (ECU) that may calculate a higher than actual vehicle speed because of an erratic rear-wheel speed sensor signal, and cause ABS activation where it is not needed, or needed ABS activation to be extended during braking as the vehicle speed drops to about 3 MPH.
Mar 09,2005	05V098000	2005 G6	Certain passenger vehicles fail to conform to the requirements of Federal Motor Vehicle Safety Standard no. 225, "Child Restraint Anchorage Systems." The child restraint lower anchors in the center rear seating position are located approximately 90mm behind the corresponding z-point of the Child Restraint Fixture (CRF). The owner's manual states that anchors are at the outboard seating positions only. Without additional and corrected owner's manual information, an owner may encounter difficulty in installing child restraints with rigid attachments in the center position.
Oct 18, 2005	05V486000	2006 Malibu	Certain vehicles fail to conform to the requirements of Federal Motor Vehicle Safety Standard No. 110, "tire selection and rims." The tire label table specifies the 15 inch spare for the front disc/rear drum brake system and it should specify a 16 inch spare tire size.
Mar 28, 2007	07V146000	2007 G6	Certain passenger vehicles fail to conform to the requirements of Federal Motor Vehicle Safety Standard no. 208, "Occupant Crash Protection." The right front passenger seats in these vehicles are built with a passenger air bag sensing system. When tested with a representative, unrestrained six year old child, the system is required to turn off the right front passenger's airbag. Interference during system calibration caused the seats to be out of specification. This can increase the risk of injury to a child seat occupant during certain crash conditions.

Recall date	Recall campaign number	Model(s) affected	Concern
Jan 28, 2009	09V036000	2005 and 2006 G6	GM is recalling certain vehicles where the brake lamps may not operate properly because of corrosion in a wiring connector. Corrosion in the connector could cause the brake lamps to illuminate when the brake pedal has not been depressed, or the lamps may not illuminate when the brake pedal is depressed. In addition, the cruise control may not engage, and greater brake pedal force may be required to shift the vehicle out of park. Brake lamps that are not operating properly may not warn a driver in a following vehicle of the braking status and could lead to a crash without prior warning.
Mar 06, 2009	08V634000	2009 Malibu	On some models, the transmission shift cable adjustment clip may not be fully engaged. If the clip is not fully engaged, the shift lever and the actual position of the transmission gear may not match. With this condition, the driver could move the shifter to PARK and remove the ignition key, but the transmission gear may not be in PARK. As a consequence, the driver may not be able to restart the vehicle and the vehicle could roll away after the driver has exited the vehicle, resulting in a possible crash without warning.
Mar 06, 2009	09V073000	2009 Aura, Malibu and G6	On some of these vehicles, the transmission shift cable adjustment clip may not be fully engaged. If the clip is not fully engaged, the shift lever and the actual position of the transmission gear may not match. With this condition, the driver could move the shifter to "Park" and remove the ignition key, but the transmission gear may not be in "Park". The driver may not be able to restart the vehicle and the vehicle could roll away after the driver has exited the vehicle, resulting in a possible crash without prior warning.
Mar 28, 2009	09E012000	2006 and 2007 Malibu	The bumper and corner lamp assemblies on certain models do not have amber side reflectors, as required by law. The reduced visibility could lead to a collision.
Sept 18, 2012	12V460000	2007, 2008, 2009 and 2010 Aura; 2008, 2009 and 2010 Malibu and G6	General Motors (GM) is recalling certain model vehicles, equipped with a 4-speed automatic transmission. On these vehicles, the tabs on the transmission shift cable end may fracture and separate. The vehicle may not be able to be restarted and the vehicle could roll away after the driver has exited the vehicle, resulting in a possible crash without warning.
Mar 31, 2014	14V153000	2004, 2005, 2006 and 2007 Aura; 2004, 2005, 2006, 2008 and 2009 Malibu; 2005, 2006, 2007, 2008 and 2009 G6	General Motors (GM) is recalling certain models where there may be a sudden loss of electric power steering (EPS) assist that could occur at any time while driving. If power steering assist is lost, greater driver effort would be required to steer the vehicle at low speeds, increasing the risk of a crash.
April 30, 2014	14V224000	2007 and 2008 Aura; 2004, 2005, 2006, 2007 and 2008 Malibu; 2005, 2006, 2007 and 2008 G6	General Motors (GM) is has increased the recall for certain model vehicles, equipped with a 4-speed automatic transmission. On these vehicles, the tabs on the transmission shift cable end may fracture and separate. The vehicle may not be able to be restarted and the vehicle could roll away after the driver has exited the vehicle, resulting in a possible crash.
May 14, 2014	14V252000	2007, 2008, 2009 and 2010 Aura; 2004, 2005, 2006, 2007, 2008, 2009, 2010, 2011 and 2012 Malibu; 2005, 2006, 2007, 2008, 2009 and 2010 G6	General Motors (GM) is recalling certain models where increased resistance in the Body Control Module (BCM) connection may result in voltage fluctuations in the Brake Apply Sensor (BAS) circuit. These fluctuations can cause one or more of these conditions: the brake lights to illuminate without the brake pedal being pushed; the brake lights to not illuminate when the pedal is pushed; difficulty disengaging the cruise control; moving the gear shifter out of the 'PARK' position without pushing the brake; and disablement of crash avoidance features such as traction control, electronic stability control, and panic braking assist features. Any of the the above failure conditions increases the risk of a crash.

Buying parts

Replacement parts are available from many sources, which generally fall into one of two categories - authorized dealer parts departments and independent retail auto parts stores. Our advice concerning these parts is as follows:

Retail auto parts stores: Good auto parts stores will stock frequently needed components which wear out relatively fast, such as clutch components, exhaust systems, brake parts, tune-up parts, etc. These stores often supply new or reconditioned parts on an exchange basis, which can save a considerable amount of money. Discount auto parts stores are often very good places to buy materials and parts needed for general vehicle maintenance such as oil, grease, filters, spark plugs, belts, touch-up paint, bulbs, etc. They also usually sell tools and general accessories, have convenient hours, charge lower prices and can often be found not far from home.

Authorized dealer parts department: This is the best source for parts which are unique to the vehicle and not generally available elsewhere (such as major engine parts, transmission parts, trim pieces, etc.).

Warranty information: If the vehicle is still covered under warranty, be sure that any replacement parts purchased - regardless of the source - do not invalidate the warranty!

To be sure of obtaining the correct parts, have engine and chassis numbers available and, if possible, take the old parts along for positive identification.

Maintenance techniques, tools and working facilities

Maintenance techniques

There are a number of techniques involved in maintenance and repair that will be referred to throughout this manual. Application of these techniques will enable the home mechanic to be more efficient, better organized and capable of performing the various tasks properly, which will ensure that the repair job is thorough and complete.

Fasteners

Fasteners are nuts, bolts, studs and screws used to hold two or more parts together. There are a few things to keep in mind when working with fasteners. Almost all of them use a locking device of some type, either a lockwasher, locknut, locking tab or thread adhesive. All threaded fasteners should be clean and straight, with undamaged threads and undamaged corners on the hex head where the wrench fits. Develop the habit of replacing all damaged nuts and bolts with new ones. Special locknuts with nylon or fiber inserts can only be used once. If they are removed, they lose their locking ability and must be replaced with new ones.

Rusted nuts and bolts should be treated with a penetrating fluid to ease removal and prevent breakage. Some mechanics use turpentine in a spout-type oil can, which works quite well. After applying the rust penetrant, let it work for a few minutes before trying to loosen the nut or bolt. Badly rusted fasteners may have to be chiseled or sawed off or removed with a special nut breaker, available at tool stores.

If a bolt or stud breaks off in an assembly, it can be drilled and removed with a special tool commonly available for this purpose. Most automotive machine shops can perform this task, as well as other repair procedures, such as the repair of threaded holes that have been stripped out.

Flat washers and lockwashers, when removed from an assembly, should always be replaced exactly as removed. Replace any damaged washers with new ones. Never use a lockwasher on any soft metal surface (such as aluminum), thin sheet metal or plastic.

Fastener sizes

For a number of reasons, automobile manufacturers are making wider and wider use of metric fasteners. Therefore, it is important to be able to tell the difference between standard (sometimes called U.S. or SAE) and metric hardware, since they cannot be interchanged.

All bolts, whether standard or metric, are sized according to diameter, thread pitch and length. For example, a standard 1/2 - 13 x 1 bolt is 1/2 inch in diameter, has 13 threads per inch and is 1 inch long. An M12 - 1.75 x 25 metric bolt is 12 mm in diameter, has a thread pitch of 1.75 mm (the distance between threads) and is 25 mm long. The two bolts are nearly identical, and easily confused, but they are not interchangeable.

In addition to the differences in diameter, thread pitch and length, metric and standard bolts can also be distinguished by examining the bolt heads. To begin with, the distance across the flats on a standard bolt head is measured in inches, while the same dimension on a metric bolt is sized in millimeters (the same is true for nuts). As a result, a standard wrench should not be used on a metric bolt and a metric wrench should not be used on a standard bolt. Also, most standard bolts have slashes radiating out from the center of the head to denote the grade or strength of the bolt, which is an indication of the amount of torque that can be applied to it. The greater the number of slashes, the greater the strength of the bolt. Grades 0 through 5 are commonly used on automobiles. Metric bolts have a property class (grade) number, rather than a slash, molded into their heads to indicate bolt strength. In this case, the higher the number, the stronger the bolt. Property class numbers 8.8, 9.8 and 10.9 are commonly used on automobiles.

Strength markings can also be used to distinguish standard hex nuts from metric hex nuts. Many standard nuts have dots stamped into one side, while metric nuts are marked with a number. The greater the number of dots, or the higher the number, the greater the strength of the nut.

Metric studs are also marked on their ends according to property class (grade). Larger studs are numbered (the same as metric bolts), while smaller studs carry a geometric code to denote grade.

It should be noted that many fasteners, especially Grades 0 through 2, have no distinguishing marks on them. When such is the case, the only way to determine whether it is standard or metric is to measure the thread pitch or compare it to a known fastener of the same size.

Standard fasteners are often referred to as SAE, as opposed to metric. However, it should be noted that SAE technically refers to a non-metric fine thread fastener only. Coarse thread non-metric fasteners are referred to as USS sizes.

Since fasteners of the same size (both standard and metric) may have different strength ratings, be sure to reinstall any bolts, studs or nuts removed from your vehicle in their original locations. Also, when replacing a fastener with a new one, make sure that the new one has a strength rating equal to or greater than the original.

Tightening sequences and procedures

Most threaded fasteners should be tightened to a specific torque value (torque is the twisting force applied to a threaded component such as a nut or bolt). Overtightening the fastener can weaken it and cause it to break, while undertightening can cause it to eventually come loose. Bolts, screws and studs, depending on the material they are made of and their thread diameters, have specific torque values, many of which are noted in the Specifications at the beginning of each Chapter. Be sure to follow the torque recommen-

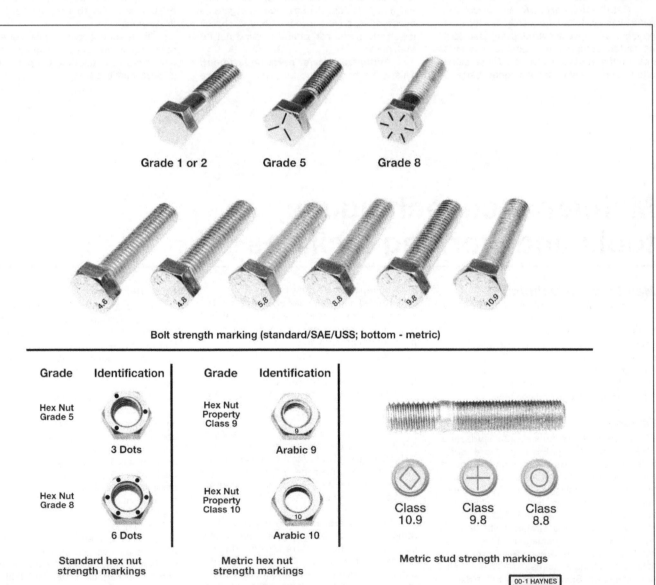

Grade 1 or 2   Grade 5   Grade 8

Bolt strength marking (standard/SAE/USS; bottom - metric)

Grade	Identification	Grade	Identification
Hex Nut Grade 5	3 Dots	Hex Nut Property Class 9	Arabic 9
Hex Nut Grade 8	6 Dots	Hex Nut Property Class 10	Arabic 10

Standard hex nut strength markings

Metric hex nut strength markings

Class 10.9   Class 9.8   Class 8.8

Metric stud strength markings

00-1 HAYNES

dations closely. For fasteners not assigned a specific torque, a general torque value chart is presented here as a guide. These torque values are for dry (unlubricated) fasteners threaded into steel or cast iron (not aluminum). As was previously mentioned, the size and grade of a fastener determine the amount of torque that can safely be applied to it. The figures listed here are approximate for Grade 2 and Grade 3 fasteners. Higher grades can tolerate higher torque values.

Fasteners laid out in a pattern, such as cylinder head bolts, oil pan bolts, differential cover bolts, etc., must be loosened or tightened in sequence to avoid warping the component. This sequence will normally be shown in the appropriate Chapter. If a specific pattern is not given, the following procedures can be used to prevent warping.

Initially, the bolts or nuts should be assembled finger-tight only. Next, they should be tightened one full turn each, in a crisscross or diagonal pattern. After each one has been tightened one full turn, return to the first one and tighten them all one-half turn, following the same pattern. Finally, tighten each of them one-quarter turn at a time until each fastener has been tightened to the proper torque. To loosen and remove the fasteners, the procedure would be reversed.

Metric thread sizes

	Ft-lbs	Nm
M-6	6 to 9	9 to 12
M-8	14 to 21	19 to 28
M-10	28 to 40	38 to 54
M-12	50 to 71	68 to 96
M-14	80 to 140	109 to 154

Pipe thread sizes

1/8	5 to 8	7 to 10
1/4	12 to 18	17 to 24
3/8	22 to 33	30 to 44
1/2	25 to 35	34 to 47

U.S. thread sizes

1/4 - 20	6 to 9	9 to 12
5/16 - 18	12 to 18	17 to 24
5/16 - 24	14 to 20	19 to 27
3/8 - 16	22 to 32	30 to 43
3/8 - 24	27 to 38	37 to 51
7/16 - 14	40 to 55	55 to 74
7/16 - 20	40 to 60	55 to 81
1/2 - 13	55 to 80	75 to 108

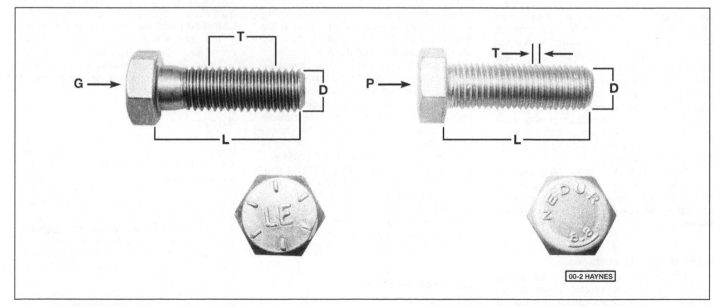

Standard (SAE and USS) bolt dimensions/grade marks

G   Grade marks (bolt strength)
L   Length (in inches)
T   Thread pitch (number of threads per inch)
D   Nominal diameter (in inches)

Metric bolt dimensions/grade marks

P   Property class (bolt strength)
L   Length (in millimeters)
T   Thread pitch (distance between threads in millimeters)
D   Diameter

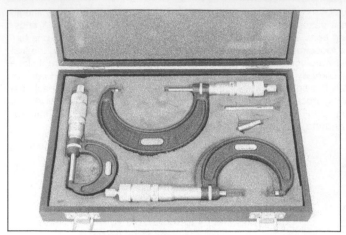

Micrometer set

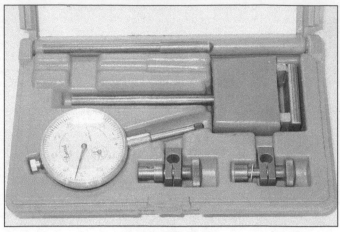

Dial indicator set

Component disassembly

Component disassembly should be done with care and purpose to help ensure that the parts go back together properly. Always keep track of the sequence in which parts are removed. Make note of special characteristics or marks on parts that can be installed more than one way, such as a grooved thrust washer on a shaft. It is a good idea to lay the disassembled parts out on a clean surface in the order that they were removed. It may also be helpful to make sketches or take instant photos of components before removal.

When removing fasteners from a component, keep track of their locations. Sometimes threading a bolt back in a part, or putting the washers and nut back on a stud, can prevent mix-ups later. If nuts and bolts cannot be returned to their original locations, they should be kept in a compartmented box or a series of small boxes. A cupcake or muffin tin is ideal for this purpose, since each cavity can hold the bolts and nuts from a particular area (i.e. oil pan bolts, valve cover bolts, engine mount bolts, etc.). A pan of this type is especially helpful when working on assemblies with very small parts, such as the carburetor, alternator, valve train or interior dash and trim pieces. The cavities can be marked with paint or tape to identify the contents.

Whenever wiring looms, harnesses or connectors are separated, it is a good idea to identify the two halves with numbered pieces of masking tape so they can be easily reconnected.

Gasket sealing surfaces

Throughout any vehicle, gaskets are used to seal the mating surfaces between two parts and keep lubricants, fluids, vacuum or pressure contained in an assembly.

Many times these gaskets are coated with a liquid or paste-type gasket sealing compound before assembly. Age, heat and pressure can sometimes cause the two parts to stick together so tightly that they are very difficult to separate. Often, the assembly can

be loosened by striking it with a soft-face hammer near the mating surfaces. A regular hammer can be used if a block of wood is placed between the hammer and the part. Do not hammer on cast parts or parts that could be easily damaged. With any particularly stubborn part, always recheck to make sure that every fastener has been removed.

Avoid using a screwdriver or bar to pry apart an assembly, as they can easily mar the gasket sealing surfaces of the parts, which must remain smooth. If prying is absolutely necessary, use an old broom handle, but keep in mind that extra clean up will be necessary if the wood splinters.

After the parts are separated, the old gasket must be carefully scraped off and the gasket surfaces cleaned. Stubborn gasket material can be soaked with rust penetrant or treated with a special chemical to soften it so it can be easily scraped off. **Caution:** *Never use gasket removal solutions or caustic chemicals on plastic or other composite components.* A scraper can be fashioned from a piece of copper tubing by flattening and sharpening one end. Copper is recommended because it is usually softer than the surfaces to be scraped, which reduces the chance of gouging the part. Some gaskets can be removed with a wire brush, but regardless of the method used, the mating surfaces must be left clean and smooth. If for some reason the gasket surface is gouged, then a gasket sealer thick enough to fill scratches will have to be used during reassembly of the components. For most applications, a non-drying (or semi-drying) gasket sealer should be used.

Hose removal tips

Warning: *If the vehicle is equipped with air conditioning, do not disconnect any of the A/C hoses without first having the system depressurized by a dealer service department or a service station.*

Hose removal precautions closely parallel gasket removal precautions. Avoid scratching or gouging the surface that the

hose mates against or the connection may leak. This is especially true for radiator hoses. Because of various chemical reactions, the rubber in hoses can bond itself to the metal spigot that the hose fits over. To remove a hose, first loosen the hose clamps that secure it to the spigot. Then, with slip-joint pliers, grab the hose at the clamp and rotate it around the spigot. Work it back and forth until it is completely free, then pull it off. Silicone or other lubricants will ease removal if they can be applied between the hose and the outside of the spigot. Apply the same lubricant to the inside of the hose and the outside of the spigot to simplify installation.

As a last resort (and if the hose is to be replaced with a new one anyway), the rubber can be slit with a knife and the hose peeled from the spigot. If this must be done, be careful that the metal connection is not damaged.

If a hose clamp is broken or damaged, do not reuse it. Wire-type clamps usually weaken with age, so it is a good idea to replace them with screw-type clamps whenever a hose is removed.

Tools

A selection of good tools is a basic requirement for anyone who plans to maintain and repair his or her own vehicle. For the owner who has few tools, the initial investment might seem high, but when compared to the spiraling costs of professional auto maintenance and repair, it is a wise one.

To help the owner decide which tools are needed to perform the tasks detailed in this manual, the following tool lists are offered: *Maintenance and minor repair, Repair/overhaul* and *Special.*

The newcomer to practical mechanics should start off with the *maintenance and minor repair* tool kit, which is adequate for the simpler jobs performed on a vehicle. Then, as confidence and experience grow, the owner can tackle more difficult tasks, buying additional tools as they are needed. Eventually the basic kit will be expanded into the *repair and overhaul* tool set. Over a period of time, the

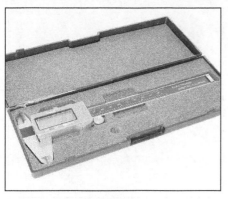

Dial caliper

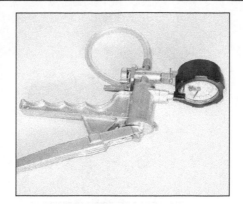

Hand-operated vacuum pump

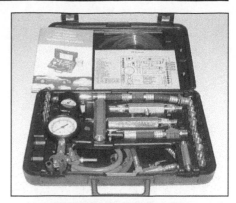

Fuel pressure gauge set

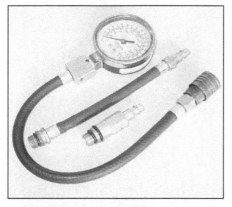

Compression gauge with spark plug hole adapter

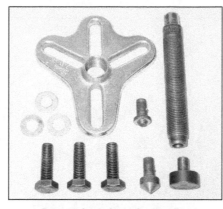

Damper/steering wheel puller

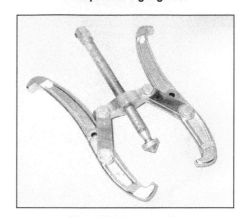

General purpose puller

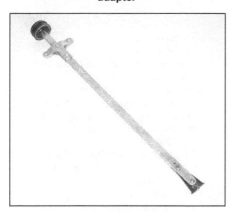

Hydraulic lifter removal tool

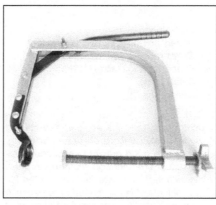

Valve spring compressor

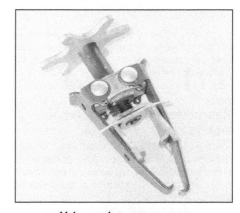

Valve spring compressor

Ridge reamer

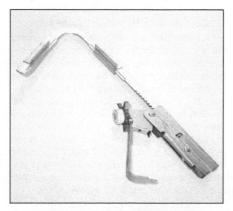

Piston ring groove cleaning tool

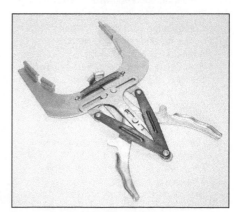

Ring removal/installation tool

Ring compressor

Cylinder hone

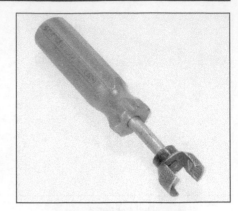

Brake hold-down spring tool

Torque angle gauge

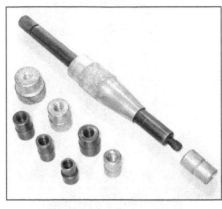

Clutch plate alignment tool

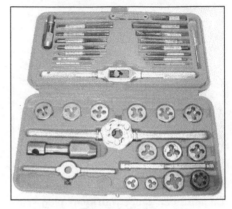

Tap and die set

experienced do-it-yourselfer will assemble a tool set complete enough for most repair and overhaul procedures and will add tools from the special category when it is felt that the expense is justified by the frequency of use.

Maintenance and minor repair tool kit

The tools in this list should be considered the minimum required for performance of routine maintenance, servicing and minor repair work. We recommend the purchase of combination wrenches (box-end and open-end combined in one wrench). While more expensive than open end wrenches, they offer the advantages of both types of wrench.

> *Combination wrench set (1/4-inch to 1 inch or 6 mm to 19 mm)*
> *Adjustable wrench, 8 inch*
> *Spark plug wrench with rubber insert*
> *Spark plug gap adjusting tool*
> *Feeler gauge set*
> *Brake bleeder wrench*
> *Standard screwdriver (5/16-inch x 6 inch)*
> *Phillips screwdriver (No. 2 x 6 inch)*
> *Combination pliers - 6 inch*
> *Hacksaw and assortment of blades*
> *Tire pressure gauge*
> *Grease gun*
> *Oil can*
> *Fine emery cloth*

> *Wire brush*
> *Battery post and cable cleaning tool*
> *Oil filter wrench*
> *Funnel (medium size)*
> *Safety goggles*
> *Jackstands (2)*
> *Drain pan*

Note: *If basic tune-ups are going to be part of routine maintenance, it will be necessary to purchase a good quality stroboscopic timing light and combination tachometer/dwell meter. Although they are included in the list of special tools, it is mentioned here because they are absolutely necessary for tuning most vehicles properly.*

Repair and overhaul tool set

These tools are essential for anyone who plans to perform major repairs and are in addition to those in the maintenance and minor repair tool kit. Included is a comprehensive set of sockets which, though expensive, are invaluable because of their versatility, especially when various extensions and drives are available. We recommend the 1/2-inch drive over the 3/8-inch drive. Although the larger drive is bulky and more expensive, it has the capacity of accepting a very wide range of large sockets. Ideally, however, the mechanic should have a 3/8-inch drive set and a 1/2-inch drive set.

> *Socket set(s)*
> *Reversible ratchet*

> *Extension - 10 inch*
> *Universal joint*
> *Torque wrench (same size drive as sockets)*
> *Ball peen hammer - 8 ounce*
> *Soft-face hammer (plastic/rubber)*
> *Standard screwdriver (1/4-inch x 6 inch)*
> *Standard screwdriver (stubby - 5/16-inch)*
> *Phillips screwdriver (No. 3 x 8 inch)*
> *Phillips screwdriver (stubby - No. 2)*
> *Pliers - vise grip*
> *Pliers - lineman's*
> *Pliers - needle nose*
> *Pliers - snap-ring (internal and external)*
> *Cold chisel - 1/2-inch*
> *Scribe*
> *Scraper (made from flattened copper tubing)*
> *Centerpunch*
> *Pin punches (1/16, 1/8, 3/16-inch)*
> *Steel rule/straightedge - 12 inch*
> *Allen wrench set (1/8 to 3/8-inch or 4 mm to 10 mm)*
> *A selection of files*
> *Wire brush (large)*
> *Jackstands (second set)*
> *Jack (scissor or hydraulic type)*

Note: *Another tool which is often useful is an electric drill with a chuck capacity of 3/8-inch and a set of good quality drill bits.*

Special tools

The tools in this list include those which are not used regularly, are expensive to buy, or which need to be used in accordance with their manufacturer's instructions. Unless these tools will be used frequently, it is not very economical to purchase many of them. A consideration would be to split the cost and use between yourself and a friend or friends. In addition, most of these tools can be obtained from a tool rental shop on a temporary basis.

This list primarily contains only those tools and instruments widely available to the public, and not those special tools produced by the vehicle manufacturer for distribution to dealer service departments. Occasionally, references to the manufacturer's special tools are included in the text of this manual. Generally, an alternative method of doing the job without the special tool is offered. However, sometimes there is no alternative to their use. Where this is the case, and the tool cannot be purchased or borrowed, the work should be turned over to the dealer service department or an automotive repair shop.

Valve spring compressor
Piston ring groove cleaning tool
Piston ring compressor
Piston ring installation tool
Cylinder compression gauge
Cylinder ridge reamer
Cylinder surfacing hone
Cylinder bore gauge
Micrometers and/or dial calipers
Hydraulic lifter removal tool
Balljoint separator
Universal-type puller
Impact screwdriver
Dial indicator set
Stroboscopic timing light (inductive pick-up)
Hand operated vacuum/pressure pump
Tachometer/dwell meter
Universal electrical multimeter
Cable hoist
Brake spring removal and installation tools
Floor jack

Buying tools

For the do-it-yourselfer who is just starting to get involved in vehicle maintenance and repair, there are a number of options available when purchasing tools. If maintenance and minor repair is the extent of the work to be done, the purchase of individual tools is satisfactory. If, on the other hand, extensive work is planned, it would be a good idea to purchase a modest tool set from one of the large retail chain stores. A set can usually be bought at a substantial savings over the individual tool prices, and they often come with a tool box. As additional tools are needed, add-on sets, individual tools and a larger tool box can be purchased to expand the tool selection. Building a tool set gradually allows the cost of the tools to be spread over a longer period of time and gives the mechanic the freedom to choose only those tools that will actually be used.

Tool stores will often be the only source of some of the special tools that are needed, but regardless of where tools are bought, try to avoid cheap ones, especially when buying screwdrivers and sockets, because they won't last very long. The expense involved in replacing cheap tools will eventually be greater than the initial cost of quality tools.

Care and maintenance of tools

Good tools are expensive, so it makes sense to treat them with respect. Keep them clean and in usable condition and store them properly when not in use. Always wipe off any dirt, grease or metal chips before putting them away. Never leave tools lying around in the work area. Upon completion of a job, always check closely under the hood for tools that may have been left there so they won't get lost during a test drive.

Some tools, such as screwdrivers, pliers, wrenches and sockets, can be hung on a panel mounted on the garage or workshop wall, while others should be kept in a tool box or tray. Measuring instruments, gauges, meters, etc. must be carefully stored where they cannot be damaged by weather or impact from other tools.

When tools are used with care and stored properly, they will last a very long time. Even with the best of care, though, tools will wear out if used frequently. When a tool is damaged or worn out, replace it. Subsequent jobs will be safer and more enjoyable if you do.

How to repair damaged threads

Sometimes, the internal threads of a nut or bolt hole can become stripped, usually from overtightening. Stripping threads is an all-too-common occurrence, especially when working with aluminum parts, because aluminum is so soft that it easily strips out.

Usually, external or internal threads are only partially stripped. After they've been cleaned up with a tap or die, they'll still work. Sometimes, however, threads are badly damaged. When this happens, you've got three choices:

1) *Drill and tap the hole to the next suitable oversize and install a larger diameter bolt, screw or stud.*
2) *Drill and tap the hole to accept a threaded plug, then drill and tap the plug to the original screw size. You can also buy a plug already threaded to the original size. Then you simply drill a hole to the specified size, then run the threaded plug into the hole with a bolt and jam nut. Once the plug is fully seated, remove the jam nut and bolt.*
3) *The third method uses a patented thread repair kit like Heli-Coil or Slimsert. These easy-to-use kits are designed to repair damaged threads in straight-through holes and blind holes. Both are available as kits which can handle a variety of sizes and thread patterns. Drill the hole, then tap it with the special included tap. Install the Heli-Coil and the hole is back to its original diameter and thread pitch.*

Regardless of which method you use, be sure to proceed calmly and carefully. A little impatience or carelessness during one of these relatively simple procedures can ruin your whole day's work and cost you a bundle if you wreck an expensive part.

Working facilities

Not to be overlooked when discussing tools is the workshop. If anything more than routine maintenance is to be carried out, some sort of suitable work area is essential.

It is understood, and appreciated, that many home mechanics do not have a good workshop or garage available, and end up removing an engine or doing major repairs outside. It is recommended, however, that the overhaul or repair be completed under the cover of a roof.

A clean, flat workbench or table of comfortable working height is an absolute necessity. The workbench should be equipped with a vise that has a jaw opening of at least four inches.

As mentioned previously, some clean, dry storage space is also required for tools, as well as the lubricants, fluids, cleaning solvents, etc. which soon become necessary.

Sometimes waste oil and fluids, drained from the engine or cooling system during normal maintenance or repairs, present a disposal problem. To avoid pouring them on the ground or into a sewage system, pour the used fluids into large containers, seal them with caps and take them to an authorized disposal site or recycling center. Plastic jugs, such as old antifreeze containers, are ideal for this purpose.

Always keep a supply of old newspapers and clean rags available. Old towels are excellent for mopping up spills. Many mechanics use rolls of paper towels for most work because they are readily available and disposable. To help keep the area under the vehicle clean, a large cardboard box can be cut open and flattened to protect the garage or shop floor.

Whenever working over a painted surface, such as when leaning over a fender to service something under the hood, always cover it with an old blanket or bedspread to protect the finish. Vinyl covered pads, made especially for this purpose, are available at auto parts stores.

Booster battery (jump) starting

Observe the following precautions when using a booster battery to start a vehicle:

a) *Before connecting the booster battery, make sure the ignition switch is in the Off position.*

b) *Turn off the lights, heater and other electrical loads.*

c) *Your eyes should be shielded. Safety goggles are a good idea.*

d) *Make sure the booster battery is the same voltage as the dead one in the vehicle.*

e) *The two vehicles MUST NOT TOUCH each other.*

f) *Make sure the transmission is in Park.*

g) *If the booster battery is not a maintenance-free type, remove the vent caps and lay a cloth over the vent holes.*

Connect the red jumper cable to the positive (+) terminals of each battery.

Connect one end of the black cable to the negative (-) terminal of the booster battery. The other end of this cable should be connected to a good ground on the engine block (see illustration) . Make sure the cable will not come into contact with the fan, drivebelts or other moving parts of the engine.

Start the engine using the booster battery, then, with the engine running at idle speed, disconnect the jumper cables in the reverse order of connection.

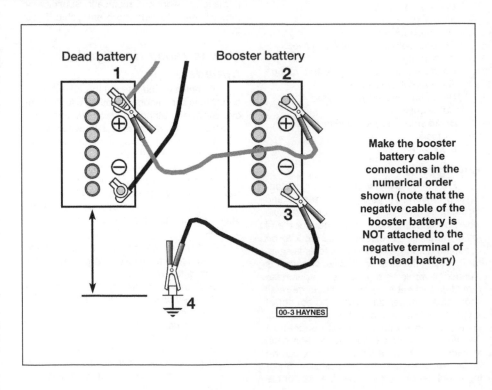

Dead battery   Booster battery

Make the booster battery cable connections in the numerical order shown (note that the negative cable of the booster battery is NOT attached to the negative terminal of the dead battery)

00-3 HAYNES

Jacking and towing

Jacking

1   The jack supplied with the vehicle should only be used for raising the vehicle for changing a tire or placing jackstands under the frame. Warning: Never crawl under the vehicle or start the engine when the jack is being used as the only means of support.
2   All vehicles are supplied with a scissors-type jack. When jacking the vehicle, it should be engaged with the notch in the rocker panel flange (see illustration).
3   The vehicle should be on level ground with the wheels blocked and the transmission in Park. Pry off the hub cap (if equipped) using the tapered end of the lug wrench. Loosen the lug nuts one-half turn and leave them in place until the wheel is raised off the ground.
4   Place the jack under the side of the vehicle in the indicated position. Use the supplied wrench to turn the jackscrew clockwise until

the wheel is raised off the ground. Remove the lug nuts, pull off the wheel and install the spare.
5   With the beveled side in, install the lug nuts and tighten them until snug. Lower the vehicle by turning the jackscrew counterclockwise. Remove the jack and tighten the nuts in a diagonal pattern to the torque listed in the Chapter 1 Specifications. If a torque wrench is not available, have the torque checked by a service station as soon as possible. Install hubcap by placing it in position and using the heel of your hand or a rubber mallet to seat it.

Towing

6   As a general rule, the vehicle should be towed with the front (drive) wheels off the ground or, preferably, on a flat bed car carrier. If the front wheels can't be raised or a carrier isn't available, place them on a dolly.

The jack fits over the rocker panel flange (there are two jacking points on each side of the vehicle)

The ignition key must be in the ACC position, since the steering lock mechanism isn't strong enough to hold the front wheels straight while towing.
Note: *Remove the "20 IBCM 1" fuse (2004 and 2005 models) or the IGN SENSOR fuse (2006 and later models) from the engine compartment fuse block to prevent the battery from draining.*
7   Some models can be towed from the front with all four wheels on the ground, provided that speeds don't exceed 55 mph.
Caution: *Malibu Sedan SS and Maxx SS models cannot be towed with all four wheels on the ground.*
8   Towing equipment specifically designed for this purpose should be used and should be attached to the main structural members of the vehicle, not the bumper or brackets.
9   Safety is a major consideration when towing and all applicable state and local laws must be obeyed. A safety chain system must be used for all towing.
10   While towing, the parking brake must be released and the transmission must be in Neutral. The steering must be unlocked (ignition switch in the Off position). Remember that power steering and power brakes will not work with the engine off.

Traction control

11   On models equipped with Traction-Control system, push in the TRAC switch anytime the vehicle is on a "rolling road" tester such as a speedometer test machine or chassis dynamometer. The TRAC OFF indicator light should illuminate when the system is turned off.

Automotive chemicals and lubricants

A number of automotive chemicals and lubricants are available for use during vehicle maintenance and repair. They include a wide variety of products ranging from cleaning solvents and degreasers to lubricants and protective sprays for rubber, plastic and vinyl.

Cleaners

Carburetor cleaner and choke cleaner is a strong solvent for gum, varnish and carbon. Most carburetor cleaners leave a dry-type lubricant film which will not harden or gum up. Because of this film it is not recommended for use on electrical components.

Brake system cleaner is used to remove brake dust, grease and brake fluid from the brake system, where clean surfaces are absolutely necessary. It leaves no residue and often eliminates brake squeal caused by contaminants.

Electrical cleaner removes oxidation, corrosion and carbon deposits from electrical contacts, restoring full current flow. It can also be used to clean spark plugs, carburetor jets, voltage regulators and other parts where an oil-free surface is desired.

Demoisturants remove water and moisture from electrical components such as alternators, voltage regulators, electrical connectors and fuse blocks. They are non-conductive and non-corrosive.

Degreasers are heavy-duty solvents used to remove grease from the outside of the engine and from chassis components. They can be sprayed or brushed on and, depending on the type, are rinsed off either with water or solvent.

Lubricants

Motor oil is the lubricant formulated for use in engines. It normally contains a wide variety of additives to prevent corrosion and reduce foaming and wear. Motor oil comes in various weights (viscosity ratings) from 0 to 50. The recommended weight of the oil depends on the season, temperature and the demands on the engine. Light oil is used in cold climates and under light load conditions. Heavy oil is used in hot climates and where high loads are encountered. Multi-viscosity oils are designed to have characteristics of both light and heavy oils and are available in a number of weights from 0W-20 to 20W-50.

Gear oil is designed to be used in differentials, manual transmissions and other areas where high-temperature lubrication is required.

Chassis and wheel bearing grease is a heavy grease used where increased loads and friction are encountered, such as for wheel bearings, balljoints, tie-rod ends and universal joints.

High-temperature wheel bearing grease is designed to withstand the extreme temperatures encountered by wheel bearings in disc brake equipped vehicles. It usually contains molybdenum disulfide (moly), which is a dry-type lubricant.

White grease is a heavy grease for metal-to-metal applications where water is a problem. White grease stays soft under both low and high temperatures (usually from -100 to +190-degrees F), and will not wash off or dilute in the presence of water.

Assembly lube is a special extreme pressure lubricant, usually containing moly, used to lubricate high-load parts (such as main and rod bearings and cam lobes) for initial start-up of a new engine. The assembly lube lubricates the parts without being squeezed out or washed away until the engine oiling system begins to function.

Silicone lubricants are used to protect rubber, plastic, vinyl and nylon parts.

Graphite lubricants are used where oils cannot be used due to contamination problems, such as in locks. The dry graphite will lubricate metal parts while remaining uncontaminated by dirt, water, oil or acids. It is electrically conductive and will not foul electrical contacts in locks such as the ignition switch.

Moly penetrants loosen and lubricate frozen, rusted and corroded fasteners and prevent future rusting or freezing.

Heat-sink grease is a special electrically non-conductive grease that is used for mounting electronic ignition modules where it is essential that heat is transferred away from the module.

Sealants

RTV sealant is one of the most widely used gasket compounds. Made from silicone, RTV is air curing, it seals, bonds, waterproofs, fills surface irregularities, remains flexible, doesn't shrink, is relatively easy to remove, and is used as a supplementary sealer with almost all low and medium temperature gaskets.

Anaerobic sealant is much like RTV in that it can be used either to seal gaskets or to form gaskets by itself. It remains flexible, is solvent resistant and fills surface imperfections. The difference between an anaerobic sealant and an RTV-type sealant is in the curing. RTV cures when exposed to air, while an anaerobic sealant cures only in the absence of air. This means that an anaerobic sealant cures only after the assembly of parts, sealing them together.

Thread and pipe sealant is used for sealing hydraulic and pneumatic fittings and vacuum lines. It is usually made from a Teflon compound, and comes in a spray, a paint-on liquid and as a wrap-around tape.

Chemicals

Anti-seize compound prevents seizing, galling, cold welding, rust and corrosion in fasteners. High-temperature ant-seize, usually made with copper and graphite lubricants, is used for exhaust system and exhaust manifold bolts.

Anaerobic locking compounds are used to keep fasteners from vibrating or working loose and cure only after installation, in the absence of air. Medium strength locking compound is used for small nuts, bolts and screws that may be removed later. High-strength locking compound is for large nuts, bolts and studs which aren't removed on a regular basis.

Oil additives range from viscosity index improvers to chemical treatments that claim to reduce internal engine friction. It should be noted that most oil manufacturers caution against using additives with their oils.

Gas additives perform several functions, depending on their chemical makeup. They usually contain solvents that help dissolve gum and varnish that build up on carburetor, fuel injection and intake parts. They also serve to break down carbon deposits that form on the inside surfaces of the combustion chambers. Some additives contain upper cylinder lubricants for valves and piston rings, and others contain chemicals to remove condensation from the gas tank.

Miscellaneous

Brake fluid is specially formulated hydraulic fluid that can withstand the heat and pressure encountered in brake systems. Care must be taken so this fluid does not come in contact with painted surfaces or plastics. An opened container should always be resealed to prevent contamination by water or dirt.

Weatherstrip adhesive is used to bond weatherstripping around doors, windows and trunk lids. It is sometimes used to attach trim pieces.

Undercoating is a petroleum-based, tar-like substance that is designed to protect metal surfaces on the underside of the vehicle from corrosion. It also acts as a sound-deadening agent by insulating the bottom of the vehicle.

Waxes and polishes are used to help protect painted and plated surfaces from the weather. Different types of paint may require the use of different types of wax and polish. Some polishes utilize a chemical or abrasive cleaner to help remove the top layer of oxidized (dull) paint on older vehicles. In recent years many non-wax polishes that contain a wide variety of chemicals such as polymers and silicones have been introduced. These non-wax polishes are usually easier to apply and last longer than conventional waxes and polishes.

Conversion factors

Length (distance)

Inches (in)	X	25.4	= Millimeters (mm)	X 0.0394	= Inches (in)
Feet (ft)	X	0.305	= Meters (m)	X 3.281	= Feet (ft)
Miles	X	1.609	= Kilometers (km)	X 0.621	= Miles

Volume (capacity)

Cubic inches (cu in; in³)	X	16.387	= Cubic centimeters (cc; cm³)	X 0.061	= Cubic inches (cu in; in³)
Imperial pints (Imp pt)	X	0.568	= Liters (l)	X 1.76	= Imperial pints (Imp pt)
Imperial quarts (Imp qt)	X	1.137	= Liters (l)	X 0.88	= Imperial quarts (Imp qt)
Imperial quarts (Imp qt)	X	1.201	= US quarts (US qt)	X 0.833	= Imperial quarts (Imp qt)
US quarts (US qt)	X	0.946	= Liters (l)	X 1.057	= US quarts (US qt)
Imperial gallons (Imp gal)	X	4.546	= Liters (l)	X 0.22	= Imperial gallons (Imp gal)
Imperial gallons (Imp gal)	X	1.201	= US gallons (US gal)	X 0.833	= Imperial gallons (Imp gal)
US gallons (US gal)	X	3.785	= Liters (l)	X 0.264	= US gallons (US gal)

Mass (weight)

Ounces (oz)	X	28.35	= Grams (g)	X 0.035	= Ounces (oz)
Pounds (lb)	X	0.454	= Kilograms (kg)	X 2.205	= Pounds (lb)

Force

Ounces-force (ozf; oz)	X	0.278	= Newtons (N)	X 3.6	= Ounces-force (ozf; oz)
Pounds-force (lbf; lb)	X	4.448	= Newtons (N)	X 0.225	= Pounds-force (lbf; lb)
Newtons (N)	X	0.1	= Kilograms-force (kgf; kg)	X 9.81	= Newtons (N)

Pressure

Pounds-force per square inch (psi; lbf/in²; lb/in²)	X	0.070	= Kilograms-force per square centimeter (kgf/cm²; kg/cm²)	X 14.223	= Pounds-force per square inch (psi; lbf/in²; lb/in²)
Pounds-force per square inch (psi; lbf/in²; lb/in²)	X	0.068	= Atmospheres (atm)	X 14.696	= Pounds-force per square inch (psi; lbf/in²; lb/in²)
Pounds-force per square inch (psi; lbf/in²; lb/in²)	X	0.069	= Bars	X 14.5	= Pounds-force per square inch (psi; lbf/in²; lb/in²)
Pounds-force per square inch (psi; lbf/in²; lb/in²)	X	6.895	= Kilopascals (kPa)	X 0.145	= Pounds-force per square inch (psi; lbf/in²; lb/in²)
Kilopascals (kPa)	X	0.01	= Kilograms-force per square centimeter (kgf/cm²; kg/cm²)	X 98.1	= Kilopascals (kPa)

Torque (moment of force)

Pounds-force inches (lbf in; lb in)	X	1.152	= Kilograms-force centimeter (kgf cm; kg cm)	X 0.868	= Pounds-force inches (lbf in; lb in)
Pounds-force inches (lbf in; lb in)	X	0.113	= Newton meters (Nm)	X 8.85	= Pounds-force inches (lbf in; lb in)
Pounds-force inches (lbf in; lb in)	X	0.083	= Pounds-force feet (lbf ft; lb ft)	X 12	= Pounds-force inches (lbf in; lb in)
Pounds-force feet (lbf ft; lb ft)	X	0.138	= Kilograms-force meters (kgf m; kg m)	X 7.233	= Pounds-force feet (lbf ft; lb ft)
Pounds-force feet (lbf ft; lb ft)	X	1.356	= Newton meters (Nm)	X 0.738	= Pounds-force feet (lbf ft; lb ft)
Newton meters (Nm)	X	0.102	= Kilograms-force meters (kgf m; kg m)	X 9.804	= Newton meters (Nm)

Vacuum

Inches mercury (in. Hg)	X	3.377	= Kilopascals (kPa)	X 0.2961	= Inches mercury
Inches mercury (in. Hg)	X	25.4	= Millimeters mercury (mm Hg)	X 0.0394	= Inches mercury

Power

Horsepower (hp)	X	745.7	= Watts (W)	X 0.0013	= Horsepower (hp)

Velocity (speed)

Miles per hour (miles/hr; mph)	X	1.609	= Kilometers per hour (km/hr; kph)	X 0.621	= Miles per hour (miles/hr; mph)

Fuel consumption*

Miles per gallon, Imperial (mpg)	X	0.354	= Kilometers per liter (km/l)	X 2.825	= Miles per gallon, Imperial (mpg)
Miles per gallon, US (mpg)	X	0.425	= Kilometers per liter (km/l)	X 2.352	= Miles per gallon, US (mpg)

Temperature

Degrees Fahrenheit = (°C x 1.8) + 32

Degrees Celsius (Degrees Centigrade; °C) = (°F - 32) x 0.56

*It is common practice to convert from miles per gallon (mpg) to liters/100 kilometers (l/100km), where mpg (Imperial) x l/100 km = 282 and mpg (US) x l/100 km = 235

DECIMALS to MILLIMETERS

Decimal	mm	Decimal	mm
0.001	0.0254	0.500	12.7000
0.002	0.0508	0.510	12.9540
0.003	0.0762	0.520	13.2080
0.004	0.1016	0.530	13.4620
0.005	0.1270	0.540	13.7160
0.006	0.1524	0.550	13.9700
0.007	0.1778	0.560	14.2240
0.008	0.2032	0.570	14.4780
0.009	0.2286	0.580	14.7320
		0.590	14.9860
0.010	0.2540		
0.020	0.5080		
0.030	0.7620		
0.040	1.0160	0.600	15.2400
0.050	1.2700	0.610	15.4940
0.060	1.5240	0.620	15.7480
0.070	1.7780	0.630	16.0020
0.080	2.0320	0.640	16.2560
0.090	2.2860	0.650	16.5100
		0.660	16.7640
0.100	2.5400	0.670	17.0180
0.110	2.7940	0.680	17.2720
0.120	3.0480	0.690	17.5260
0.130	3.3020		
0.140	3.5560		
0.150	3.8100		
0.160	4.0640	0.700	17.7800
0.170	4.3180	0.710	18.0340
0.180	4.5720	0.720	18.2880
0.190	4.8260	0.730	18.5420
		0.740	18.7960
0.200	5.0800	0.750	19.0500
0.210	5.3340	0.760	19.3040
0.220	5.5880	0.770	19.5580
0.230	5.8420	0.780	19.8120
0.240	6.0960	0.790	20.0660
0.250	6.3500		
0.260	6.6040		
0.270	6.8580	0.800	20.3200
0.280	7.1120	0.810	20.5740
0.290	7.3660	0.820	21.8280
		0.830	21.0820
0.300	7.6200	0.840	21.3360
0.310	7.8740	0.850	21.5900
0.320	8.1280	0.860	21.8440
0.330	8.3820	0.870	22.0980
0.340	8.6360	0.880	22.3520
0.350	8.8900	0.890	22.6060
0.360	9.1440		
0.370	9.3980		
0.380	9.6520		
0.390	9.9060	0.900	22.8600
0.400	10.1600	0.910	23.1140
0.410	10.4140	0.920	23.3680
0.420	10.6680	0.930	23.6220
0.430	10.9220	0.940	23.8760
0.440	11.1760	0.950	24.1300
0.450	11.4300	0.960	24.3840
0.460	11.6840	0.970	24.6380
0.470	11.9380	0.980	24.8920
0.480	12.1920	0.990	25.1460
0.490	12.4460	1.000	25.4000

FRACTIONS to DECIMALS to MILLIMETERS

Fraction	Decimal	mm	Fraction	Decimal	mm
1/64	0.0156	0.3969	33/64	0.5156	13.0969
1/32	0.0312	0.7938	17/32	0.5312	13.4938
3/64	0.0469	1.1906	35/64	0.5469	13.8906
1/16	0.0625	1.5875	9/16	0.5625	14.2875
5/64	0.0781	1.9844	37/64	0.5781	14.6844
3/32	0.0938	2.3812	19/32	0.5938	15.0812
7/64	0.1094	2.7781	39/64	0.6094	15.4781
1/8	0.1250	3.1750	5/8	0.6250	15.8750
9/64	0.1406	3.5719	41/64	0.6406	16.2719
5/32	0.1562	3.9688	21/32	0.6562	16.6688
11/64	0.1719	4.3656	43/64	0.6719	17.0656
3/16	0.1875	4.7625	11/16	0.6875	17.4625
13/64	0.2031	5.1594	45/64	0.7031	17.8594
7/32	0.2188	5.5562	23/32	0.7188	18.2562
15/64	0.2344	5.9531	47/64	0.7344	18.6531
1/4	0.2500	6.3500	3/4	0.7500	19.0500
17/64	0.2656	6.7469	49/64	0.7656	19.4469
9/32	0.2812	7.1438	25/32	0.7812	19.8438
19/64	0.2969	7.5406	51/64	0.7969	20.2406
5/16	0.3125	7.9375	13/16	0.8125	20.6375
21/64	0.3281	8.3344	53/64	0.8281	21.0344
11/32	0.3438	8.7312	27/32	0.8438	21.4312
23/64	0.3594	9.1281	55/64	0.8594	21.8281
3/8	0.3750	9.5250	7/8	0.8750	22.2250
25/64	0.3906	9.9219	57/64	0.8906	22.6219
13/32	0.4062	10.3188	29/32	0.9062	23.0188
27/64	0.4219	10.7156	59/64	0.9219	23.4156
7/16	0.4375	11.1125	15/16	0.9375	23.8125
29/64	0.4531	11.5094	61/64	0.9531	24.2094
15/32	0.4688	11.9062	31/32	0.9688	24.6062
31/64	0.4844	12.3031	63/64	0.9844	25.0031
1/2	0.5000	12.7000	1	1.0000	25.4000

Safety first!

Regardless of how enthusiastic you may be about getting on with the job at hand, take the time to ensure that your safety is not jeopardized. A moment's lack of attention can result in an accident, as can failure to observe certain simple safety precautions. The possibility of an accident will always exist, and the following points should not be considered a comprehensive list of all dangers. Rather, they are intended to make you aware of the risks and to encourage a safety conscious approach to all work you carry out on your vehicle.

Essential DOs and DON'Ts

DON'T rely on a jack when working under the vehicle. Always use approved jackstands to support the weight of the vehicle and place them under the recommended lift or support points.

DON'T attempt to loosen extremely tight fasteners (i.e. wheel lug nuts) while the vehicle is on a jack - it may fall.

DON'T start the engine without first making sure that the transmission is in Neutral (or Park where applicable) and the parking brake is set.

DON'T remove the radiator cap from a hot cooling system - let it cool or cover it with a cloth and release the pressure gradually.

DON'T attempt to drain the engine oil until you are sure it has cooled to the point that it will not burn you.

DON'T touch any part of the engine or exhaust system until it has cooled sufficiently to avoid burns.

DON'T siphon toxic liquids such as gasoline, antifreeze and brake fluid by mouth, or allow them to remain on your skin.

DON'T inhale brake lining dust - it is potentially hazardous (see *Asbestos* below).

DON'T allow spilled oil or grease to remain on the floor - wipe it up before someone slips on it.

DON'T use loose fitting wrenches or other tools which may slip and cause injury.

DON'T push on wrenches when loosening or tightening nuts or bolts. Always try to pull the wrench toward you. If the situation calls for pushing the wrench away, push with an open hand to avoid scraped knuckles if the wrench should slip.

DON'T attempt to lift a heavy component alone - get someone to help you.

DON'T rush or take unsafe shortcuts to finish a job.

DON'T allow children or animals in or around the vehicle while you are working on it.

DO wear eye protection when using power tools such as a drill, sander, bench grinder, etc. and when working under a vehicle.

DO keep loose clothing and long hair well out of the way of moving parts.

DO make sure that any hoist used has a safe working load rating adequate for the job.

DO get someone to check on you periodically when working alone on a vehicle.

DO carry out work in a logical sequence and make sure that everything is correctly assembled and tightened.

DO keep chemicals and fluids tightly capped and out of the reach of children and pets.

DO remember that your vehicle's safety affects that of yourself and others. If in doubt on any point, get professional advice.

Steering, suspension and brakes

These systems are essential to driving safety, so make sure you have a qualified shop or individual check your work. Also, compressed suspension springs can cause injury if released suddenly - be sure to use a spring compressor.

Airbags

Airbags are explosive devices that can **CAUSE** injury if they deploy while you're working on the vehicle. Follow the manufacturer's instructions to disable the airbag whenever you're working in the vicinity of airbag components.

Asbestos

Certain friction, insulating, sealing, and other products - such as brake linings, brake bands, clutch linings, torque converters, gaskets, etc. - may contain asbestos or other hazardous friction material. Extreme care must be taken to avoid inhalation of dust from such products, since it is hazardous to health. If in doubt, assume that they do contain asbestos.

Fire

Remember at all times that gasoline is highly flammable. Never smoke or have any kind of open flame around when working on a vehicle. But the risk does not end there. A spark caused by an electrical short circuit, by two metal surfaces contacting each other, or even by static electricity built up in your body under certain conditions, can ignite gasoline vapors, which in a confined space are highly explosive. Do not, under any circumstances, use gasoline for cleaning parts. Use an approved safety solvent.

Always disconnect the battery ground (-) cable at the battery before working on any part of the fuel system or electrical system. Never risk spilling fuel on a hot engine or exhaust component. It is strongly recommended that a fire extinguisher suitable for use on fuel and electrical fires be kept handy in the garage or workshop at all times. Never try to extinguish a fuel or electrical fire with water.

Fumes

Certain fumes are highly toxic and can quickly cause unconsciousness and even death if inhaled to any extent. Gasoline vapor falls into this category, as do the vapors from some cleaning solvents. Any draining or pouring of such volatile fluids should be done in a well ventilated area.

When using cleaning fluids and solvents, read the instructions on the container carefully. Never use materials from unmarked containers.

Never run the engine in an enclosed space, such as a garage. Exhaust fumes contain carbon monoxide, which is extremely poisonous. If you need to run the engine, always do so in the open air, or at least have the rear of the vehicle outside the work area.

The battery

Never create a spark or allow a bare light bulb near a battery. They normally give off a certain amount of hydrogen gas, which is highly explosive.

Always disconnect the battery ground (-) cable at the battery before working on the fuel or electrical systems.

If possible, loosen the filler caps or cover when charging the battery from an external source (this does not apply to sealed or maintenance-free batteries). Do not charge at an excessive rate or the battery may burst.

Take care when adding water to a non maintenance-free battery and when carrying a battery. The electrolyte, even when diluted, is very corrosive and should not be allowed to contact clothing or skin.

Always wear eye protection when cleaning the battery to prevent the caustic deposits from entering your eyes.

Household current

When using an electric power tool, inspection light, etc., which operates on household current, always make sure that the tool is correctly connected to its plug and that, where necessary, it is properly grounded. Do not use such items in damp conditions and, again, do not create a spark or apply excessive heat in the vicinity of fuel or fuel vapor.

Secondary ignition system voltage

A severe electric shock can result from touching certain parts of the ignition system (such as the spark plug wires) when the engine is running or being cranked, particularly if components are damp or the insulation is defective. In the case of an electronic ignition system, the secondary system voltage is much higher and could prove fatal.

Hydrofluoric acid

This extremely corrosive acid is formed when certain types of synthetic rubber, found in some O-rings, oil seals, fuel hoses, etc. are exposed to temperatures above 750-degrees F (400-degrees C). The rubber changes into a charred or sticky substance containing the acid. *Once formed, the acid remains dangerous for years. If it gets onto the skin, it may be necessary to amputate the limb concerned.*

When dealing with a vehicle which has suffered a fire, or with components salvaged from such a vehicle, wear protective gloves and discard them after use.

Troubleshooting

Contents

This section provides an easy reference guide to the more common problems which may occur during the operation of your vehicle. Various symptoms and their possible causes are grouped under headings denoting components or systems, such as Engine, Cooling system, etc. They also refer to the Chapter and/or Section that deals with the problem.

Remember that successful troubleshooting isn't a mysterious art practiced only by professional mechanics. It's simply the result of knowledge combined with an intelligent, systematic approach to a problem. Always use a process of elimination, starting with the simplest solution and working through to the most complex - and never overlook the obvious. Anyone can run the gas tank dry or leave the lights on overnight, so don't assume that you're exempt from such oversights.

Finally, always establish a clear idea why a problem has occurred and take steps to ensure that it doesn't happen again. If the electrical system fails because of a poor connection, check all other connections in the system to make sure they don't fail as well. If a particular fuse continues to blow, find out why - don't just go on replacing fuses. Remember, failure of a small component can often be indicative of potential failure or incorrect functioning of a more important component or system.

Engine and performance

1 Engine will not rotate when attempting to start

1   Battery terminal connections loose or corroded (Chapter 1).
2   Battery discharged or faulty (Chapter 1).
3   Automatic transaxle not completely engaged in Park (Chapter 7B).
4   Broken, loose or disconnected wiring in the starting circuit (Chapters 5 and 12).
5   Starter motor pinion jammed in flywheel ring gear (Chapter 5).
6   Starter solenoid faulty (Chapter 5).
7   Starter motor faulty (Chapter 5).
8   Ignition switch faulty (Chapter 12).
9   Transaxle range switch faulty (Chapter 6).
10   Starter pinion or driveplate teeth worn or broken (Chapter 5).

2 Engine rotates but will not start

1   Fuel tank empty.
2   Battery discharged (engine rotates slowly) (Chapter 5).
3   Battery terminal connections loose or corroded (Chapter 1).
4   Leaking fuel injector(s), fuel pump, pressure regulator, etc. (Chapter 4).
5   Fuel not reaching fuel injection system (Chapter 4).
6   Ignition components damp or damaged (Chapter 5).
7   Worn, faulty or incorrectly gapped spark plugs (Chapter 1).
8   Broken, loose or disconnected wires at the ignition coil(s) or faulty coil(s) (Chapter 5).

3 Engine hard to start when cold

1   Battery discharged or low (Chapter 1).
2   Fuel system malfunctioning (Chapter 4).
3   Emissions or engine control system malfunctioning (Chapter 6).

4 Engine hard to start when hot

1   Air filter clogged (Chapter 1).
2   Fuel not reaching the fuel injection system (Chapter 4).
3   Corroded battery connections, especially ground (Chapter 1).
4   Emissions or engine control system malfunctioning (Chapter 6).

5 Starter motor noisy or excessively rough in engagement

1   Pinion or driveplate gear teeth worn or broken (Chapter 5).
2   Starter motor mounting bolts loose or missing (Chapter 5).

6 Engine starts but stops immediately

1   Loose or faulty electrical connections at coil pack or alternator (Chapter 5).
2   Insufficient fuel reaching the fuel injectors (Chapter 4).
3   Vacuum leak at the gasket between the intake manifold/plenum and throttle body (Chapters 1 and 4).
4   Restricted exhaust system (most likely the catalytic converter) (Chapters 4 and 6).
5   Vehicle anti-theft system.

7 Oil puddle under engine

1   Oil pan gasket and/or oil pan drain bolt seal leaking (Chapters 1 and 2).
2   Oil pressure sending unit leaking (Chapter 2A).
3   Valve cover gaskets leaking (Chapter 2A).
4   Engine oil seals leaking (Chapter 2A).

8 Engine lopes while idling or idles erratically

1   Vacuum leakage (Chapter 4).
2   Leaking EGR valve or plugged PCV system (Chapter 6).
3   Air filter clogged (Chapter 1).
4   Fuel pump not delivering sufficient fuel to the fuel injection system (Chapter 4).
5   Leaking head gasket (Chapter 2A).
6   Camshaft lobes worn (Chapter 2A).

9 Engine misses at idle speed

1   Spark plugs worn or not gapped properly (Chapter 1).
2   Faulty spark plug wires (Chapter 1).
3   Vacuum leaks (Chapter 1 and Chapter 4).
4   Uneven or low compression (Chapter 2C).

10 Engine misses throughout driving speed range

1   Fuel filter clogged and/or impurities in the fuel system (Chapters 1 and 4).
2   Low fuel output at the injector (Chapter 4).
3   Faulty or incorrectly gapped spark plugs (Chapter 1).
4   Leaking spark plug wires (Chapter 1) or faulty ignition coil (see Chapter 5).
5   Faulty emission system components (Chapter 6).
6   Low or uneven cylinder compression pressures (Chapter 2A).
7   Weak or faulty ignition system (Chapter 5).
8   Vacuum leak in fuel injection system, intake manifold or vacuum hoses (Chapter 4).

11 Engine stumbles on acceleration

1   Spark plugs fouled (Chapter 1).
2   Fuel injection system malfunctioning (Chapter 4).
3   Fuel filter clogged (Chapter 1).
4   Intake manifold air leak (Chapter 4).

12 Engine surges while holding accelerator steady

1   Intake air leak (Chapter 4).
2   Fuel pump faulty (Chapter 4).
3   Defective Throttle Position (TP) sensor (Chapter 6).
4   Defective PCM (Chapter 6).

13 Engine stalls

1   Idle speed incorrect (Chapters 1 and 4).
2   Fuel filter clogged and/or water and impurities in the fuel system (Chapters 1 and 4).
3   Ignition components damp or damaged (Chapter 5).
4   Faulty emissions system components (Chapter 6).
5   Faulty or incorrectly gapped spark plugs (Chapter 1).
6   Faulty spark plug wires (Chapter 1).
7   Vacuum leak in the intake manifold or vacuum hoses (Chapter 4).
8   Throttle body dirty (Chapter 4).

14 Engine lacks power

1   Faulty or incorrectly gapped spark plugs (Chapter 1).
2   Restricted exhaust system (most likely the catalytic converter (Chapters 4 and 6).
3   Fuel injection system malfunctioning (Chapter 4).
4   Faulty coil(s) (Chapter 5).
5   Brakes binding (Chapter 1).
6   Automatic transaxle fluid level incorrect (Chapter 1).
7   Fuel filter clogged and/or impurities in the fuel system (Chapter 1).
8   Emission control system not functioning properly (Chapter 6).
9   Low or uneven cylinder compression pressures (Chapter 2A).

15 Engine backfires

1   Emissions system not functioning properly (Chapter 6).
2   Fuel injection system malfunctioning (Chapter 4).
3   Vacuum leak at fuel injectors, intake manifold or vacuum hoses (Chapter 4).
4   Valves sticking (Chapter 2A).

16 Pinging or knocking engine sounds during acceleration or uphill

1   Incorrect grade of fuel.
2   Fuel injection system malfunctioning (Chapter 4).
3   Improper or damaged spark plugs or wires (Chapter 1).
4   Worn or damaged ignition components (Chapter 5).
5   Faulty emissions system (Chapter 6).
6   Vacuum leak (Chapter 4).

17 Engine runs with oil pressure light on

1   Low oil level (Chapter 1).
2   Short in wiring circuit (Chapter 12).
3   Faulty oil pressure sender (Chapter 2A).
4   Oil viscosity too low or oil diluted.
5   Worn engine bearings and/or oil pump (Chapter 2A).

18 Engine diesels (continues to run) after switching off

1   Excessive engine operating temperature (Chapter 3).
2   Excessive carbon deposits on valves and pistons.
3   Leaking fuel injectors (Chapter 4).

Engine electrical system

19 Battery will not hold a charge

1   Alternator drivebelt defective or not adjusted properly (Chapter 1).
2   Battery terminals loose or corroded (Chapter 1).
3   Alternator not charging properly (Chapter 5).
4   Loose, broken or faulty wiring in the charging circuit (Chapter 5).
5   Short in vehicle wiring (Chapter 5 and Chapter 12).
6   Internally defective battery (Chapter 1 and Chapter 5).

20 Voltage warning light fails to go out

1   Faulty alternator or charging circuit (Chapter 5).
2   Alternator drivebelt defective or out of adjustment (Chapter 1).
3   Alternator voltage regulator inoperative (Chapter 5).

21 Voltage warning light fails to come on when key is turned on

1   Warning light bulb defective (Chapter 12).
2   Fault in the printed circuit, dash wiring or bulb holder (Chapter 12).

Fuel system

22 Excessive fuel consumption

1   Dirty or clogged air filter element (Chapter 1).
2   Emissions system not functioning properly (Chapter 6).
3   Fuel injection system malfunctioning (Chapter 4).
4   Low tire pressure or incorrect tire size (Chapter 1).

23 Fuel leakage and/or fuel odor

1   Leak in a fuel feed or vent line (Chapter 4).
2   Tank overfilled.
3   Evaporative emissions control canister defective (Chapters 1 and 6).
4   Fuel injector seals faulty (Chapter 4).

Cooling system

24 Overheating

1   Insufficient coolant in system (Chapter 1).
2   Water pump drivebelt defective or out of adjustment (Chapter 1).
3   Radiator core blocked or grille restricted (Chapter 3).
4   Thermostat faulty (Chapter 3).
5   Electric cooling fan blades broken or cracked (Chapter 3).
6   Expansion tank cap not maintaining proper pressure.

25 Overcooling

Incorrect (opening temperature too low) or faulty thermostat (Chapter 3).

26 External coolant leakage

1   Deteriorated/damaged hoses or loose clamps (Chapters 1 and 3).
2   Water pump seal defective (Chapters 1 and 3).
3   Leakage from radiator core (Chapter 3).
4   Engine drain or water jacket core plugs leaking (Chapter 2A).

27 Internal coolant leakage

1 Leaking cylinder head gasket (Chapter 2A).
2 Cracked cylinder bore or cylinder head (Chapter 2A).

28 Coolant loss

1 Too much coolant in system (Chapter 1).
2 Coolant boiling away because of overheating (Chapter 3).
3 Internal or external leakage (Chapter 3).
4 Faulty expansion tank cap.

29 Poor coolant circulation

1 Inoperative water pump (Chapter 3).
2 Restriction in cooling system (Chapter 1 and Chapter 3).
3 Water pump drivebelt defective or out of adjustment (Chapter 1).
4 Thermostat sticking (Chapter 3).

Automatic transaxle

30 Fluid leakage

1 Automatic transmission fluid is a deep red color. Fluid leaks should not be confused with engine oil, which can easily be blown by airflow to the transaxle.
2 To pinpoint a leak, first remove all built-up dirt and grime from the transaxle housing with degreasing agents and/or steam cleaning. Drive the vehicle at low speeds so air flow will not blow the leak far from its source. Raise the vehicle and determine where the leak is coming from. Common areas of leakage are:
a) *Fluid pan*
b) *Fill plug (Chapter 1)*
c) *Fluid cooler lines (Chapter 7)*
d) *Vehicle Speed Sensor (Chapter 6)*

31 Transaxle fluid brown or has a burned smell

Transaxle overheated. Change fluid (Chapter 1).

32 General shift mechanism problems

1 Chapter 7B deals with checking and adjusting the shift cable on automatic transax

les. Common problems which may be attributed to a poorly adjusted cable are:
a) *Engine starting in gears other than Park or Neutral.*
b) *Indicator on shifter pointing to a gear other than the one actually being used.*
c) *Vehicle moves when in Park.*
2 Refer to Chapter 7B for the shift cable adjustment procedure.

33 Engine will start in gears other than Park or Neutral

Transmission range switch malfunctioning (Chapter 6).

34 Transaxle slips, shifts roughly, is noisy or has no drive in forward or reverse gears

1 There are many probable causes for the above problems, but the home mechanic should be concerned with only one possibility - fluid level. Before taking the vehicle to a repair shop, check the level and condition of the fluid as described in Chapter 1.
2 Correct the fluid level as necessary or change the fluid and filter if needed. If the problem persists, have a professional diagnose the probable cause.

Driveaxles

35 Clicking noise in turns

Worn or damaged outer CV joint. Check for cut or damaged boots (Chapter 1). Repair as necessary (Chapter 8).

36 Knock or clunk when accelerating after coasting

Worn or damaged CV joint. Check for cut or damaged boots (Chapter 1). Repair as necessary (Chapter 8).

37 Shudder or vibration during acceleration

1 Worn or damaged CV joints. Repair or replace as necessary (Chapter 8).
2 Sticking inner joint assembly. Correct or replace as necessary (Chapter 8).

Brakes

38 Vehicle pulls to one side during braking

1 Incorrect tire pressures (Chapter 1).
2 Front end out of alignment (have the front end aligned).
3 Unmatched tires on same axle.
4 Restricted brake lines or hoses (Chapter 9).
5 Sticking caliper or wheel cylinder piston (Chapter 9).
6 Loose suspension parts (Chapter 10).
7 Contaminated brake pad or shoe material (Chapter 9).

39 Noise (grinding or high-pitched squeal) when the brakes are applied

1 Disc brake pads worn out. Replace pads with new ones immediately (Chapter 9).
2 Drum brake shoes worn out. Replace the shoes immediately (Chapter 9).

40 Brake roughness or chatter (pedal pulsates)

1 Excessive brake disc lateral runout or brake drum out-of-round (Chapter 9).
2 Parallelism of disc not within specifications (Chapter 9).
3 Uneven pad wear caused by caliper not sliding due to improper clearance or dirt (Chapter 9).
4 Defective brake disc (Chapter 9).

41 Excessive pedal effort required to stop vehicle

1 Malfunctioning power brake booster (Chapter 9).
2 Partial system failure (Chapter 9).
3 Excessively worn pads (Chapter 9).
4 One or more caliper or wheel cylinder pistons seized or sticking (Chapter 9).
5 Brake pads contaminated with oil or grease (Chapter 9).
6 New pads or shoes installed and not yet seated. It will take a while for the new material to seat.

42 Excessive brake pedal travel

1 Partial brake system failure (Chapter 9).
2 Insufficient fluid in master cylinder (Chapter 1 and Chapter 9).
3 Air trapped in system (Chapter 9).
4 Faulty master cylinder (Chapter 9).

43   Dragging brakes

1   Master cylinder pistons not returning correctly (Chapter 9).
2   Restricted brake lines or hoses (Chapter 1 and Chapter 9).
3   Incorrect parking brake adjustment (Chapter 9).
4   Defective brake calipers (Chapter 9).

44   Grabbing or uneven braking action

1   Malfunction of proportioning valve (Chapter 9).
2   Malfunction of power brake booster unit (Chapter 9).
3   Binding brake pedal mechanism (Chapter 9).
4   Contaminated brake linings (Chapter 9).

45   Brake pedal feels spongy when depressed

1   Air in hydraulic lines (Chapter 9).
2   Master cylinder mounting bolts loose (Chapter 9).
3   Master cylinder defective (Chapter 9).

46   Brake pedal travels to the floor with little resistance

Little or no fluid in the master cylinder reservoir caused by leaking caliper, or loose, damaged or disconnected brake lines (Chapter 9).

47   Parking brake does not hold

Parking brake cables improperly adjusted (Chapter 9).

Suspension and steering systems

48   Vehicle pulls to one side

1   Mismatched or uneven tires (Chapter 10).
2   Broken or sagging springs (Chapter 10).
3   Wheel alignment incorrect (Chapter 10).
4   Front brakes dragging (Chapter 9).

49   Abnormal or excessive tire wear

1   Front wheel alignment incorrect (Chapter 10).
2   Sagging or broken springs (Chapter 10).
3   Tire out-of-balance (Chapter 10).
4   Worn strut or shock absorber (Chapter 10).
5   Overloaded vehicle.
6   Tires not rotated regularly.

50   Wheel makes a "thumping" noise

1   Blister or bump on tire (Chapter 1).
2   Improper strut or shock absorber action (Chapter 10).

51   Shimmy, shake or vibration

1   Tire or wheel out-of-balance or out-of-round (Chapter 10).
2   Loose or worn wheel bearings (Chapter 10).
3   Worn tie-rod ends (Chapter 10).
4   Worn balljoints (Chapter 10).
5   Excessive wheel runout (Chapter 10).
6   Blister or bump on tire (Chapter 1).

52   Hard steering

1   Lack of lubrication at balljoints, tie-rod ends and steering gear assembly (Chapter 10).
2   Front wheel alignment incorrect (Chapter 10).
3   Low tire pressure (Chapter 1).

53   Steering wheel does not return to center position correctly

1   Lack of lubrication at balljoints and tie-rod ends (Chapter 1 and Chapter 10).
2   Binding in steering column (Chapter 10).
3   Defective rack-and-pinion assembly (Chapter 10).
4   Front wheel alignment problem (Chapter 10).

54   Abnormal noise at the front end

1   Lack of lubrication at balljoints and tie-rod ends (Chapter 1).
2   Loose upper strut mount (Chapter 10).
3   Worn tie-rod ends (Chapter 10).
4   Loose stabilizer bar (Chapter 10).
5   Loose wheel lug nuts (Chapter 1).

6   Loose suspension bolts (Chapter 10).

55   Wander or poor steering stability

1   Mismatched or uneven tires (Chapter 10).
2   Lack of lubrication at balljoints or tie-rod ends (Chapter 1 and Chapter 10).
3   Worn struts or shock absorbers (Chapter 10).
4   Loose stabilizer bar (Chapter 10).
5   Broken or sagging springs (Chapter 10).
6   Front wheel alignment incorrect.
7   Loose steering gear mounting fasteners (Chapter 10).

56   Erratic steering when braking

1   Wheel bearings worn (Chapter 10).
2   Broken or sagging springs (Chapter 10).
3   Leaking caliper (Chapter 9).
4   Warped brake discs (Chapter 9).
5   Worn steering gear clamp bushing (Chapter 10).
6   Wheel alignment incorrect.

57   Excessive pitching and/or rolling around corners or during braking

1   Loose stabilizer bar (Chapter 10).
2   Worn struts/shock absorbers or mounts (Chapter 10).
3   Broken or sagging springs (Chapter 10).
4   Overloaded vehicle.

58   Suspension bottoms

1   Overloaded vehicle.
2   Worn struts or shock absorbers (Chapter 10).
3   Incorrect, broken or sagging springs (Chapter 10).

59   Cupped tires

1   Front wheel alignment incorrect (Chapter 10).
2   Worn struts or shock absorbers (Chapter 10).
3   Wheel bearings worn (Chapter 10).
4   Excessive tire or wheel runout (Chapter 10).
5   Worn balljoints (Chapter 10).

60 Excessive tire wear on outside edge

1   Inflation pressures incorrect (Chapter 1).
2   Excessive speed in turns.
3   Wheel alignment incorrect (excessive toe-in or positive camber). Have professionally aligned.
4   Suspension arm bent or twisted (Chapter 10).

61 Excessive tire wear on inside edge

1   Inflation pressures incorrect (Chapter 1).

2   Wheel alignment incorrect (toe-out or excessive negative camber). Have professionally aligned.
3   Loose or damaged steering components (Chapter 10).

62 Tire tread worn in one place

1   Tires out-of-balance.
2   Damaged or buckled wheel. Inspect and replace if necessary.
3   Defective tire (Chapter 1).

63 Excessive play or looseness in steering system

1   Wheel bearings worn (Chapter 10).
2   Tie-rod end loose or worn (Chapter 10).
3   Steering gear loose (Chapter 10).

64 Rattling or clicking noise in steering gear

1   Steering gear mounting bolts loose (Chapter 10).
2   Steering gear defective (Chapter 10).

Notes

Chapter 1
Tune-up and routine maintenance

Contents

Specifications

Recommended lubricants and fluids

Note: *Listed here are manufacturer recommendations at the time this manual was written. Manufacturers occasionally upgrade their fluid and lubricant specifications, so check with your local auto parts store for current recommendations.*

Engine oil	
Type	API "certified for gasoline engines"
Viscosity	SAE 5W-30
Fuel	Unleaded gasoline, 87 Octane minimum
Automatic transaxle fluid	
2005 and earlier models	DEXRON ® III
2006 and later models	DEXRON ® VI
Manual transaxle lubricant	GM part no. 21018899 or equivalent
Brake fluid	DOT 3 brake fluid
Engine coolant	50/50 mixture of DEX-COOL® and distilled water
Power steering system (if equipped)	GM power steering fluid

Capacities*

Engine oil (including filter)
    Four-cylinder engine .... 5.0 quarts
    V6 engine
        3.5L and 3.9L V6 engines .... 4.0 quarts
        3.6L V6 engine .... 5.5 quarts
Coolant
    Four-cylinder engine
        2005 and earlier models .... 6.9 quarts
        2006 and later models .... 7.5 quarts
    V6 engine
        2005 and earlier models .... 10.1 quarts
        2006 and later models .... 9.8 quarts
Automatic transaxle
    4T40-E/4T45-E .... 6.9 quarts
    4T65-E .... 7.4 quarts
    GT70 .... 9.5 quarts
Manual transaxle .... 3.1 quarts

Note: *This is an initial-fill specification for routine fluid replacement. Be sure to follow the fluid check procedure in Section 24, as additional fluid may be required.*
*All capacities approximate. Add as necessary to bring up to appropriate level.

Ignition system

Spark plug type and gap
    Type
        Four-cylinder engine
            2006 and earlier models.. AC41-981 or equivalent
            2007 and later models.. AC41-103 or equivalent
        V6 engines
            3.5L V6
                2006 and earlier models .... AC41-101 or equivalent
                2007 and later models .... AC41-100 or equivalent
            3.6L V6 .... AC41-990 or equivalent
            3.9L V6 .... AC41-100 or equivalent
    Gap
        Four-cylinder engine
            2006 and earlier models .... 0.042 inch
            2007 and later models... 0.040 inch
        V6 engines
            3.5L V6
                 2006 and earlier models .... 0.060 inch
                2007 and later models .... 0.040 inch
            3.6L V6 .... 0.043 inch
            3.9L V6
                2006 G6 models.. 0.043 inch
                2007 and later models .... 0.040 inch
Engine firing order
    Four-cylinder engine 1-3-4-2
    V6 engine 1-2-3-4-5-6

**Cylinder locations -
four-cylinder engine**

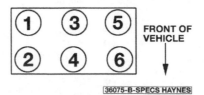

Cylinder locations - 3.6L V6 engines

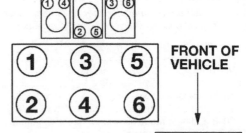

**Cylinder locations and coil terminal identification
diagram - 3.5L and 3.9L V6 engines**

Brakes

Disc brake pad lining thickness (minimum)..	1/8 inch
Drum brake shoe lining thickness (minimum)	1/16 inch

Torque specifications   **Ft-lbs (unless otherwise indicated)**

Note: *One foot-pound (ft-lb) of torque is equivalent to 12 inch-pounds (in-lbs) of torque. Torque values below approximately 15 ft-lbs are expressed in inch-pounds, since most foot-pound torque wrenches are not accurate at these smaller values.*

Engine oil drain plug
Four-cylinder engine	18
V6 engines	19

Automatic transaxle
Fluid pan bolts...	108 in-lbs
Fluid level check plug..	108 in-lbs

Spark plugs
Four-cylinder engine	15
V6 engines	132 in-lbs

Drivebelt tensioner bolt
Four-cylinder engine	33
V6 engine	37

Water pump drain bolt	15
Wheel lug nuts...	100

Engine compartment layout (four-cylinder engine)

1	Coolant expansion tank	5	Windshield washer fluid reservoir	9	Engine oil filler cap
2	Brake fluid reservoir	6	Upper radiator hose	10	Lower radiator hose
3	Battery	7	Engine oil dipstick	11	Air filter housing
4	Underhood fuse/relay block	8	Oil filter cover		

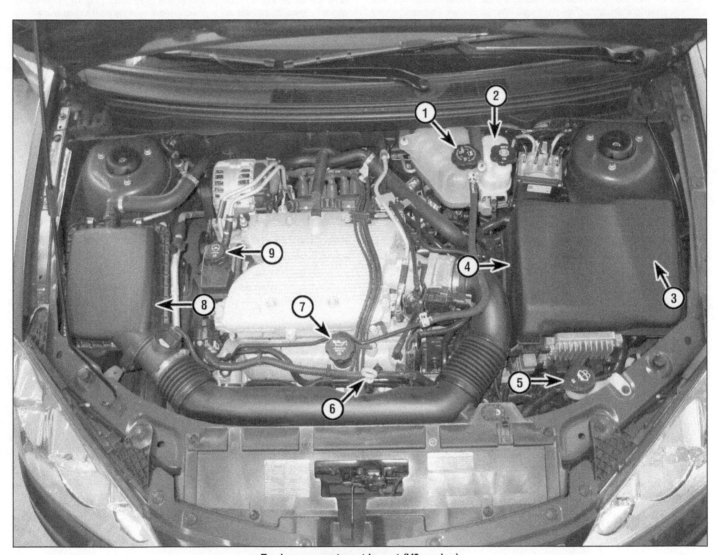

Engine compartment layout (V6 engine)

1	Coolant expansion tank	4	Underhood fuse/relay block	7	Engine oil filler cap
2	Brake fluid reservoir	5	Windshield washer fluid reservoir	8	Air filter housing
3	Battery	6	Engine oil dipstick	9	Power steering fluid reservoir

1 Maintenance schedule

The maintenance intervals in this manual are provided with the assumption that you, not the dealer, will be doing the work. These are the minimum maintenance intervals recommended by the factory for vehicles that are driven daily. If you wish to keep your vehicle in peak condition at all times, you may wish to perform some of these procedures even more often. Because frequent maintenance enhances the efficiency, performance and resale value of your car, we encourage you to do so. If you drive in dusty areas, tow a trailer, idle or drive at low speeds for extended periods or drive for short distances (less than four miles) in below freezing temperatures, shorter intervals are also recommended.

When your vehicle is new, it should be serviced by a factory authorized dealer service department to protect the factory warranty. In many cases, the initial maintenance check is done at no cost to the owner.

Every 250 miles or weekly, whichever comes first

Check the engine oil level (Section 4)
Check the engine coolant level (Section 4)
Check the brake fluid level (Section 4)
Check the windshield washer fluid level (Section 4)
Check the power steering fluid level (Section 4)
Check the tires and tire pressures (Section 5)

Every 3000 miles or 3 months, whichever comes first

All items listed above plus:
Change the engine oil and oil filter (Section 6)

Every 6000 miles or 6 months, whichever comes first

All items listed above plus:
Inspect (and replace, if necessary) the windshield wiper-blades (Section 7)
Check and service the battery (Section 8)
Check the cooling system (Section 9)
Rotate the tires (Section 10)
Check the seat belts (Section 11)
Inspect the brake system (Section 12)

Every 15,000 miles or 12 months, whichever comes first

All items listed above plus:
Check all underhood hoses (Section 13)
Inspect the suspension and steering components (Section 14)
Inspect the driveaxle boots (Section 15)
Check the exhaust system (Section 16)
Check the fuel system (Section 17)
Check the engine drivebelt (Section 18)

Every 30,000 miles or 24 months, whichever comes first

All items listed above plus:
Check (and replace, if necessary) the air filter (Section 19)*
Change the brake fluid (Section 20)
Replace the spark plugs (non-platinum type spark plugs) (Section 21)
Inspect the spark plug wires (Section 22)

Every 60,000 miles or 48 months, whichever comes first

All items listed above plus:
Replace the spark plugs (Platinum type spark plugs) (Section 21)

Every 100,000 miles

Replace the automatic transaxle fluid (Section 24)**
Service the cooling system (drain, flush and refill) (Section 23)

This item is affected by "severe" operating conditions as described below. If your vehicle is operated under "severe" conditions, perform all maintenance indicated with an asterisk (*) at 3000 mile/3 month intervals. Severe conditions are indicated if you mainly operate your vehicle under one or more of the following conditions:
Operating in dusty areas
Towing a trailer
Idling for extended periods and/or low speed operation

** If operated under one or more of the following conditions, change the automatic transaxle fluid lubricant every 50,000 miles:*
In heavy city traffic where the outside temperature regularly reaches 90-degrees F (32-degrees C) or higher
In hilly or mountainous terrain
Frequent towing of a trailer

2 Introduction

1   This Chapter is designed to help the home mechanic maintain the Aura, Malibu and G6 with the goals of maximum performance, economy, safety and reliability in mind.

2   Included is a master maintenance schedule, followed by procedures dealing specifically with each item on the schedule. Visual checks, adjustments, component replacement and other helpful items are included. Refer to the accompanying illustrations of the engine compartment and the underside of the vehicle for the locations of various components.

3   Servicing the vehicle, in accordance with the mileage/time maintenance schedule and the step-by-step procedures will result in a planned maintenance program that should produce a long and reliable service life. Keep in mind that it is a comprehensive plan, so maintaining some items but not others at the specified intervals will not produce the same results.

4   As you service the vehicle, you will discover that many of the procedures can - and should - be grouped together because of the nature of the particular procedure you're performing or because of the close proximity of two otherwise unrelated components to one another.

5   For example, if the vehicle is raised for chassis lubrication, you should inspect the exhaust, suspension, steering and fuel systems while you're under the vehicle. When you're rotating the tires, it makes good sense to check the brakes since the wheels are already removed. Finally, let's suppose you have to borrow or rent a torque wrench. Even if you only need it to tighten the spark plugs, you might as well check the torque of as many critical fasteners as time allows.

6   The first step in this maintenance program is to prepare yourself before the actual work begins. Read through all the procedures you're planning to do, then gather up all the parts and tools needed. If it looks like you might run into problems during a particular job, seek advice from a mechanic or an experienced do-it-yourselfer.

Owner's Manual and VECI label information

7   Your vehicle owner's manual was written for your year and model and contains very specific information on component locations, specifications, fuse ratings, part numbers, etc. The owner's manual is an important resource for the do-it-yourselfer to have; if one was not supplied with your vehicle, it can generally be ordered from a dealer parts department.

8   Among other important information, the Vehicle Emissions Control Information (VECI) label contains specifications and procedures for applicable tune-up adjustments and, in some instances, spark plugs (see Chapter 6 for more information on the VECI label). The information on this label is the exact maintenance data recommended by the manufac-

turer. This data often varies by intended operating altitude, local emissions regulations, month of manufacture, etc.

9   This Chapter contains procedural details, safety information and more ambitious maintenance intervals than you might find in manufacturer's literature. However, you may also find procedures or specifications in your owner's manual or VECI label that differ with what's printed here. In these cases, the owner's manual or VECI label can be considered correct, since it is specific to your particular vehicle.

3 Tune-up general information

1   The term tune-up is used in this manual to represent a combination of individual operations rather than one specific procedure.

2   If, from the time the vehicle is new, the routine maintenance schedule is followed closely and frequent checks are made of fluid levels and high wear items, as suggested throughout this manual, the engine will be kept in relatively good running condition and the need for additional work will be minimized.

3   More likely than not, however, there will be times when the engine is running poorly due to lack of regular maintenance. This is even more likely if a used vehicle, which has not received regular and frequent maintenance checks, is purchased. In such cases, an engine tune-up will be needed outside of the regular routine maintenance intervals.

4   The first step in any tune-up or diagnostic procedure to help correct a poor running engine is a cylinder compression check. A compression check (see Chapter 2D) will help determine the condition of internal engine components and should be used as a guide for tune-up and repair procedures. If, for instance, a compression check indicates serious internal engine wear, a conventional tune-up will not improve the performance of the engine and would be a waste of time and money. Because of its importance, the com-

pression check should be done by someone with the right equipment and the knowledge to use it properly.

5   The following procedures are those most often needed to bring a generally poor running engine back into a proper state of tune.

Minor tune-up

Check all engine related fluids (Section 4)
Clean, inspect and test the battery (Section 8)
Check the cooling system (Section 9)
Check all underhood hoses (Section 13)
Check the fuel system (Section 17)
Check the air filter (Section 19)

Major tune-up

All items listed under Minor tune-up, plus . . .
Check the drivebelt (Section 18)
Replace the air filter (Section 19)
Replace the spark plugs (Section 21)

4 Fluid level checks (every 250 miles or weekly)

1   Fluids are an essential part of the lubrication, cooling, brake and windshield washer systems. Because the fluids gradually become depleted and/or contaminated during normal operation of the vehicle, they must be periodically replenished. See Recommended lubricants and fluids at the beginning of this Chapter before adding fluid to any of the following components.

Note: *The vehicle must be on level ground when fluid levels are checked.*

Engine oil

2   The oil level is checked with a dipstick, which is attached to the engine block (see illustrations). The dipstick extends through a metal tube down into the oil pan.

3   The oil level should be checked before the vehicle has been driven, or about 5 minutes after the engine has been shut off. If the oil is checked immediately after driving

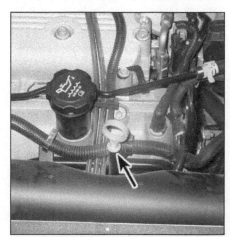

4.2a Engine oil dipstick location - V6 engines

4.2b Engine oil dipstick location - four-cylinder engine

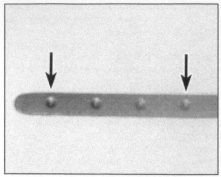

4.4 The oil level should be in the cross-hatched area - if it's below the lower dimple, add enough oil to bring it up to or near the upper dimple

4.6 The oil filler cap is located on the valve cover - always make sure the area around the opening is clean before unscrewing the cap to prevent dirt from contaminating the engine (four-cylinder engine shown)

4.8 The cooling system expansion tank is located at the rear of the engine compartment

the vehicle, some of the oil will remain in the upper part of the engine, resulting in an inaccurate reading on the dipstick.

4   Pull the dipstick out of the tube and wipe all the oil from the end with a clean rag or paper towel. Insert the clean dipstick all the way back into the tube and pull it out again. Note the oil at the end of the dipstick. At its highest point, the level should be between the MIN and MAX marks on the dipstick (see illustration).

5   It takes one quart of oil to raise the level from the MIN mark to the MAX mark on the dipstick. Do not allow the level to drop below the MIN mark or oil starvation may cause engine damage. Conversely, overfilling the engine (adding oil above the MAX mark) may cause oil fouled spark plugs, oil leaks or oil seal failures. Maintaining the oil level above the MAX mark can cause excessive oil consumption.

6   To add oil, remove the filler cap from the valve cover (see illustration). After adding oil, wait a few minutes to allow the level to stabilize, then pull out the dipstick and check the level again. Add more oil if required. Install the filler cap and tighten it by hand only.

7   Checking the oil level is an important preventive maintenance step. A consistently low oil level indicates oil leakage through damaged seals, defective gaskets or past worn rings or valve guides. If the oil looks milky in color or has water droplets in it, the cylinder head gasket(s) may be blown or the head(s) or block may be cracked. The engine should be checked immediately. The condition of the

oil should also be checked. Whenever you check the oil level, slide your thumb and index finger up the dipstick before wiping off the oil. If you see small dirt or metal particles clinging to the dipstick, the oil should be changed (see Section 6).

Engine coolant

Warning: *Do not allow antifreeze to come in contact with your skin or painted surfaces of the vehicle. Flush contaminated areas immediately with plenty of water. Don't store new coolant or leave old coolant lying around where it's accessible to children or pets - they're attracted by its sweet smell. Ingestion of even a small amount of coolant can be fatal! Wipe up garage floor and drip pan spills immediately. Keep antifreeze containers covered and repair cooling system leaks as soon as they're noticed.*

8   All vehicles covered by this manual are equipped with a pressurized coolant recovery system. A plastic expansion tank located at the rear of the engine compartment is connected by a hose to the radiator (see illustration). As the engine heats up during operation, the expanding coolant fills the tank.

9   The coolant level in the tank should be checked regularly.

Warning: *Do not remove the expansion tank cap to check the coolant level when the engine is warm!*

The level in the tank varies with the temperature of the engine. When the engine is cold,

the coolant level should be at the COLD mark on the reservoir (see illustration). If it isn't, remove the cap from the tank and add a 50/50 mixture of ethylene glycol based antifreeze and water.

10   Drive the vehicle, let the engine cool completely then recheck the coolant level. Don't use rust inhibitors or additives. If only a small amount of coolant is required to bring the system up to the proper level, water can be used. However, repeated additions of water will dilute the antifreeze and water solution. In order to maintain the proper ratio of antifreeze and water, always top up the coolant level with the correct mixture. An empty plastic milk jug or bleach bottle makes an excellent container for mixing coolant.

11   If the coolant level drops consistently, there may be a leak in the system. Inspect the radiator, hoses, filler cap, drain plugs and water pump (see Section 9). If no leaks are noted, have the expansion tank cap pressure tested by a service station.

12   If you have to remove the expansion tank cap wait until the engine has cooled completely, then wrap a thick cloth around the cap and unscrew it slowly, stopping if you hear a hissing noise. If coolant or steam escapes, let the engine cool down longer, then remove the cap.

13   Check the condition of the coolant as well. It should be relatively clear. If it's brown or rust colored, the system should be drained, flushed and refilled. Even if the coolant appears to be normal, the corrosion inhibitors wear out, so it must be replaced at the specified intervals.

Brake fluid

14   The brake master cylinder is mounted on the front of the power booster unit in the engine compartment.

15   To check the fluid level of brake system, simply look at the MAX and MIN marks on the brake fluid reservoir (see illustration).

16   If the level is low, wipe the top of the reservoir cover with a clean rag to prevent contamination of the brake system before lifting the cover.

17   Add only the specified brake fluid to the reservoir (refer to *Recommended lubricants and fluids* at the front of this Chapter or to your owner's manual). Mixing different types

4.9 When the engine is cold, the engine coolant level should be at the COLD mark (expansion tank seam)

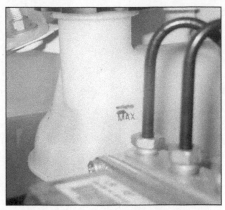

4.15 The brake fluid level should be kept between the MIN and MAX marks on the translucent plastic reservoir; the same reservoir contains the clutch fluid and is connected to the clutch master cylinder by a hose

of brake fluid can damage the system. Fill the brake master cylinder reservoir only to the MAX line.

Warning: *Use caution when filling the reservoir - brake fluid can harm your eyes and damage painted surfaces. Do not use brake fluid that is more than one year old or has been left open. Brake fluid absorbs moisture from the air. Excess moisture can cause a dangerous loss of braking.*

18   While the reservoir cap is removed, inspect the master cylinder reservoir for contamination. If deposits, dirt particles or water droplets are present, the system should be drained and refilled.

19   After filling the reservoir to the proper level, make sure the cap is properly seated to prevent fluid leakage.

20   The fluid in the brake master cylinder will drop slightly as the brake pads at each wheel wear down during normal operation. If the master cylinder requires repeated replenishing to keep it at the proper level, this is an indication of leakage in the brake system, which should be corrected immediately. If the brake system shows an indication of leakage check all brake lines and connections, along with the calipers, wheel cylinders and booster (see Section 12 for more information).

21   If, upon checking the brake master cylinder fluid level, you discover the reservoir empty or nearly empty, the system should be checked, repaired and bled (see Chapter 9).

Windshield washer fluid

22   Fluid for the windshield washer system is stored in a plastic reservoir located at the left front of the engine compartment (see illustration).

23   In milder climates, plain water can be used in the reservoir, but it should be kept no more than 2/3 full to allow for expansion if the water freezes. In colder climates, use windshield washer system antifreeze, available at any auto parts store, to lower the freezing point of the fluid. Mix the antifreeze with water

4.22 The windshield washer fluid reservoir is located in the left front corner of the engine compartment

in accordance with the manufacturer's directions on the container.

Caution: *Do not use cooling system antifreeze - it will damage the vehicle's paint.*

Power steering fluid

Note: *Only 2006 3.9L V6 models and all 2007 V6 models are equipped with hydraulically assisted power steering.*

24   Check the power steering fluid level periodically to avoid steering system problems, such as damage to the pump.

Caution: *DO NOT hold the steering wheel against either stop (extreme left or right turn) for more than five seconds. If you do, the power steering pump could be damaged.*

25   The fluid reservoir for the power steering system is mounted to the power steering pump near the alternator engine.

26   The fluid level can be checked by removing the dipstick. With the engine cool, the fluid level should be kept between the MIN and MAX marks on the dipstick (see illustration).

27   Add small amounts of fluid until the level is correct.

Caution: *Do not overfill the reservoir. If too much fluid is added, remove the excess with a*

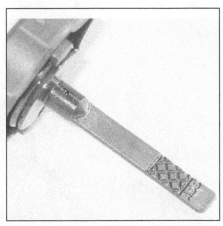

4.26 The power steering fluid level should be kept in the cross-hatched area of the dipstick

clean syringe or suction pump.

28   If the reservoir requires frequent fluid additions, all power steering hoses, hose connections, the power steering pump and the steering gear assembly should be carefully checked for leaks.

Manual transaxle lubricant

Note: *It isn't necessary to check this lubricant weekly; every 30,000 miles or 24 months will be adequate.*

29   The manual transaxle does not have a dipstick. To check the fluid level, raise the vehicle and support it securely on jackstands. On the side of the transaxle housing near the left driveaxle you will see a check plug (the upper one) and a drain plug (the lower one). Remove the check plug. If the lubricant level is correct, it should be up to the lower edge of the hole.

30   If the transaxle needs more lubricant (if the level is not up to the hole), open the fill plug at the top of the transaxle and add more (be sure to use the proper type, listed in this Chapter's Specifications). Stop filling the transaxle when the lubricant begins to run out the check hole on the side of the transaxle.

31   Install the plugs and tighten them securely. Drive the vehicle a short distance, then check for leaks.

5   Tire and tire pressure checks (every 250 miles or weekly)

1   Periodic inspection of the tires may spare you the inconvenience of being stranded with a flat tire. It can also provide you with vital information regarding possible problems in the steering and suspension systems before major damage occurs.

2   The original tires on this vehicle are equipped with 1/2-inch wide bands that will appear when tread depth reaches 1/16-inch, at which point they can be considered worn out. Tread wear can be monitored with a simple, inexpensive device known as a tread depth indicator (see illustration).

5.2 A tire tread depth indicator should be used to monitor tire wear - they are available at auto parts stores and service stations and cost very little

UNDERINFLATION

CUPPING

Cupping may be caused by:

• Underinflation and/or mechanical irregularities such as out-of-balance condition of wheel and/or tire, and bent or damaged wheel.
• Loose or worn steering tie-rod or steering idler arm.
• Loose, damaged or worn front suspension parts.

OVERINFLATION

INCORRECT TOE-IN OR EXTREME CAMBER

FEATHERING DUE TO MISALIGNMENT

5.3 This chart will help you determine the condition of your tires, the probable cause(s) of abnormal wear and the corrective action necessary

3   Note any abnormal tread wear (see illustration). Tread pattern irregularities such as cupping, flat spots and more wear on one side than the other are indications of front end alignment and/or balance problems. If any of these conditions are noted, take the vehicle to a tire shop or service station to correct the problem.
4   Look closely for cuts, punctures and embedded nails or tacks. Sometimes a tire will hold air pressure for a short time or leak down very slowly after a nail has embedded itself in the tread. If a slow leak persists, check the valve stem core to make sure it is tight (see illustration). Examine the tread for an object that may have embedded itself in the tire or for a "plug" that may have begun to leak (radial tire punctures are repaired with a plug that is installed in a puncture). If a puncture is suspected, it can be easily verified by spraying a solution of soapy water onto the puncture area (see illustration). The soapy solution will bubble if there is a leak. Unless the puncture is unusually large, a tire shop or service station can usually repair the tire.
5   Carefully inspect the inner sidewall of each tire for evidence of brake fluid leakage. If

5.4a If a tire loses air on a steady basis, check the valve core first to make sure it's snug (special inexpensive wrenches are commonly available at auto parts stores)

5.4b If the valve core is tight, raise the corner of the vehicle with the low tire and spray a soapy water solution onto the tread as the tire is turned slowly - slow leaks will cause small bubbles to appear

5.8 To extend the life of your tires, check the air pressure at least once a week with an accurate gauge (don't forget the spare!)

6.7 Use a proper size box-end wrench or socket to remove the oil drain plug and avoid rounding it off

6.11a Unscrew the cap to access the oil filter (four-cylinder models)

you see any, inspect the brakes immediately.

6   Correct air pressure adds miles to the life span of the tires, improves mileage and enhances overall ride quality. Tire pressure cannot be accurately estimated by looking at a tire, especially if it's a radial. A tire pressure gauge is essential. Keep an accurate gauge in the glove compartment. The pressure gauges attached to the nozzles of air hoses at gas stations are often inaccurate.

7   Always check tire pressure when the tires are cold. Cold, in this case, means the vehicle has not been driven over a mile in the three hours preceding a tire pressure check. A pressure rise of four to eight pounds is not uncommon once the tires are warm.

8   Unscrew the valve cap protruding from the wheel or hubcap and push the gauge firmly onto the valve stem (see illustration). Note the reading on the gauge and compare the figure to the recommended tire pressure shown on the tire placard on the driver's side door. Be sure to reinstall the valve cap to keep dirt and moisture out of the valve stem mechanism. Check all four tires and, if necessary, add enough air to bring them up to the recommended pressure.

9   Don't forget to keep the spare tire inflated to the specified pressure (refer to the pressure molded into the tire sidewall).

6 Engine oil and filter change (every 3000 miles or 3 months)

Note: *These vehicles are equipped with an oil life indicator system that illuminates a light or message on the instrument panel when the system deems it necessary to change the oil. A number of factors are taken into consideration to determine when the oil should be considered "worn out." Generally, this system will allow the vehicle to accumulate more miles between oil changes than the tra-*

ditional 3000 mile interval, but we believe that frequent oil changes are "cheap insurance" and will prolong engine life. If you do decide not to change your oil every 3000 miles and rely on the oil life indicator instead, make sure you don't exceed 10,000 miles before the oil is changed, regardless of what the oil life indicator shows.

1   Frequent oil changes are the most important preventive maintenance procedures that can be done by the home mechanic. As engine oil ages, it becomes diluted and contaminated, which leads to premature engine wear.

2   Make sure that you have all the necessary tools before you begin this procedure. You should also have plenty of rags or newspapers handy for mopping up oil spills.

3   Access to the oil drain plug and filter will be improved if the vehicle can be lifted on a hoist, driven onto ramps or supported by jackstands. Warning: Do not work under a vehicle supported only by a jack - always use jackstands!

4   If you haven't changed the oil on this vehicle before, get under it and locate the oil drain plug and the oil filter. The exhaust components will be warm as you work, so note how they are routed to avoid touching them when you are under the vehicle.

5   Start the engine and allow it to reach normal operating temperature - oil and sludge will flow out more easily when warm. If new oil, a filter or tools are needed, use the vehicle to go get them and warm up the engine/oil at the same time. Park on a level surface and shut off the engine when it's warmed up. Remove the oil filler cap from the valve cover.

6   Raise the vehicle and support it on jackstands. Make sure it is safely supported!

7   Being careful not to touch the hot exhaust components, position a drain pan under the plug in the bottom of the engine, then remove the plug (see illustration). It's a good idea to

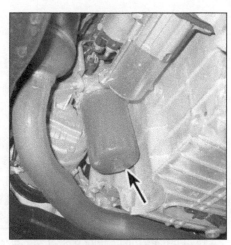

6.11b The oil filter is located at the front of the engine and is accessed from under the vehicle (V6 models)

wear a rubber glove while unscrewing the plug the final few turns to avoid being scalded by hot oil.

8   It may be necessary to move the drain pan slightly as oil flow slows to a trickle. Inspect the old oil for the presence of metal particles.

9   After all the oil has drained, wipe off the drain plug with a clean rag. Any small metal particles clinging to the plug would immediately contaminate the new oil.

10   Clean the area around the drain plug opening, reinstall the plug and tighten it securely, but don't strip the threads.

11   Move the drain pan into position under the oil filter. On four-cylinder engines, the canister-type oil filter is located at the front left side of the engine (see illustration). On V6 models, the filter is located at the front of the engine and is accessed from under the vehicle (see illustration).

6.12a Remove the filter cartridge . . .

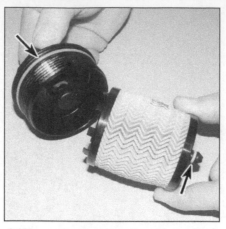

6.12b . . . then separate the element from the cap

6.12c The oil filter is usually on very tight and will require a special wrench for removal - DO NOT use the wrench to tighten the new filter!

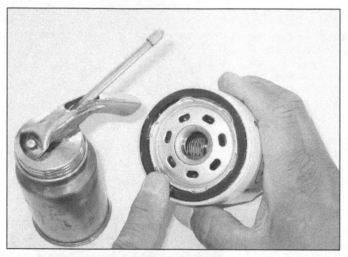

6.15 Lubricate the oil filter gasket with clean engine oil before installing the filter on the engine

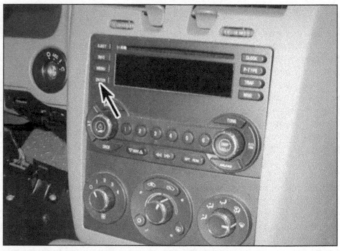

6.20 With "Oil Life Reset" displayed on the Driver's Information center, press and hold the enter button for at least one second

12   On four-cylinder models, unscrew the oil filter cap and withdraw it, together with the element (see illustrations). On V6 models, loosen the oil filter (see illustration) by turning it counterclockwise with the filter wrench. Any standard filter wrench should work. Once the filter is loose, use your hands to unscrew it from the block.
Warning: *The exhaust pipes may still be hot, so be careful.*
13   On four-cylinder models, use a clean rag to remove all oil, dirt and sludge from the oil filter housing and cap. On V6 models, wipe off the mounting surface on the block. If a residue of old oil is allowed to remain, it will smoke when the block is heated up. It will also prevent the new filter from seating properly. Also make sure that the none of the old gasket remains stuck to the mounting surface. It can be removed with a scraper if necessary.
14   On four-cylinder models, install a new O-ring seal in the groove on the retaining cap,

then install the new element in the cap and insert them both in the filter housing. Screw on the cap and tighten it securely.
15   On V6 models, compare the old filter with the new one to make sure they are the same type. Smear some engine oil on the rubber gasket of the new filter and screw it into place (see illustration). Because over-tightening the filter will damage the gasket, do not use a filter wrench to tighten the filter. Tighten it by hand until the gasket contacts the seating surface. Then seat the filter by giving it an additional 3/4-turn.
16   Remove all tools and materials from under the vehicle, being careful not to spill the oil in the drain pan, then lower the vehicle.
17   Add new oil to the engine through the oil filler cap. Use a funnel to prevent oil from spilling onto the top of the engine. Pour four quarts of fresh oil into the engine. Wait a few minutes to allow the oil to drain into the pan, then check the level on the dipstick (see Section 4 if necessary). If the oil level is in the OK

range, install the filler cap.
18   Start the engine and run it for about a minute. While the engine is running, look under the vehicle and check for leaks at the oil pan drain plug and around the oil filter. If either one is leaking, stop the engine and tighten the plug or filter slightly.
19   Wait a few minutes, then recheck the level on the dipstick. Add oil as necessary to bring the level into the OK range.
20   Be sure to reset the Change Engine Oil light. On all models, first turn the ignition to the Run position with the engine Off. On 2007 and earlier models, display "Oil Life Reset" on the Driver's Information Center, then hold down the enter button for at least one second (see illustration). On 2008 and later models, press the accelerator pedal to the floor and release it three times within 5 seconds. Several beeps sound. If the CHANGE/OIL message comes back on when you start the engine, the engine oil life system has not been reset. Repeat the procedure.

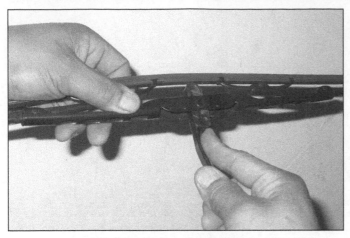

7.4a Depress the release lever . . .

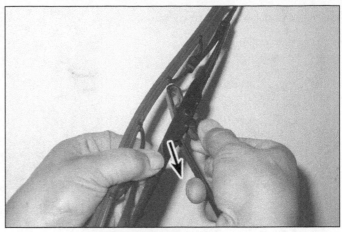

7.4b . . . and slide the wiper assembly down the wiper arm and out of the hook in the end of the arm

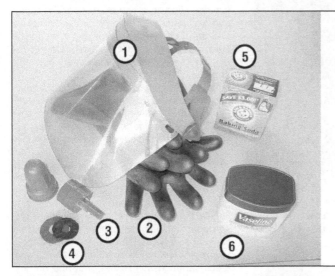

8.1 Tools and materials required for battery maintenance

1   *Face shield/safety goggles - When removing corrosion with a brush, the acidic particles can easily fly up into your eyes*
2   *Rubber gloves - Another safety item to consider when servicing the battery - remember that's acid inside the battery!*
3   *Battery terminal/cable cleaner - This wire brush cleaning tool will remove all traces of corrosion from the battery posts and cable clamps*
4   *Treated felt washers - Placing one of these on each post, directly under the cable clamps, will help prevent corrosion (be sure to get the correct type for side terminal batteries)*
5   *Baking soda - A solution of baking soda and water can be used to neutralize corrosion*
6   *Petroleum jelly - A layer of this on the battery posts will help prevent corrosion*

21   During the first few trips after an oil change, make it a point to check frequently for leaks and proper oil level.
22   The old oil drained from the engine cannot be reused in its present state and should be disposed of. Check with your local auto parts store, disposal facility or environmental agency to see if they will accept the oil for recycling. After the oil has cooled it can be drained into a container (capped plastic jugs, topped bottles, milk cartons, etc.) for transport to one of these disposal sites. Don't dispose of the oil by pouring it on the ground or down a drain!

7   Windshield wiper blade inspection and replacement (every 6000 miles or 6 months)

1   The windshield wiper and blade assembly should be inspected periodically for damage, loose components and cracked or worn blade elements.
2   Road film can build up on the wiper blades and affect their efficiency, so they should be washed regularly with a mild detergent solution.
3   If the wiper blade elements are cracked, worn or warped, or no longer clean adequately, they should be replaced with new ones.
4   Lift the arm assembly away from the glass for clearance, pry up on the release lever, then slide the wiper blade assembly out of the hook in the end of the arm (see illustrations).
5   Attach the new wiper to the arm. Connection can be confirmed by an audible click.

8   Battery check, maintenance and charging (every 6000 miles or 6 months)

Warning: *Certain precautions must be followed when checking and servicing the battery. Hydrogen gas, which is highly flammable, is always present in the battery cells, so keep lighted tobacco and all other open flames and sparks away from the battery. The electrolyte inside the battery is actually dilute sulfuric acid,* which will cause injury if splashed on your skin or in your eyes. It will also ruin clothes and painted surfaces. When removing the battery cables, always detach the negative cable first and hook it up last!

1   A routine preventive maintenance program for the battery in your vehicle is the only way to ensure quick and reliable starts. But before performing any battery maintenance, make sure that you have the proper equipment necessary to work safely around the battery (see illustration).
2   There are also several precautions that should be taken whenever battery maintenance is performed. Before servicing the battery, always turn the engine and all accessories off and disconnect the cable from the negative terminal of the battery.
3   The battery produces hydrogen gas, which is both flammable and explosive. Never create a spark, smoke or light a match around the battery. Always charge the battery in a ventilated area.
4   Electrolyte contains poisonous and corrosive sulfuric acid. Do not allow it to get in your eyes, on your skin or on your clothes. Never ingest it. Wear protective safety glasses

Terminal end corrosion or damage.

Insulation cracks.

Chafed insulation
or exposed wires.

Burned or melted insulation.

8.5 Typical battery cable problems

8.7a A tool like this one (available at auto parts stores) is used to clean the side-terminal type battery-cable contact area

8.7c Regardless of the type of tool used on the battery and cables, a clean, shiny surface should be the result

8.7b Use the brush side of the tool to finish the job

when working near the battery. Keep children away from the battery.

5   Note the external condition of the battery. If the positive terminal and cable clamp on your vehicle's battery is equipped with a rubber protector, make sure that it's not torn or damaged. It should completely cover the terminal. Look for any corroded or loose connections, cracks in the case or cover or loose hold-down clamps. Also check the entire length of each cable for cracks and frayed conductors (see illustration).

6   If corrosion, which looks like white, fluffy deposits is evident, particularly around the terminals, the battery should be removed for cleaning. Loosen the cable bolts with a wrench, being careful to remove the ground cable first, and slide them off the terminals. Then disconnect the hold-down clamp bolt and nut, remove the clamp and lift the battery from the engine compartment.

7   Clean the cable ends thoroughly with a battery brush or a terminal cleaner and a solution of warm water and baking soda. Wash the terminals and the side of the battery case with the same solution but make sure that the solution doesn't get into the battery. When cleaning the cables, terminals and battery

case, wear safety goggles and rubber gloves to prevent any solution from coming in contact with your eyes or hands. Wear old clothes too - even diluted, sulfuric acid splashed onto clothes will burn holes in them. If the terminals have been corroded, clean them up with a terminal cleaner (see illustrations). Thoroughly wash all cleaned areas with plain water.

8   Make sure that the battery tray is in good condition and the hold-down clamp bolts are tight. If the battery is removed from the tray, make sure no parts remain in the bottom of the tray when the battery is reinstalled. When reinstalling the hold-down clamp bolts, do not overtighten them.

9   Any metal parts of the vehicle damaged by corrosion should be covered with a zinc-based primer, then painted.

10   Information on removing and installing the battery can be found in Chapter 5. Information on jump starting can be found at the front of this manual. For more detailed battery checking procedures, refer to the *Haynes Automotive Electrical Manual*.

Charging

Warning: *When batteries are being charged, hydrogen gas, which is very explosive and*

flammable, is produced. Do not smoke or allow open flames near a charging or a recently charged battery. Wear eye protection when near the battery during charging. Also, make sure the charger is unplugged before connecting or disconnecting the battery from the charger.
Note: *The manufacturer recommends the battery be removed from the vehicle for charging because the gas that escapes during this procedure can damage the paint. Fast charging with the battery cables connected can result in damage to the electrical system.*

11   Slow-rate charging is the best way to restore a battery that's discharged to the point where it will not start the engine. It's also a good way to maintain the battery charge in a vehicle that's only driven a few miles between starts. Maintaining the battery charge is particularly important in the winter when the battery must work harder to start the engine and electrical accessories that drain the battery are in greater use.

12   It's best to use a one or two-amp battery charger (sometimes called a "trickle" charger). They are the safest and put the least strain on the battery. They are also the least expensive. For a faster charge, you can use a higher amperage charger, but don't use one rated more than 1/10th the amp/hour rating of the battery. Rapid boost charges that claim to restore the power of the battery in one to two hours are hardest on the battery and can damage batteries not in good condition. This type of charging should only be used in emergency situations.

13   The average time necessary to charge a battery should be listed in the instructions that come with the charger. As a general rule, a trickle charger will charge a battery in 12 to 16 hours.

14   Remove all the cell caps (if equipped) and cover the holes with a clean cloth to prevent spattering electrolyte. Disconnect the negative battery cable and hook the battery charger cable clamps up to the battery posts (positive to positive, negative to negative), then plug in the charger. Make sure it is set at 12-volts if it has a selector switch.

Check for a chafed area that could fail prematurely.

Check for a soft area indicating the hose has deteriorated inside.

Overtightening the clamp on a hardened hose will damage the hose and cause a leak.

Check each hose for swelling and oil-soaked ends. Cracks and breaks can be located by squeezing the hose

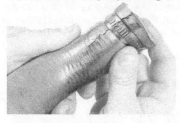

9.4 Hoses, like drivebelts, have a habit of failing at the worst possible time - to prevent the inconvenience of a blown radiator or heater hose, inspect them carefully as shown here

15   If you're using a charger with a rate higher than two amps, check the battery regularly during charging to make sure it doesn't overheat. If you're using a trickle charger, you can safely let the battery charge overnight after you've checked it regularly for the first couple of hours.

16   If the battery has removable cell caps, measure the specific gravity with a hydrometer every hour during the last few hours of the charging cycle. Hydrometers are available inexpensively from auto parts stores - follow the instructions that come with the hydrometer. Consider the battery charged when there's no change in the specific gravity reading for two hours and the electrolyte in the cells is gassing (bubbling) freely. The specific gravity reading from each cell should be very close to the others. If not, the battery

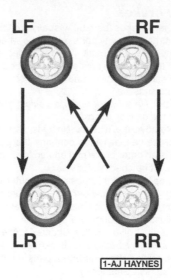

10.2 Four-tire rotation pattern

probably has a bad cell(s).

17   Some batteries with sealed tops have built-in hydrometers on the top that indicate the state of charge by the color displayed in the hydrometer window. Normally, a bright-colored hydrometer indicates a full charge and a dark hydrometer indicates the battery still needs charging.

18   If the battery has a sealed top and no built-in hydrometer, you can hook up a digital voltmeter across the battery terminals to check the charge. A fully charged battery should read 12.5 volts or higher.

19   Further information on the battery and jump-starting can be found in Chapter 5 and at the front of this manual.

9   Cooling system check (every 6000 miles or 6 months)

1   Many major engine failures can be caused by a faulty cooling system.

2   The engine must be cold for the cooling system check, so perform the following procedure before the vehicle is driven for the day or after it has been shut off for at least three hours.

3   Remove the pressure relief cap from the expansion tank. Clean the cap thoroughly, inside and out, with clean water. The presence of rust or corrosion in the expansion tank means the coolant should be changed (see Section 23). The coolant inside the expansion tank should be relatively clean and transparent. If it's rust colored, drain the system and refill it with new coolant.

4   Carefully check the radiator hoses and the smaller diameter heater hoses (see illustrations in Chapter 3). Inspect each coolant hose along its entire length, replacing any hose which is cracked, swollen or deteriorated (see illustration). Cracks will show up better if the

hose is squeezed. Pay close attention to hose clamps that secure the hoses to cooling system components. Hose clamps can pinch and puncture hoses, resulting in coolant leaks.

5   Make sure that all hose connections are tight. A leak in the cooling system will usually show up as white or rust colored deposits on the area adjoining the leak. If wire-type clamps are used on the hoses, it may be a good idea to replace them with screw-type clamps.

6   Clean the front of the radiator and air conditioning condenser with compressed air, if available, or a soft brush. Remove all bugs, leaves, etc. embedded in the radiator fins. Be extremely careful not to damage the cooling fins or cut your fingers on them.

7   If the coolant level has been dropping consistently and no leaks are detectable, have the expansion tank cap and cooling system pressure checked at a service station.

10   Tire rotation (every 6000 miles or 6 months)

1   The tires should be rotated at the specified intervals and whenever uneven wear is noticed. Since the vehicle will be raised and the tires removed anyway, check the brakes also (see Section 12).

2   Radial tires must be rotated in a specific pattern (see illustration). If your vehicle has a compact spare tire, don't include it in the rotation pattern.

3   Refer to the information in *Jacking and towing* at the front of this manual for the proper procedure to follow when raising the vehicle and changing a tire. If the brakes must be checked, don't apply the parking brake as stated.

4   The vehicle must be raised on a hoist or supported on jackstands to get all four wheels off the ground. Make sure the vehicle is safely supported!

5   After the rotation procedure is finished, check and adjust the tire pressures as necessary and be sure to check the lug nut tightness.

11   Seat belt check (every 6000 miles or 6 months)

1   Check seat belts, buckles, latch plates and guide loops for obvious damage and signs of wear.

2   See if the seat belt reminder light comes on when the key is turned to the Run or Start position. A chime should also sound.

3   The seat belts are designed to lock up during a sudden stop or impact, yet allow free movement during normal driving. Make sure the retractors return the belt against your chest while driving and rewind the belt fully when the buckle is unlatched.

4   If any of the above checks reveal problems with the seat belt system, replace parts as necessary.

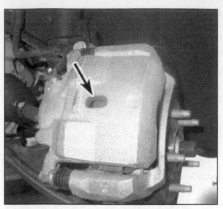

12.6 You will find an inspection hole like this in each caliper through which you can view the thickness of remaining friction material for the inner pad

12 Brake check (every 6000 miles or 6 months)

Warning: *The dust created by the brake system is harmful to your health. Never blow it out with compressed air and don't inhale any of it. An approved filtering mask should be worn when working on the brakes. Do not, under any circumstances, use petroleum-based solvents to clean brake parts. Use brake system cleaner only!*

Note: *For detailed photographs of the brake system, refer to Chapter 9.*

1   In addition to the specified intervals, the brakes should be inspected every time the wheels are removed or whenever a defect is suspected.

2   Any of the following symptoms could indicate a potential brake system defect: The vehicle pulls to one side when the brake pedal is depressed; the brakes make squealing or dragging noises when applied; brake pedal travel is excessive; the pedal pulsates; or brake fluid leaks, usually onto the inside of the tire or wheel.

Disc brakes

3   Disc brakes can be visually checked without removing any parts except the wheels. Remove the hub caps (if applicable) and loosen the wheel lug nuts a quarter turn each.

4   Raise the vehicle and place it securely on jackstands.

Warning: *Never work under a vehicle that is supported only by a jack!*

5   Remove the wheels. Now visible is the disc brake caliper which contains the pads. There is an outer brake pad and an inner pad. Both must be checked for wear.

6   Measure the thickness of the outer pad at each end of the caliper and the inner pad through the inspection hole in the caliper body (see illustration). Compare the measurement with the limit given in this Chapter's Specifications; if any brake pad thickness is less than specified, then all brake pads must be replaced (see Chapter 9).

7   If you're in doubt as to the exact pad thickness or quality, remove them for measurement and further inspection (see Chapter 9).

8   Check the disc for score marks, wear and burned spots. If any of these conditions exist, the disc should be removed for servicing or replacement (see Chapter 9).

9   Before installing the wheels, check all the brake lines and hoses for damage, wear, deformation, cracks, corrosion, leakage, bends and twists, particularly in the vicinity of the rubber hoses and calipers.

10   Install the wheels, lower the vehicle and tighten the wheel lug nuts to the torque given in this Chapter's Specifications.

Drum brakes

11   On models with rear drum brakes, make sure the parking brake is off, then tap on the outside of the drum with a rubber mallet to loosen it.

12   Remove the brake drums. If the drum still won't come off, refer to Chapter 9

13   With the drums removed, carefully clean the brake assembly with brake system cleaner.

Warning: *Don't blow the dust out with compressed air and don't inhale any of it (it is harmful to your health).*

14   Note the thickness of the lining material on both the front and rear brake shoes (see illustration). Compare the measurement with the limit given in this Chapter's Specifications; if any lining thickness is less than specified, then all of the brake shoes must be replaced (see Chapter 9). The shoes should also be replaced if they're cracked, glazed (shiny areas), or covered with brake fluid.

15   Make sure all the brake assembly springs are connected and in good condition.

16   Check the brake components for signs of fluid leakage. With your finger or a small screwdriver, carefully pry back the rubber cups on the wheel cylinder located at the top of the brake shoes (see illustration). Any leakage here is an indication that the wheel cylinders should be replaced immediately (see Chapter 9). Also, check all hoses and connections for signs of leakage.

17   Wipe the inside of the drum with a clean rag and denatured alcohol or brake cleaner. Again, be careful not to breathe the dangerous brake dust.

18   Check the inside of the drum for cracks, score marks, deep scratches and "hard spots" which will appear as small discolored areas. If imperfections cannot be removed with fine emery cloth, the drum must be taken to an automotive machine shop for resurfacing.

19   Repeat the procedure for the remaining wheel. If the inspection reveals that all parts are in good condition, reinstall the brake drums, install the wheels and lower the vehicle to the ground.

Brake booster check

20   Sit in the driver's seat and perform the following sequence of tests.

21   With the brake fully depressed, start the engine - the pedal should move down a little when the engine starts.

22   With the engine running, depress the brake pedal several times - the travel distance

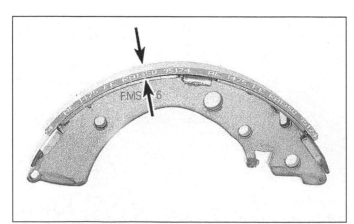

12.14 If the lining is bonded to the brake shoe, measure the lining thickness from the outer surface to the metal shoe, as shown here; if the lining is riveted to the shoe, measure from the lining outer surface to the rivet head

12.16 Carefully peel back the wheel cylinder boot and check for leaking fluid, indicating that the cylinder must be replaced

should not change.

23   Depress the brake, stop the engine and hold the pedal in for about 30 seconds - the pedal should neither sink nor rise.

24   Restart the engine, run it for about a minute and turn it off. Then, firmly depress the brake several times - the pedal travel should decrease with each application.

25   If your brakes do not operate as described, the brake booster has failed. Refer to Chapter 9 for the replacement procedure.

13   Underhood hose check and replacement (every 15,000 miles or 12 months)

Warning: *Replacement of air conditioning hoses must be left to a dealer service department or air conditioning shop that has the equipment to depressurize the system safely. Never remove air conditioning components or hoses until the system has been depressurized.*

General

1   High temperatures under the hood can cause deterioration of the rubber and plastic hoses used for engine, accessory and emission systems operation. Periodic inspection should be made for cracks, loose clamps, material hardening and leaks.

2   Information specific to the cooling system hoses can be found in Section 9.

3   Most (but not all) hoses are secured to the fittings with clamps. Where clamps are used, check to be sure they haven't lost their tension, allowing the hose to leak. If clamps aren't used, make sure the hose has not expanded and/or hardened where it slips over the fitting, allowing it to leak.

PCV system hose

4   To reduce hydrocarbon emissions, crankcase blow-by gas is vented through the PCV valve in the rocker arm cover to the intake manifold via a rubber hose on most models. The blow-by gases mix with incoming air in the intake manifold before being burned in the combustion chambers.

5   Check the PCV hose for cracks, leaks and other damage. Disconnect it from the valve cover and the intake manifold and check the inside for obstructions. If it's clogged, clean it out with solvent.

Vacuum hoses

6   It's quite common for vacuum hoses, especially those in the emissions system, to be color coded or identified by colored stripes molded into them. Various systems require hoses with different wall thickness, collapse resistance and temperature resistance. When replacing hoses, be sure the new ones are made of the same material.

7   Often the only effective way to check a hose is to remove it completely from the vehicle. If more than one hose is removed, be sure to label the hoses and fittings to ensure

correct installation.

8   When checking vacuum hoses, be sure to include any plastic T-fittings in the check. Inspect the fittings for cracks and the hose where it fits over each fitting for distortion, which could cause leakage.

9   A small piece of vacuum hose (1/4-inch inside diameter) can be used as a stethoscope to detect vacuum leaks. Hold one end of the hose to your ear and probe around vacuum hoses and fittings, listening for the "hissing" sound characteristic of a vacuum leak. **Warning:** *When probing with the vacuum hose stethoscope, be careful not to come into contact with moving engine components such as drivebelts, the cooling fan, etc.*

Fuel hose

Warning: *Gasoline is flammable, so take extra precautions when you work on any part of the fuel system. Don't smoke or allow open flames or bare light bulbs near the work area, and don't work in a garage where a gas-type appliance (such as a water heater or clothes dryer) is present. Since fuel is carcinogenic, wear latex gloves when there's a possibility of being exposed to fuel, and, if you spill any fuel on your skin, rinse it off immediately with soap and water. Mop up any spills immediately and do not store fuel-soaked rags where they could ignite. The fuel system is under constant pressure, so, if any fuel lines are to be disconnected, the fuel pressure in the system must be relieved first (see Chapter 4 for more information). When you perform any kind of work on the fuel system, wear safety glasses and have a Class B type fire extinguisher on hand.*

10   The fuel lines are usually under pressure, so if any fuel lines are to be disconnected be prepared to catch spilled fuel. Warning: Your vehicle is equipped with fuel injection and you must relieve the fuel system pressure before servicing the fuel lines. Refer to Chapter 4 for the fuel system pressure relief procedure.

11   Check all flexible fuel lines for deterioration and chafing. Check especially for cracks in areas where the hose bends and just before fittings, such as where a hose attaches to the fuel pump, fuel filter and fuel injection unit.

12   When replacing a hose, use only hose that is specifically designed for your fuel injection system.

13   Spring-type clamps are sometimes used on fuel return or vapor lines. These clamps often lose their tension over a period of time, and can be "sprung" during removal. Replace all spring-type clamps with screw clamps whenever a hose is replaced. Some fuel lines use spring-lock type couplings, which require a special tool to disconnect. See Chapter 4 for more information on this type of coupling.

Metal lines

14   Sections of metal line are often used for fuel line between the fuel pump and the fuel injection unit. Check carefully to make sure the line isn't bent, crimped or cracked.

15   If a section of metal fuel line must be

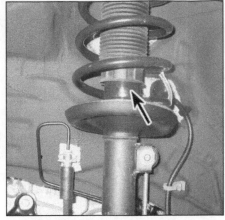

14.4 Check the shocks for leakage at the indicated area

replaced, use seamless steel tubing only, since copper and aluminum tubing do not have the strength necessary to withstand vibration caused by the engine.

16   Check the metal brake lines where they enter the master cylinder and brake proportioning unit (if used) for cracks in the lines and loose fittings. Any sign of brake fluid leakage calls for an immediate thorough inspection of the brake system.

14   Steering and suspension check (every 15,000 miles or 12 months)

Note: *For detailed illustrations of the steering and suspension components, refer to Chapter 10.*

With the wheels on the ground

1   With the vehicle stopped and the front wheels pointed straight ahead, rock the steering wheel gently back and forth. If freeplay is excessive, a front wheel bearing, steering shaft universal joint or lower arm balljoint is worn or the steering gear is out of adjustment or broken. Refer to Chapter 10 for the appropriate repair procedure.

2   Other symptoms, such as excessive vehicle body movement over rough roads, swaying (leaning) around corners and binding as the steering wheel is turned, may indicate faulty steering and/or suspension components.

3   Check the shock absorbers by pushing down and releasing the vehicle several times at each corner. If the vehicle does not come back to a level position within one or two bounces, the shocks/struts are worn and must be replaced. When bouncing the vehicle up and down, listen for squeaks and noises from the suspension components.

4   Check the struts and shock absorbers for evidence of fluid leakage (see illustration). A light film of fluid is no cause for concern. Make sure that any fluid noted is from the struts/shocks and not from some other

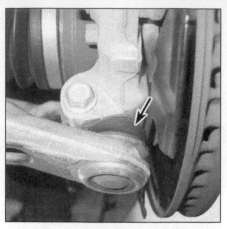

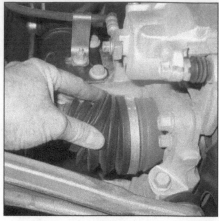

14.10 To check a balljoint for wear, try to pry the control arm up and down to make sure there is no play in the balljoint (if there is, replace it)

14.11 Check the balljoint boot for damage

15.2 Flex the driveaxle boots by hand to check for cracks and/or leaking grease

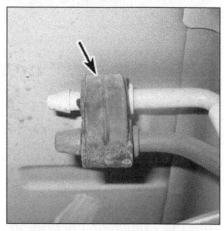

16.2 Be sure to check each exhaust system rubber hanger for damage

source. If leakage is noted, replace the struts/shocks as a set.

5   Check the struts and shocks to be sure they are securely mounted and undamaged. Check the upper mounts for damage and wear. If damage or wear is noted, replace the shocks as a set (front and rear).

6   If the shocks must be replaced, refer to Chapter 10 for the procedure.

Under the vehicle

7   Raise the vehicle with a floor jack and support it securely on jackstands. See *Jacking and towing* at the front of this book for the proper jacking points.

8   Check the tires for irregular wear patterns and proper inflation. See Section 5 in this Chapter for information regarding tire wear and Chapter 10 for information on wheel bearing replacement.

9   Inspect the universal joint between the steering shaft and the steering gear housing. Check the steering gear housing for lubri-

cant leakage. Make sure that the dust seals and boots are not damaged and that the boot clamps are not loose. Check the steering linkage for looseness or damage. Check the tie-rod ends for excessive play. Look for loose bolts, broken or disconnected parts and deteriorated rubber bushings on all suspension and steering components. While an assistant turns the steering wheel from side to side, check the steering components for free movement, chafing and binding. If the steering components do not seem to be reacting with the movement of the steering wheel, try to determine where the slack is located.

10   Check the balljoints for wear by trying to move each control arm up and down with a prybar (see illustration) to ensure that its balljoint has no play. If any balljoint does have play, replace it. See Chapter 10 for the balljoint replacement procedure.

11   Inspect the balljoint boots for damage and leaking grease (see illustration). Replace the balljoints with new ones if they are damaged (see Chapter 10).

12   At the rear of the vehicle, inspect the suspension arm bushings for deterioration. Additional information on suspension components can be found in Chapter 10.

15   Driveaxle boot check (every 15,000 miles or 12 months)

Note: *For detailed illustrations of the driveaxles, refer to Chapter 8.*

1   The driveaxle boots are very important because they prevent dirt, water and foreign material from entering and damaging the constant velocity (CV) joints. Oil and grease can cause the boot material to deteriorate prematurely, so it's a good idea to wash the boots with soap and water. Because it constantly pivots back and forth following the steering action of the front hub, the outer CV boot

wears out sooner and should be inspected.

2   Inspect the boots for tears and cracks as well as loose clamps (see illustration). If there is any evidence of cracks or leaking lubricant, they must be replaced as described in Chapter 8.

16   Exhaust system check (every 15,000 miles or 12 months)

1   With the engine cold (at least three hours after the vehicle has been driven), check the complete exhaust system from the engine to the end of the tailpipe. Ideally, the inspection should be done with the vehicle on a hoist to permit unrestricted access. If a hoist isn't available, raise the vehicle and support it securely on jackstands.

2   Check the exhaust pipes and connections for evidence of leaks, severe corrosion and damage. Make sure that all brackets and hangers are in good condition and tight (see illustration).

3   At the same time, inspect the underside of the body for holes, corrosion, open seams, etc. which may allow exhaust gases to enter the passenger compartment. Seal all body openings with silicone or body putty.

4   Rattles and other noises can often be traced to the exhaust system, especially the mounts and hangers. Try to move the pipes, muffler and catalytic converter. If the components can come in contact with the body or suspension parts, secure the exhaust system with new mounts.

5   Check the running condition of the engine by inspecting inside the end of the tailpipe. The exhaust deposits here are an indication of engine state-of-tune. If the pipe is black and sooty or coated with white deposits, the engine may need a tune-up, including a thorough fuel system inspection.

17 Fuel system check (every 15,000 miles or 12 months)

Warning: *Gasoline is flammable, so take extra precautions when you work on any part of the fuel system. Don't smoke or allow open flames or bare light bulbs near the work area, and don't work in a garage where a gas-type appliance (such as a water heater or clothes dryer) is present. Since fuel is carcinogenic, wear fuel-resistant gloves when there's a possibility of being exposed to fuel, and, if you spill any fuel on your skin, rinse it off immediately with soap and water. Mop up any spills immediately and do not store fuel-soaked rags where they could ignite. When you perform any kind of work on the fuel system, wear safety glasses and have a Class B type fire extinguisher on hand. The fuel system is under constant pressure, so, before any lines are disconnected, the fuel system pressure must be relieved (see Chapter 4).*

1   If you smell gasoline while driving or after the vehicle has been sitting in the sun, inspect the fuel system immediately.

2   Remove the fuel filler cap and inspect it for damage and corrosion. The gasket should have an unbroken sealing imprint. If the gasket is damaged or corroded, install a new cap.

3   Inspect the fuel feed line for cracks. Make sure that the connections between the fuel lines and the fuel injection system and between the fuel lines and the in-line fuel filter are tight.

Warning: *Your vehicle is fuel injected, so you must relieve the fuel system pressure before servicing fuel system components. The fuel system pressure relief procedure is outlined in Chapter 4.*

4   Since some components of the fuel system - the fuel tank and part of the fuel feed line, for example - are underneath the vehicle, they can be inspected more easily with the vehicle raised on a hoist. If that's not possible, raise the vehicle and support it on jackstands.

5   With the vehicle raised and safely supported, inspect the gas tank and filler neck for punctures, cracks and other damage. The connection between the filler neck and the tank is particularly critical. Sometimes a rubber filler neck will leak because of loose clamps or deteriorated rubber. Inspect all fuel tank mounting brackets and straps to be sure that the tank is securely attached to the vehicle.

Warning: *Do not, under any circumstances, try to repair a fuel tank (except rubber components). A welding torch or any open flame can easily cause fuel vapors inside the tank to explode.*

6   Carefully check all rubber hoses and metal lines leading away from the fuel tank. Check for loose connections, deteriorated hoses, crimped lines and other damage. Repair or replace damaged sections as necessary.

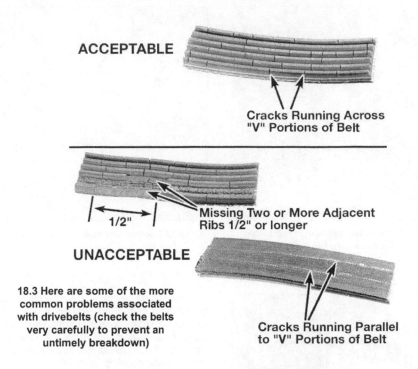

ACCEPTABLE

Cracks Running Across "V" Portions of Belt

|← 1/2" →| Missing Two or More Adjacent Ribs 1/2" or longer

UNACCEPTABLE

Cracks Running Parallel to "V" Portions of Belt

18.3 Here are some of the more common problems associated with drivebelts (check the belts very carefully to prevent an untimely breakdown)

18 Drivebelt check and replacement (every 15,000 miles or 12 months)

1   A single serpentine drivebelt is located at the front of the engine and plays an important role in the overall operation of the engine and its components. Due to its function and material make up, the belt is prone to wear and should be periodically inspected. The serpentine belt drives the alternator and air conditioning compressor. Although the belt should be inspected at the recommended intervals, replacement may not be necessary for more than 100,000 miles.

Check

2   With the engine stopped, inspect the full length of the drivebelt for cracks and separation of the belt plies. It will be necessary to turn the engine (using a wrench or socket and bar on the crankshaft pulley bolt) in order to move the belt from the pulleys so that the belt can be inspected thoroughly. Twist the belt between the pulleys so that both sides can be viewed. Also check for fraying, and glazing which gives the belt a shiny appearance. Check the pulleys for nicks, cracks, distortion and corrosion.

3   Note that it is not unusual for a ribbed belt to exhibit small cracks in the edges of the belt ribs, and unless these are extensive or very deep, belt replacement is not essential (see illustration).

Replacement

4   On four-cylinder models, loosen the right front wheel lug nuts, then raise the front of the vehicle and support it on jackstands. Remove

18.6 Rotate the tensioner arm to relieve belt tension

the right front wheel and remove the lower splash shield from the underbody.

5   If you're working on a V6 engine, remove the air filter housing (see Chapter 4), and the engine mount strut (see Chapter 2A).

6   Insert a 3/8-inch drive ratchet or breaker bar into the tensioner hole and pull the handle counterclockwise to release the drivebelt tension (see illustration). Note how the drivebelt is routed, then remove the belt from the pulleys.

7   Fit the new drivebelt onto the crankshaft, alternator, power steering pump, and air conditioning compressor pulleys, as applicable, then turn the release the tensioner and locate the drivebelt on the pulley. Make sure that the drivebelt is correctly seated in all of the pulley grooves, then release the tensioner.

8   Installation is the reverse of removal.

18.10 Remove the tensioner mounting bolt, then remove the tensioner

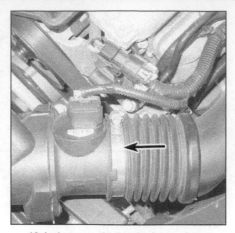

19.1a Loosen the intake hose clamp

19.1b Release the clips from the air filter housing lid . . .

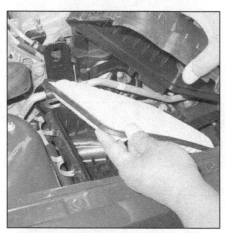

19.1c . . . then remove the lid and pull out the filter

Tensioner replacement

9   Remove the drivebelt as described previously.
10   Remove the bolt securing the tensioner to the engine (see illustration).
11   Installation is the reverse of removal. Be sure to tighten the tensioner bolt to the torque listed in this Chapter's Specifications.

19 Air filter check and replacement (every 30,000 miles or 24 months)

1   The air filter is located inside a housing at the right (passenger's) side of the engine compartment. To remove the air filter, loosen the clamp securing the inlet tube to the air filter cover, remove the screws that secure the two halves of the housing together, then separate the cover halves and remove the air filter element (see illustrations).
2   Inspect the outer surface of the filter element. If it is dirty, replace it. If it is only moderately dusty, it can be reused by blowing it clean from the back to the front surface with

compressed air. Because it is a pleated paper type filter, it cannot be washed or oiled. If it cannot be cleaned satisfactorily with compressed air, discard and replace it. While the cover is off, be careful not to drop anything down into the housing.
Caution: *Never drive the vehicle with the air filter removed. Excessive engine wear could result and backfiring could even cause a fire under the hood.*
3   Wipe out the inside of the housing.
4   Place the new filter into the housing, making sure it seats properly.
5   Installation of the housing is the reverse of removal.

20 Brake fluid change (every 30,000 miles or 24 months)

Warning: *Brake fluid can harm your eyes and damage painted surfaces, so use extreme caution when handling or pouring it. Do not use brake fluid that has been standing open or is more than one year old. Brake fluid absorbs moisture from the air. Excess moisture can cause a dangerous loss of braking effectiveness.*
1   At the specified intervals, the brake fluid should be drained and replaced. Since the brake fluid may drip or splash when pouring it, place plenty of rags around the master cylinder to protect any surrounding painted surfaces.
2   Before beginning work, purchase the specified brake fluid (see *Recommended lubricants and fluids* at the beginning of this Chapter).
3   Remove the cap from the master cylinder reservoir.
4   Using a hand suction pump or similar device, withdraw the fluid from the master cylinder reservoir.
5   Add new fluid to the master cylinder until it rises to the base of the filler neck.
6   Bleed the brake system as described in Chapter 9 at all four brakes until new and

uncontaminated fluid is expelled from the bleeder screw. Be sure to maintain the fluid level in the master cylinder as you perform the bleeding process. If you allow the master cylinder to run dry, air will enter the system.
7   Refill the master cylinder with fluid and check the operation of the brakes. The pedal should feel solid when depressed, with no sponginess.
Warning: *Do not operate the vehicle if you are in doubt about the effectiveness of the brake system.*

21 Spark plug check and replacement (see Maintenance schedule for service intervals)

1   The spark plugs are located in the cylinder head(s).
2   In most cases, the tools necessary for spark plug replacement include a spark plug socket which fits onto a ratchet (spark plug sockets are padded inside to prevent damage to the porcelain insulators on the new plugs), various extensions and a gap gauge to check and adjust the gaps on the new plugs (see illustration). A torque wrench should be used to tighten the new plugs.
3   The best approach when replacing the spark plugs is to purchase the new ones in advance, adjust them to the proper gap and replace the plugs one at a time. When buying the new spark plugs, be sure to obtain the correct plug type for your particular engine. This information can be found in the Specifications Section at the beginning of this Chapter or in your owner's manual.
4   Allow the engine to cool completely before attempting to remove any of the plugs. These engines are equipped with aluminum cylinder heads, which can be damaged if the spark plugs are removed when the engine is hot. While you are waiting for the engine to cool, check the new plugs for defects and adjust the gaps.

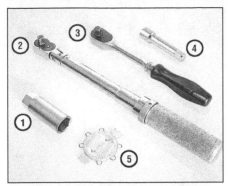

21.2 Tools required for changing spark plugs

1 **Spark plug socket** - *This will have special padding inside to protect the spark plug's porcelain insulator*
2 **Torque wrench** - *Although not mandatory, using this tool is the best way to ensure the plugs are tightened properly*
3 **Ratchet** - *Standard hand tool to fit the spark plug socket*
4 **Extension** - *Depending on model and accessories, you may need special extensions and universal joints to reach one or more of the plugs*
5 **Spark plug gap gauge** - *This gauge for checking the gap comes in a variety of styles. Make sure the gap for your engine is included*

21.5a The manufacturer recommends using a wire-type thickness gauge when checking the gap - if the wire does not slide between the electrodes with a slight drag, adjustment is required

21.5b To change the gap, bend the side electrode only, and be very careful not to crack or chip the porcelain insulator surrounding the center electrode

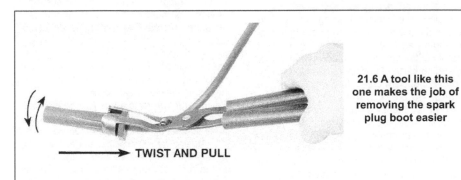

TWIST AND PULL

21.6 A tool like this one makes the job of removing the spark plug boot easier

5   The gap is checked by inserting the proper-thickness gauge between the electrodes at the tip of the plug (see illustration). The gap between the electrodes should be the same as the one specified on the Vehicle Emissions Control Information label or in this Chapter's Specifications. The gauge should just slide between the electrodes with a slight amount of drag. If the gap is incorrect, use the adjuster on the gauge body to bend the curved side electrode slightly until the proper gap is obtained (see illustration). If the side electrode is not exactly over the center electrode, bend it with the adjuster until it is. Check for cracks in the porcelain insulator (if any are found, the plug should not be used).
Note: *We recommend using a wire-type thickness gauge when checking platinum- or iridium-type spark plugs. Other types of gauges may scrape the thin coating from the electrodes, thus dramatically shortening the life of the plugs.*
6   Four-cylinder and 3.6L V6 engines are equipped with individual ignition coils which must be removed first to access the spark plugs (see Chapter 5). On 3.5L and 3.9L V6 engines, remove the spark plug wire from one spark plug. Pull only on the boot at the end of the wire - do not pull on the wire. A plug wire removal tool should be used if available (see illustration).
7   If compressed air is available, use it to blow any dirt or foreign material away from the spark plug hole. The idea here is to eliminate

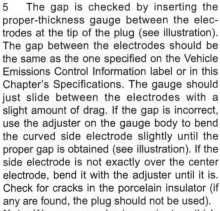

21.8 Use a ratchet and extension to remove the spark plugs

the possibility of debris falling into the cylinder as the spark plug is removed.
8   Place the spark plug socket over the plug and remove it from the engine by turning it in a counterclockwise direction (see illustration).
9   Compare the spark plug to those shown in the photos located on the inside back cover to get an indication of the general running

21.10 A length of snug-fitting rubber hose will save time and prevent damaged threads when installing the spark plugs

condition of the engine.
10   Install one of the new plugs into the hole until you can no longer turn it with your fingers, then tighten it with a torque wrench (if available) or the ratchet. It is a good idea to slip a short length of rubber hose over the end of the plug to use as a tool to thread it into place (see illustration). The hose will grip the

plug well enough to turn it, but will start to slip if the plug begins to cross-thread in the hole - this will prevent damaged threads and the accompanying repair costs.

11   Before pushing the ignition coil onto the end of the plug, inspect the ignition coil following the procedures outlined in Chapter 5. On V6 engines, inspect the plug wire following the procedures outlined in Section 22.

12   Repeat the procedure for the remaining spark plugs.

22   Spark plug wire check and replacement (V6 engines) (see Maintenance schedule for service intervals)

1   The spark plug wires should be checked at the recommended intervals or whenever new spark plugs are installed.

2   Begin this procedure by making a visual check of the spark plug wires while the engine is running. In a darkened garage (make sure there is adequate ventilation) or at night, start the engine and observe each plug wire. Be careful not to come into contact with any moving engine parts. If possible, use an insulated or non-conductive object to wiggle each wire. If there is a break in the wire, you will see arcing or a small blue spark coming from the damaged area. Secondary ignition voltage increases with engine speed and sometimes a damaged wire will not produce an arc at idle speed. Have an assistant press the accelerator pedal to raise the engine speed to approximately 2000 rpm. Check the spark plug wires for arcing as stated previously. If arcing is noticed, replace all spark plug wires.

23   Cooling system servicing (draining, flushing and refilling) (see Maintenance schedule for service intervals)

Warning: *Do not allow antifreeze to come in contact with your skin or painted surfaces of the vehicle. Rinse off spills immediately with plenty of water. Antifreeze is highly toxic if ingested. Never leave antifreeze lying around in an open container or in puddles on the floor; children and pets are attracted by its sweet smell and may drink it. Check with local authorities on disposing of used antifreeze. Many communities have collection centers that will see that antifreeze is disposed of safely.* **Note:** *Non-toxic antifreeze is now manufactured and available at local auto parts stores, but even this type should be disposed of properly.*

Draining

1   Periodically, the cooling system should be drained, flushed and refilled to replenish the antifreeze mixture and prevent formation of rust and corrosion, which can impair the performance of the cooling system and cause

23.3a If the radiator drain valve is stuck, you'll have to use pliers to open it

engine damage. When the cooling system is serviced, all hoses and the expansion tank cap should be checked and replaced if necessary.

2   Apply the parking brake and block the wheels. Raise the front of the vehicle and support it securely on jackstands, then remove the under-vehicle splash shield.

Warning: *If the vehicle has just been driven, wait several hours to allow the engine to cool down before beginning this procedure.*

3   Move a large container under the radiator drain to catch the coolant. Turn the knob on the radiator drain valve. Remove the cap from the coolant expansion tank and allow the coolant to drain. If you're working on a four-cylinder model, also unscrew the water pump drain bolt and allow the coolant to drain into a container (see illustrations).

4   While the coolant is draining, check the condition of the radiator hoses, heater hoses and clamps (refer to Section 9 if necessary).

5   Replace any damaged clamps or hoses. Close the radiator drain valve and, if you're working on a four-cylinder model, tighten the water pump drain bolt to the torque listed in this Chapter's Specifications.

Flushing

6   Fill the cooling system with clean water, following the *Refilling* procedure (see Step 12).

7   Start the engine and allow it to reach normal operating temperature, then rev up the engine a few times.

8   Turn the engine off and allow it to cool completely, then drain the system as described earlier.

9   Repeat Steps 6 through 8 until the water being drained is free of contaminants.

10   In severe cases of contamination or clogging of the radiator, remove the radiator (see Chapter 3) and have a radiator repair facility clean and repair it if necessary.

11   Many deposits can be removed by the chemical action of a cleaner available at auto parts stores. Follow the procedure outlined in

23.3b Location of the water pump drain bolt (four-cylinder models)

the manufacturer's instructions.

Note: *When the coolant is regularly drained and the system refilled with the correct antifreeze/water mixture, there should be no need to use chemical cleaners or descalers.*

Refilling

12   Close and tighten the radiator drain.

13   Place the heater temperature control in the maximum heat position.

14   Slowly add new coolant (a 50/50 mixture of water and antifreeze) to the expansion tank until the level is at the COLD mark on the expansion tank.

15   Install the expansion tank cap and run the engine in a well-ventilated area until the engine cooling fan turns on.

16   Turn the engine off and let it cool. If necessary, add more coolant mixture to bring the level to the COLD mark on the expansion tank.

17   Squeeze the upper radiator hose to expel air, then add more coolant mixture if necessary. Replace the expansion tank cap.

18   Start the engine, allow it to reach normal operating temperature and check for leaks. Also, set the heater and blower controls to the maximum setting and check to see that the heater output from the air ducts is warm. This is a good indication that all air has been purged from the cooling system.

24   Automatic transaxle fluid change (see Maintenance schedule for service intervals)

Four-speed transaxles with a transaxle pan

1   At the specified time intervals, the transaxle fluid should be drained and replaced. Since the fluid will remain hot long after driving, perform this procedure only after everything has cooled down completely.

2   Before beginning work, purchase the

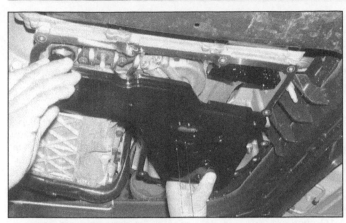

24.10a Pull the transaxle filter straight down and out of the transaxle - there are no fasteners (four-speed transaxles)

24.10b Pry out the old seal, being careful not to damage the aluminum housing (four-speed transaxles)

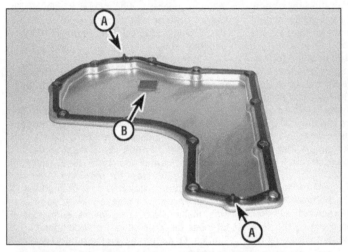

24.12 Place the gasket on the pan, aligning the plastic pins (A) with the holes in the pan for alignment (they also help align the pan to the transaxle) - make sure the magnet (B) is clean and in place before installing the pan (four-speed transaxles)

24.17 Location of the transaxle fluid check plug (four-speed transaxles)

specified transaxle fluid (see *Recommended lubricants and fluids* at the front of this Chapter) and a new filter.

3   Other tools necessary for this job include jackstands to support the vehicle in a raised position, a drain pan capable of holding several quarts, newspapers and clean rags.

4   Raise and support the vehicle on jackstands.

5   With a drain pan in place, remove the front and side transaxle pan mounting bolts.

6   Loosen the rear pan bolts one turn.

7   Carefully pry the transaxle pan loose with a screwdriver, allowing the fluid to drain.

8   Remove the remaining bolts, pan and gasket. Carefully clean the gasket surface of the transaxle to remove all traces of the old gasket and sealant.

9   Drain the fluid from the transaxle pan, clean the pan with solvent and dry it with compressed air. Be careful not to lose the magnet.

10   Remove the filter and pry out the seal (see illustrations).

11   Push a new filter seal fully into its bore, then install the new filter.

12   Make sure the gasket surface on the transaxle pan is clean, then install the new gasket (see illustration). Put the pan in place against the transaxle and install the bolts. Working around the pan, tighten each bolt a little at a time until the final torque figure is reached.

13   Lower the vehicle and add the specified amount of automatic transmission fluid through the vent/fill cap and check the fluid level (see below).

14   Check under the vehicle for leaks during the first few trips.

Fluid level check

15   The manufacturer states that routine checks of the automatic transaxle fluid are not necessary; this procedure should only be used when refilling the transaxle after the fluid has been drained, unless an obvious leak has been detected. Low fluid level can lead to slipping or loss of drive, while overfilling can

cause foaming and loss of fluid.

Warning: *This procedure is potentially dangerous and is best left to a professional shop with a safe lifting apparatus. The vehicle must be kept level while being safely raised high enough for access to the check plug on the transaxle.*

16   With the vehicle raised and safely supported, start the engine, then move the shift lever through all the gear ranges, ending in Park.

Note: *Incorrect fluid level readings will result if the vehicle has just been driven at high speeds for an extended period, in hot weather in city traffic, or if it has been pulling a trailer. If any of these conditions apply, wait until the fluid has cooled (about 30 minutes).*

17   Remove the vent/fill cap (see illustration 24.19). With the engine running and the transaxle at normal operating temperature (having idled for 3 to 5 minutes), locate the check plug on the transaxle. The check plug is located near the pan, adjacent to the engine oil drain plug (see illustration).

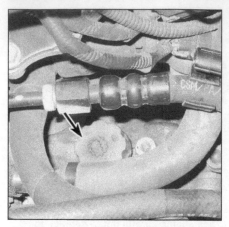

24.19 Location of the vent/fill plug on the automatic transaxle (four-speed transaxles)

18   Place a container under the check plug and remove it. Observe the fluid as it drips into the pan, indicating correct fluid level.

19   The fluid level should be at the bottom of the check hole. If fluid pours out excessively, the transaxle may have been overfilled. Double-check to make sure the vehicle is level. If no fluid drips from the check hole, add small amounts of fluid through the vent/fill cap at the top of the transaxle until the level is at the bottom of the check hole (see illustration). A long-necked funnel will be necessary to add fluid.

20   The condition of the fluid should also be checked along with the level. If the fluid in the drain pan is a dark reddish-brown color, or if the fluid has a burned smell, the fluid should be changed (see above). If you're in doubt about the condition of the fluid, purchase some new fluid and compare the two for color and smell.

21   Be sure to install the check plug and tighten it securely when you're done.

Later five- and six-speed transaxles with no transaxle pan

22   Place the drain pan underneath the transaxle drain plug at the bottom of the transaxle. Remove the drain plug and allow the fluid to drain.

23   Install the drain plug and tighten it securely.

24   Remove the transaxle fluid dipstick located on the front side of the transaxle.

25   Add approximately 4-1/2 quarts of the specified type of automatic transmission fluid through the dipstick hole. Reinstall the dipstick.

26   Start the engine. With the brake pedal depressed, slowly shift through each gear range, then place the shifter in Park. Let the engine idle for one minute.

27   The transaxle should be between 180 and 200-degrees F. If it is not up to 180-degrees F, drive the vehicle until it is. If it's above 200-degrees F, wait until the fluid has cooled back down and falls within the temperature range.

Note: *Some scan tools are capable of displaying transmission fluid temperatures.*

28   Remove the dipstick from the filler tube. Wipe the fluid from the dipstick with a clean rag, then reinsert it.

29   Pull the dipstick out again and note the fluid level; it should be in the crosshatched range. If additional fluid is required, add fluid a little at a time and keep checking the level until it's correct. It is important not to overfill the transmission.

25   Manual transaxle lubricant change

1   Raise the vehicle and support it securely on jackstands.

2   Move a drain pan, rags, newspapers and wrenches under the transaxle.

3   On the side of the transaxle housing near the left driveaxle you will see a check plug (the upper one) and a drain plug (on the bottom side of the transaxle). Remove the transaxle check plug and the drain plug from the case., then allow the lubricant to drain into the pan.

4   After the lubricant has drained completely, reinstall the drain plug and tighten it securely.

5   Remove the fill plug at the top of the transaxle, then using a funnel, fill the transaxle with the specified lubricant until it is level with the lower edge of the check hole. Reinstall the fill plug and check plug and tighten them securely.

6   Lower the vehicle.

7   Drive the vehicle for a short distance, then check the drain and fill plugs for leakage.

8   The old lubricant drained from the transaxle cannot be reused in its present state and should be disposed of. Check with your local auto parts store, disposal facility or environmental agency to see if they will accept the lubricant for recycling. After the lubricant has cooled it can be drained into a container (capped plastic jugs, topped bottles, milk cartons, etc.) for transport to one of these disposal sites. Don't dispose of the lubricant by pouring it on the ground or down a drain!

Chapter 2 Part A
Four-cylinder engines

Contents

Specifications

General

Displacement
    F engine code models .... 134 cubic inches (2.2 liters)
    All other models .... 146 cubic inches (2.4 liters)
Bore
    2.2L models .... 3.385 to 3.386 inches
    2.4L models .... 3.466 to 3.467 inches
Stroke
    2.2L models .... 3.73 inches
    2.4L models .... 3.86 inches
Firing order 1-3-4-2
Compression ratio 10:1
Compression pressure .... See Chapter 2D
Oil pressure .... See Chapter 2D

FRONT OF VEHICLE → **1 2 3 4**

1-3-4-2   67020-B-SPECS HAYNES

Cylinder locations and firing order

Timing chain tensioner

Timing chain tensioner compressed length .... 2.83 inches

Hydraulic lash adjuster

Lash adjuster bore diameter .... 0.4730 to 0.4739 inch
Lash adjuster diameter .... 0.4723 to 0.4728 inch
Lash adjuster-to-bore clearance .... 0.0005 to 0.0020 inch

Camshafts

Lobe lift (intake and exhaust) .... See Section 11
Allowable lobe lift loss .... See Section 11
Endplay .... 0.0016 to 0.0057 inch
Journal diameter (all) .... 1.0604 to 1.0614 inches
Bearing inside diameter (all) .... Not available
Journal-to-bearing (oil) clearance .... Not available (but a typical clearance would be 0.0015 to 0.0034 inch)

Torque specifications   Ft-lbs (unless otherwise indicated)

Note: *One foot-pound (ft-lb) of torque is equivalent to 12 inch-pounds (in-lbs) of torque. Torque values below approximately 15 ft-lbs are expressed in inch-pounds, since most foot-pound torque wrenches are not accurate at these smaller values.*

Camshaft sprocket bolts*
    2.2L engine
        Step 1 63
        Step 2 .... Tighten an additional 30-degrees
    2.4L engine
        Step 1 22
        Step 2 .... Tighten an additional 100-degrees

* Bolt(s) must be replaced.

Torque specifications (continued)   Ft-lbs (unless otherwise indicated)

Note: *One foot-pound (ft-lb) of torque is equivalent to 12 inch-pounds (in-lbs) of torque. Torque values below approximately 15 ft-lbs are expressed in inch-pounds, since most foot-pound torque wrenches are not accurate at these smaller values.*

Camshaft bearing cap bolts	
Intake camshaft rear cap bolts	18
All other camshaft cap bolts	89 in-lbs
Crankshaft pulley bolt*	
Step 1	74
Step 2	Tighten an additional 125-degrees
Cylinder head bolts* (in sequence - see illustration 12.16)	
Step 1 Main bolts (1 through 10)	22
Step 2 Main bolts (1 through 10)	Tighten an additional 155-degrees
Step 3 Front bolts *(11 through 14)	26
Drivebelt tensioner bolt	33
Driveplate bolts	
Step 1	39
Step 2	Tighten an additional 25-degrees
Exhaust manifold nuts	124 in-lbs
Exhaust manifold heat shield bolts	17
Exhaust pipe-to-manifold nuts	22
Engine front cover perimeter bolts	18
Engine front cover water pump bolt	18
Drivebelt tensioner bolt	33
Intake manifold bolts/nuts	89 in-lbs
Oil pump cover-to-engine front cover bolts	53 in-lbs
Oil pump pressure relief valve plug	30
Oil pan-to-transaxle bolts	18
Balance shaft retaining bolts	89 in-lbs
Balance shaft chain tensioner	89 in-lbs
Balance shaft chain guides	
Adjustable balance shaft chain guide bolts	89 in-lbs
Small balance shaft chain guide bolts	
2004 and 2005	89 in-lbs
2006 and later	132 in-lbs
Upper balance shaft guide bolts	
2004 and 2005	89 in-lbs
2006 and later	132 in-lbs
Balance shaft sprocket bolt	
2.2L engine	37
2.4L engine	
Step 1	71 in-lbs
Step 2	Tighten an additional 30-degrees
Timing chain tensioner	55
Timing chain guides	
Adjustable timing chain guide bolts	89 in-lbs
Fixed timing chain guide bolts	
2004 and 2005	89 in-lbs
2006 and later	132 in-lbs
Upper timing chain guide bolts	89 in-lbs
Timing chain oiling nozzle bolt	89 in-lbs
Timing chain guide access hole plug	66
Valve cover bolts	89 in-lbs
Valve cover ground strap bolt	89 in-lbs
Water pump bolts	18
Water pump drain bolt	15

Bolt(s) must be replaced.

---

1 General Information

1   This Part of Chapter 2A is devoted to in-vehicle repair procedures for the 2.2L and 2.4L DOHC (Double Overhead Camshaft) engine. Information concerning engine removal and installation and engine block overhaul can be found in Chapter 2D.

2   This engine is equipped with a single timing chain to drive the camshafts. The balance shaft chain drives the two balance shafts and the water pump sprocket. The balance shaft chain is mounted directly behind the camshaft timing chain.

2 Repair operations possible with the engine in the vehicle

1   Many major repair operations can be accomplished without removing the engine from the vehicle.

2   Clean the engine compartment and the exterior of the engine with some type of degreaser before any work is done. It will make the job easier and help keep dirt out of the internal areas of the engine.

3   Depending on the components involved, it may be helpful to remove the hood to improve access to the engine as repairs are performed (refer to Chapter 11 if necessary). Cover the fenders to prevent damage to the paint. Special pads are available, but an old bedspread or blanket will also work.

4   If vacuum, exhaust, oil or coolant leaks develop, indicating a need for gasket or seal replacement, the repairs can generally be made with the engine in the vehicle. The intake and exhaust manifold gaskets, oil pan gasket, crankshaft oil seals and cylinder head gasket are all accessible with the engine in place.

5   Exterior engine components, such as the intake and exhaust manifolds, the oil pan, the oil pump, the water pump, the starter motor, the alternator and the fuel system components can be removed for repair with the engine in place.

6   Since the cylinder head can be removed without pulling the engine, camshaft and valve component servicing can also be accomplished with the engine in the vehicle. Replacement of the timing chain and sprockets is also possible with the engine in the vehicle.

7   In extreme cases caused by a lack of necessary equipment, repair or replacement of piston rings, pistons, connecting rods and rod bearings is possible with the engine in the vehicle. However, this practice is not recommended because of the cleaning and preparation work that must be done to the components involved.

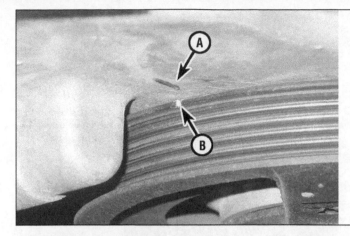

3.8 Timing marks - align the pointer on the engine front cover (A) with the notch in the crankshaft pulley (B)

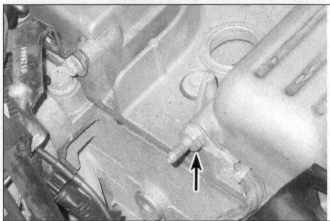

4.7 Location of the ground strap bolt

3   Top Dead Center (TDC) for No. 1 piston - locating

1   Top Dead Center (TDC) is the highest point in the cylinder that each piston reaches as it travels up the cylinder bore. Each piston reaches TDC on the compression stroke and again on the exhaust stroke, but TDC generally refers to piston position on the compression stroke.

2   Positioning the piston(s) at TDC is an essential part of certain repair procedures discussed in this manual.

3   Before beginning this procedure, be sure to place the transmission in Park and apply the parking brake or block the rear wheels. Remove the spark plugs (see Chapter 1). If in the next step you plan on using the starter motor to rotate the engine, disable the fuel pump by removing the fuel pump fuse (see Chapter 4, Section 2).

4   In order to bring any piston to TDC, the crankshaft must be turned using one of the methods outlined below. When looking at the front of the engine, normal crankshaft rotation is clockwise.

a) *The preferred method is to turn the crankshaft with a socket and ratchet attached to the bolt threaded into the front of the crankshaft. Turn the bolt in a clockwise direction only.*

b) *A remote starter switch, which may save some time, can also be used. Follow the instructions included with the switch. Once the piston is close to TDC, use a socket and ratchet as described in the previous paragraph.*

c) *If an assistant is available to turn the ignition switch to the Start position in*

short bursts, you can get the piston close to TDC without a remote starter switch. Make sure your assistant is out of the vehicle, away from the ignition switch, then use a socket and ratchet as described in Paragraph a) to complete the procedure.

5   Install a compression gauge in the No. 1 spark plug hole. It should be a gauge with a screw-in type fitting and a hose at least six inches long.

6   Rotate the crankshaft using one of the methods described above while observing for pressure on the compression gauge. The moment the gauge shows pressure indicates that the No. 1 cylinder has begun the compression stroke.

7   Once the compression stroke has begun, TDC for the compression stroke is reached by bringing the piston to the top of the cylinder.

8   Continue turning the crankshaft until the TDC notch in the crankshaft damper is aligned with the TDC mark on the timing chain cover (see illustration). At this point, the No. 1 cylinder is at TDC on the compression stroke. If the marks were aligned and there was no compression, the piston was on the exhaust stroke. Continue rotating the crankshaft 360 degrees (1-turn) and realign the marks.

9   After the No. 1 piston has been positioned at TDC on the compression stroke, TDC for any of the remaining pistons can be located by turning the crankshaft and follow-

ing the firing order. Divide the crankshaft pulley into two equal sections with chalk marks at each point, each indicating 180 degrees of crankshaft rotation. Rotating the engine past TDC no. 1 to the next mark will place the engine at TDC for cylinder no. 3.

4   Valve cover - removal and installation

Removal

1   On a 2007 and later models, relieve the fuel system pressure (see Chapter 4).

2   Lift the battery box cover up and off of the ball studs, if equipped then disconnect the cable from the negative battery terminal (see Chapter 5, Section 1).

3   Remove the air intake duct from the air filter housing (see Chapter 4). On 2007 and later models, also detach the fuel feed line from the fuel rail (see Chapter 4).

4   Remove the fuel line bracket.

5   Remove the ignition coil/module assembly (2004 through 2006 models) or the ignition coils (2007 and later models) (see Chapter 5).

6   Detach the PCV hose from the valve cover.

7   Remove the ground strap from the cylinder head (see illustration).

8   Detach the engine wiring harness clips from the valve cover, on models so equipped.

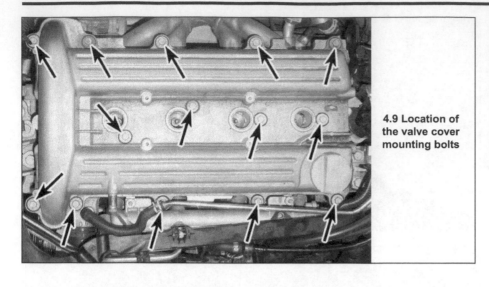

4.9 Location of the valve cover mounting bolts

4.11 Install the gasket into the grooved recess in the valve cover

4.12a Remove the spark plug seal and the O-ring from the valve cover

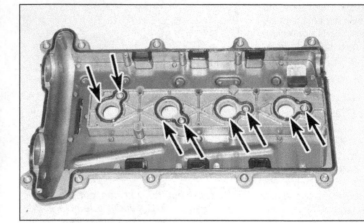

4.12b Be sure to change all the spark plug seals and O-rings in the valve cover

9   Remove the valve cover bolts (see illustration). Lift the valve cover off. Tap gently with a soft-face hammer if necessary to break the gasket seal.

Installation

10   Clean the gasket surfaces on the intake manifold, cylinder head and valve cover. Use a shop rag, lacquer thinner or acetone to wipe off all residue and gasket material from the sealing surfaces.

11   Insert a new valve cover gasket into the grooved recess in the valve cover. Make sure the gasket is positioned properly inside the valve cover groove (see illustration).

12   Install new O-rings and spark plug seals in the valve cover (see illustrations).

13   The remainder of installation is the reverse of the removal Steps. Tighten the valve cover bolts evenly, starting with the center bolts and working out, to the torque listed in this Chapter's Specifications.

14   Reconnect the battery (see Chapter 5, Section 1).

5   Intake manifold - removal and installation

Warning: *The engine must be completely cool before beginning this procedure, otherwise the intake manifold may become damaged.*

Removal

1   Relieve the fuel system pressure (see Chapter 4).

2   Disconnect the cable from the negative terminal of the battery (see Chapter 5).

3   Remove the throttle body (see Chapter 4).

4   Remove the EVAP tube from the intake manifold and the canister purge solenoid valve (see Chapter 6).

5   On 2006 and earlier models disconnect the PCV hose from the intake manifold (see Chapter 1).

6   Remove the fuel rail (see Chapter 4).

7   Disconnect the brake booster vacuum hose and any other vacuum hoses from the intake manifold. Mark each hose with tape

to insure correct reassembly. On 2008 and later models, remove the upper radiator hose.

8   Detach the knock sensor electrical connector from the intake manifold (2006 and earlier models) or from the dipstick tube (2007 models). Remove the dipstick tube mounting bolt and position the dipstick to the side.

9   On 2.4L (LAT) engine models, remove the alternator (see Chapter 2D).

10   Detach the engine wiring harness from the clips on the intake manifold.

11   Remove the intake manifold mounting bolts and nuts (see illustration).

12   Detach the intake manifold from the cylinder head.

Installation

13   Clean away all traces of old gasket material. Remove oil and dirt with a cloth and solvent.

14   Install a new gasket, if necessary.

Note: *The intake manifold gasket does not need to be replaced unless it is deteriorated or has become damaged during the removal process.*

5.11 Intake manifold mounting bolts and nuts

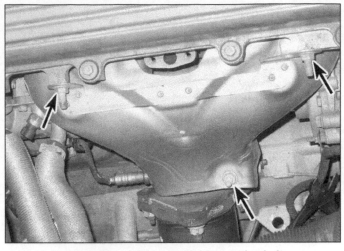

6.3 Location of the heat shield mounting bolts

15   Install the manifold over the studs on the cylinder head. Install the bolts and nuts.
16   Tighten the bolts and nuts to the torque listed in this Chapter's Specifications, starting with the center bolts and working towards the ends.
17   The remainder of installation is the reverse of the removal steps.
18   Reconnect the battery (see Chapter 5, Section 1).
19   Refill the cooling system, if drained (see Chapter 1).
20   Run the engine and check for vacuum and fuel leaks.

6   Exhaust manifold/catalytic converter assembly - removal and installation

Warning: *The engine must be completely cool before beginning this procedure.*

Removal

1   Disconnect the cable from the negative terminal of the battery (see Chapter 5, Section 1).
2   If equipped with an engine block heater, drain the cooling system (see Chapter 1), then disconnect the engine block heater and remove the heater from the engine.
3   Remove the exhaust manifold heat shield (see illustration).
4   Remove the upstream oxygen sensor (see Chapter 6).
5   Raise the vehicle and support it securely on jackstands.
6   Detach the exhaust pipe from the manifold/catalytic converter assembly. On 2007 models, unbolt and remove the exhaust manifold brace.
7   Unbolt the exhaust manifold from the cylinder head (see illustration).

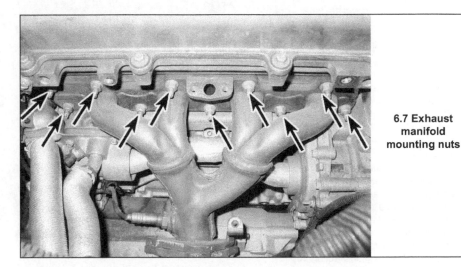

6.7 Exhaust manifold mounting nuts

Installation

8   Using a scraper, thoroughly clean the mating surfaces on the cylinder head, manifold and exhaust pipe. Remove the residue with solvent.
9   Check that the mating surfaces are perfectly flat and not damaged in any way. Warped or damaged manifolds may require machining. Install the new gasket to the cylinder head studs and place the manifold on the cylinder head. Tighten the nuts evenly, working from the center outward, to the torque listed in this Chapter's Specifications.
10   Connect the exhaust pipe to the manifold/catalytic converter assembly and tighten the nuts to the torque listed in this Chapter's Specifications.
11   The remainder of installation is the reverse of the removal steps.
12   Reconnect the battery (see Chapter 5, Section 1).
13   Refill the cooling system, if drained (see Chapter 1).

14   Run the engine and check for exhaust leaks.

7   Front cover - removal and installation

Removal

1   Disconnect the cable from the negative battery terminal (see Chapter 5).
2   Drain the engine oil (see Chapter 1).
3   Remove the drivebelt (see Chapter 1).
4   Remove the drivebelt tensioner from the front cover (see Chapter 1).
5   Remove the crankshaft pulley (see Section 10).
6   Loosen the right front wheel lug nuts, raise the front of the vehicle and support it securely on jackstands. Remove the right front wheel. On 2008 and later models, remove the right engine mount, being careful to support the engine with a hydraulic floor jack. Use a piece of wood between the floor jack and the

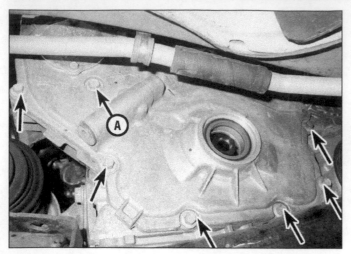

7.7a The front cover mounting bolts can be accessed from below . . .

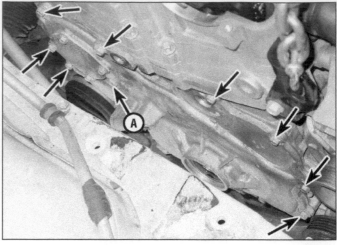

7.7b . . . and from above the engine compartment - don't forget the front cover/water pump bolt (A)

oil pan (see Section 17)

7   Loosen the front cover fasteners gradually and evenly, then remove the fasteners (see illustrations).

Note: *Draw a sketch of the front cover and cover fasteners. Identify the location of all bolts for installation in their original locations.*

8   Remove the water pump bolt from the engine front cover (see illustration 7.7b).

9   Remove the front cover.

10   Remove the front cover-to-block gasket.

Installation

11   Inspect and clean all sealing surfaces of the front cover and the block.

Caution: *Be very careful when scraping on aluminum engine parts. Aluminum is soft and gouges easily. Severely gouged parts may require replacement.*

12   If necessary, replace the crankshaft oil seal in the front cover (see Section 10).

13   Install the front cover gasket on the engine block.

Caution: *The engine cover gasket is reusable. Make sure the gasket has not been damaged. Install a new gasket if necessary.*

14   Install the front cover and cover fasteners. Make sure the hub on the inner rotor is aligned with the flats on the crankshaft and the front cover fasteners are in their original locations. Tighten the fasteners by hand until the cover is contacting the block around its entire periphery.

15   Install the long water pump bolt.

16   Tighten the bolts to the torque listed in this Chapter's Specifications.

17   Install the drivebelt and tensioner. Tighten the drivebelt tensioner to the torque listed in this Chapter's Specifications.

18   Install the crankshaft pulley (see Section 10).

19   Connect the wiring harness connectors. Secure the wiring harnesses with the clamps.

20   Reinstall the remaining parts in the reverse order of removal.

21   Fill the crankcase with the recommended

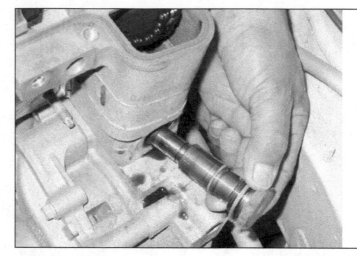

8.8 Remove the timing chain tensioner from the cylinder head

oil (see Chapter 1).

22   Reconnect the battery (see Chapter 5, Section 1).

23   Start the engine and check for leaks. Check all fluid levels.

---

8   Timing chain and sprockets - removal, inspection and installation

---

Removal

Note: *The procedure for 2008 and later vehicles is similar to prior years. It is important in Step 2, that 2008 and later vehicles are set to TDC on the exhaust stroke - not the compression stroke. This can be accomplished easily by finding TDC (see Section 3) then rotating the crankshaft one full revolution.*

Note: *The camshaft sprockets can be removed without the removal of the timing chain if GM special tool J44217 or an equivalent is used to securely hold the timing chain in position while the sprockets are removed. This tool is available from GM dealers and special-*

ty automotive tool suppliers. Make sure that when you use a timing chain holding tool you tighten it fully and engage the hooks fully into the chain.

Note: *Some illustrations show a model without a variable camshaft timing sprocket. All procedures are the same regardless of the type of sprocket.*

1   Disconnect the cable from the negative battery terminal (see Chapter 5).

2   Set the engine to TDC compression for cylinder number 1 (see Section 3). On 2007 and earlier models, refer to Section 3 and set number 1 cylinder to TDC on the compression stroke. On 2008 and later models, rotate the engine exactly one additional full turn so that number 1 cylinder is at TDC on the exhaust stroke.

3   Drain the engine oil (see Chapter 1).

4   Remove the drivebelt (see Chapter 1).

5   Remove the drivebelt tensioner from the front cover.

6   Remove the front cover (see Section 7).

7   Remove the valve cover (see Section 4).

8   Remove the timing chain tensioner (see illustration).

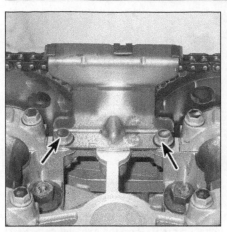

8.9 Remove the upper timing chain guide mounting bolts

8.10 Use a wrench on the hex drive on the camshaft to prevent the camshaft from turning while loosening the sprocket bolt

8.11a Remove the adjustable timing chain guide mounting bolt . . .

8.11b . . . then lift the guide out through the top of the cylinder head

8.12 Access plug for the fixed timing chain guide upper mounting bolt

8.14 Use a wrench on the hex drive on the camshaft when loosening the camshaft sprocket bolt

9   Remove the upper timing chain guide (see illustration).
10   Remove the exhaust camshaft sprocket bolt (see illustration). Be sure to discard the bolt and install a new bolt on reassembly.
11   Remove the adjustable timing chain guide (see illustrations).
12   Unscrew the access bolt and remove the fixed timing chain guide upper mounting bolt (see illustration).
13   Remove the fixed timing chain guide lower mounting bolt and lift the guide out.
14   Remove the intake camshaft sprocket bolt (see illustration). Be sure to discard the bolt and install a new bolt on reassembly.
15   Remove the timing chain through the top of the cylinder head (see illustration).
16   Remove the timing chain guide sprocket and slide the timing chain oiling nozzle off the engine block.

Inspection

17   Clean all parts with clean solvent and dry with compressed air, if available.

18   Inspect the chain tensioner for excessive wear or other damage. Be sure to drain all the oil out of the chain tensioner if it is to be reused.
19   Inspect the timing chain guides for deep grooves, excessive wear, or other damage.
20   Inspect the timing chain for excessive wear or damage.
21   Inspect the crankshaft and camshaft sprockets for chipped or broken teeth, excessive wear, or damage.
22   Replace any component that is in questionable condition.

Installation

Caution: *Before starting the engine, carefully rotate the crankshaft by hand through at least two full revolutions (use a socket and breaker bar on the crankshaft pulley center bolt). If you feel any resistance, STOP! There is something wrong - most likely, valves are contacting the pistons. You must find the problem before proceeding.*

8.15 Carefully remove the timing chain and the sprocket through the top of the cylinder head

8.24 The round dot (alignment mark) on the sprocket should be in the 5 o'clock position. When installing the chain, one of the silver plated links must be aligned with this dot

8.25 The copper link must align with the INT on the intake camshaft and the silver link with the EXH on the exhaust camshaft

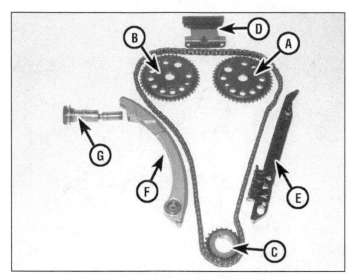

8.29 Timing chain component details

A	Intake camshaft sprocket	E	Fixed timing chain guide
B	Exhaust camshaft sprocket	F	Adjustable timing
C	Crankshaft sprocket		chain guide
D	Upper timing chain guide	G	Timing chain tensioner

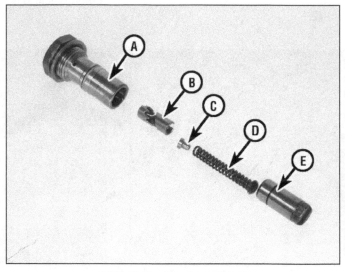

8.33a Timing tensioner details

A	Timing chain	C	Spring adjuster
	tensioner body	D	Spring
B	Ratchet cylinder	E	Piston

23   If the crankshaft has been rotated during this procedure, make sure the No. 1 piston is at the top of its stroke (TDC) (see Section 3). The crankshaft key should be at the 12 o'clock position.

24   Be sure to check that the notch on the intake camshaft is in the 5 o'clock position (see illustration) and the exhaust camshaft notch is in the 7 o'clock position, then install the intake camshaft sprocket onto the camshaft. Be sure to install a new bolt. Tighten the intake camshaft sprocket bolt lightly, finger tight at this time.

Caution: *Do not turn the camshaft more than 1/2 turn to avoid any valve/piston contact. The camshafts should be positioned correctly before the timing chain is installed.*

25   Install the timing chain by lowering it from the top through the opening. Be sure the timing chain drops down around both sides of the cylinder block bosses. Be sure the bright plated link (copper) on the chain is aligned

with the INT designation on the camshaft sprocket (see illustration).

Note: *The copper link will be installed at the intake camshaft sprocket (front) while the silver links will be installed at the crankshaft sprocket and the exhaust camshaft sprocket (rear).*

26   Drape the timing chain over the crankshaft sprocket and engage the chain over the plated link (silver) on the crankshaft sprocket timing mark located in the 5 o'clock position (see illustration 8.24).

27   Install the adjustable timing chain guide. Install the bolts and tighten them to the torque listed in this Chapter's Specifications.

28   Install the exhaust camshaft sprocket onto the camshaft. Be sure to install a new bolt. Be sure the plated link (silver) on the chain is aligned with the EXH designation on the camshaft sprocket (see illustration 8.25). Tighten the exhaust camshaft sprocket bolt lightly, finger-tight at this time.

29   Install the fixed timing chain guide (see illustration). Tighten the bolts to the torque listed in this Chapter's Specifications.

30   Install the upper timing chain guide (see illustration 8.9). Tighten the bolts to the torque listed in this Chapter's Specifications.

31   Hold the intake camshaft with a wrench on its hex as a back-up, and tighten the intake camshaft bolt to the torque listed in this Chapter's Specifications.

32   Hold the exhaust camshaft with a wrench on its hex as a back-up, and tighten the exhaust camshaft bolt to the torque listed in this Chapter's Specifications.

33   Install the timing chain tensioner.

Caution: *The timing chain tensioner must be installed in the compressed state. Do not install a tensioner in its released state. Damage to the tensioner and timing chain will occur.*

a)   *Completely disassemble the tensioner and drain all the oil (see illustration).*

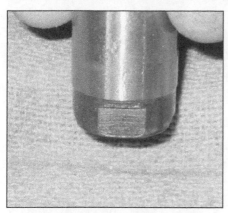

8.33b Install the piston with the flats locked into the jaws of the vise

8.33c Align the groove in the ratchet cylinder with the pin in the piston

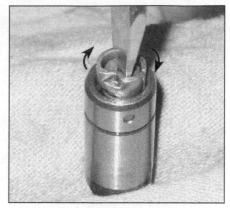

8.33d Using a flat-bladed screwdriver, drive the ratchet cylinder down to the bottom and rotate it clockwise to lock it into position

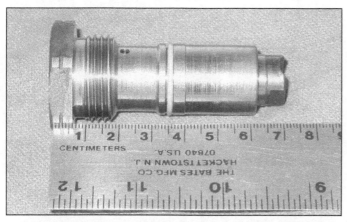

8.33e The tensioner should measure the correct length in its compressed state or it must be replaced with a new tensioner

9.5 Location of the balance shaft chain tensioner mounting bolts

b) Inspect the tensioner body, the piston and all components for scoring or damage. If necessary, replace the tensioner with a new one.
c) Install the tensioner piston into the vise with the flats seated in the jaws of the vise (see illustration).
d) Install the ratchet cylinder into the piston, aligning the groove with the locating pin (see illustration).
e) Drive the ratchet cylinder into the piston with a flat-bladed screwdriver. Rotate the ratchet cylinder clockwise when it reaches the bottom (see illustration). The ratchet cylinder should be locked into position.
f) The tensioner must measure 2.83 inches (72 mm) from end-to-end (see illustration).

34 Install the timing chain oiling nozzle. Tighten the bolt to the torque listed in this Chapter's Specifications.
35 Apply a small amount of RTV sealant to the threads and install the timing chain guide access plug. Tighten the bolt to the torque listed in this Chapter's Specifications.
36 Use a wrench on the crankshaft bolt to

rotate the engine several complete revolutions. This will be easier with the spark plugs removed. If the engine does not rotate freely, then the timing marks may not have been aligned correctly. If this is the case, stop and check your work before proceeding.
37 Install the valve cover (see Section 4).
38 Install the engine front cover (see Section 7).
39 The remainder of installation is the reverse of removal.
40 Reconnect the battery (see Chapter 5, Section 1).
41 Refill the crankcase with oil and change the oil filter (see Chapter). Run the engine and check for leaks.

---

9 Balance shaft chain and balance shafts - removal, inspection and installation

Note: This procedure covers removal of the balance shaft chain and balance shafts, but take note that the shafts themselves can only be removed from the engine block af-

ter the engine has been removed from the vehicle. If there is a problem with the balance shafts that does warrant their removal, the engine would have to be removed anyway, since replacement of the balance shaft bushings is a job that must be left to an automotive machine shop. If you're just removing or replacing the chain, ignore the steps that don't apply.

Removal

1 Disconnect the cable from the negative battery terminal (see Chapter 5).
2 Drain the engine oil (see Chapter 1).
3 Remove the timing chain, timing chain guides and sprockets (see Section 8).
4 Check to make sure the engine is positioned at TDC for cylinder number 1 (see Section 3).
Caution: Do not rotate the engine to find TDC number 1 when the timing chain is removed unless the engine has been rotated accidentally. If the engine is not positioned at TDC number 1, the camshafts must be removed to prevent damage to the valves (see Section 11).
5 Remove the balance shaft chain tensioner (see illustration).

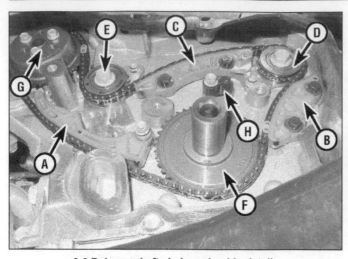

9.6 Balance shaft chain and guide details

A   Adjustable balance shaft chain guide
B   Small balance shaft chain guide
C   Upper balance shaft chain guide
D   Intake side (front) balance shaft sprocket
E   Exhaust side (rear) balance shaft sprocket
F   Crankshaft/balance shaft sprocket
G   Water pump sprocket
H   Timing chain oiling nozzle

9.9 Balance shaft sprocket/chain alignment marks (A)
and retainer bolts (B)

6   Remove the adjustable balance shaft chain guide (see illustration).
7   Remove the small balance shaft chain guide (see illustration 9.6).
8   Remove the upper balance shaft chain guide (see illustration 9.6).
9   Remove the balance shaft drive chain (see illustration).
Note: *To aid in removal, gather all the slack in the chain between the water pump sprocket and the crankshaft sprocket.*
10   If you're removing the balance shafts (engine removed from the vehicle), remove the balance shaft retainer bolts.
11   Remove the balance shafts from the engine block.
Caution: *Mark each balance shaft to insure correct reassembly. The balance shafts are not interchangeable. Do not install the balance shaft into the wrong bore or extreme engine vibration will occur.*

Inspection

12   Clean all parts with clean solvent and dry with compressed air, if available.
13   Inspect the chain tensioners for excessive wear or other damage.
14   Inspect the balance shaft chain guides for deep grooves, excessive wear, or other damage.

15   Inspect the balance shaft chain for excessive wear or damage.
16   Inspect the crankshaft and water pump sprockets for chipped or broken teeth, excessive wear, or damage.
17   Replace any component that is damaged.

Installation

18   Before installing the balance shaft chain, make sure the crankshaft timing mark (round dot) is pointing to the 6 o'clock position (see illustration).
Caution: *Do not rotate the engine to find TDC number 1 after the timing chain has been removed unless the engine has been rotated accidentally. If the engine is not positioned at TDC number 1, the camshafts must be removed to prevent damage to the valves (see Section 11).*
19   Install the balance shafts into the bores and tighten the balance shaft retainer bolts to the torque listed in this Chapter's Specifications.
20   Align the balance shaft sprockets before installing the balance shaft chain. Starting with the intake side balance shaft, place the alignment arrow pointing up, then temporarily install a drill bit into the alignment hole and the sprocket teeth to lock the balance shaft sprocket in place (see illustration).
21   Now position the exhaust side (rear) balance shaft sprocket with the arrow pointing down and aligned with the cutout in the retainer, then install a drill bit into the alignment hole to hold the sprocket (see illustrations).
22   Install the balance shaft chain onto the

9.18 The timing mark (round dot) on the crankshaft sprocket should point to the 6 o'clock position (approximately)

9.20 With the arrow on the intake side balance shaft sprocket pointing up (and aligned with the cutout on the balance shaft retainer, not visible in this photo, but similar to the one shown in illustration

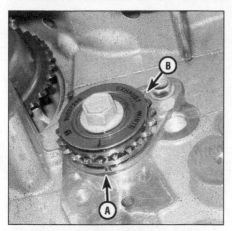

9.21a Location of the alignment notch for the sprocket arrow (A) and the alignment hole (B) on the exhaust side balance shaft sprocket

9.21b Install the drill bit into the exhaust balance shaft retainer to lock it into position

9.26a Rotate the plunger 90-degrees, align the holes in the body and piston . . .

9.26b . . . then install a drill bit to retain the piston in the locked position

9.28 After the tensioner is installed and the bolts tightened, remove the drill bit

10.5 A pin spanner can be used to prevent the pulley from rotating

balance shaft/crankshaft sprocket and the balance shafts. Align the colored links with the alignment marks on each sprocket. Position the copper-colored link onto the intake side balance shaft, aligning the mark with the colored link at approximately the 12 o'clock position (see illustration 9.9).

Note: *The copper link will be installed at the intake balance shaft sprocket (front) while the silver links will be installed at the crankshaft sprocket and the exhaust balance shaft sprocket (rear).*

23 Working clockwise, position the second colored link (silver) on the crankshaft/balance shaft sprocket, aligning the mark on the sprocket with the colored link at the 6 o'clock position (see illustration 9.18).

24 Finally, pass the chain over the water pump sprocket, under the exhaust balance shaft sprocket and into position. Align the third colored link (silver) on the exhaust balance shaft sprocket, aligning the mark on the sprocket with the colored link at the 6 o'clock position.

25 Install the balance shaft chain guides (see illustration 9.6). Tighten the bolts to the torque listed in this Chapter's Specifications.

26 Reset the balance shaft chain tensioner. Turn the tensioner plunger 90-degrees in the bore and compress the tensioner plunger (see illustration). Rotate the plunger back to the original position at 12 o'clock and install a paper clip through the hole in the body into the plunger (see illustration).

27 Install the balance shaft chain tensioner and tighten the bolts to the torque listed in this Chapter's Specifications.

28 Remove the drill bit to release the plunger (see illustration).

29 Recheck all the balance shaft chain timing marks.

30 Install the timing chain (see Section 8) and all components removed previously.

31 Reconnect the battery (see Chapter 5, Section 1).

32 Install a new oil filter and refill the crankcase with oil (see Chapter 1).

33 Run the engine and check for leaks.

10 Crankshaft pulley and front oil seal - removal and installation

Removal

1 Disconnect the cable from the negative battery terminal (see Chapter 5).

2 Remove the drivebelt (see Chapter 1).

3 Raise the vehicle and support it securely on jackstands.

4 Remove the splash shield from below the engine compartment (see Chapter 11).

5 Use a breaker bar and socket to remove the crankshaft pulley center bolt (see illustration). Discard the bolt and obtain a new one for installation.

Note: *It will be necessary to lock the pulley in position using a strap wrench or special tool. Be sure to wrap a length of old drivebelt around the pulley if you are using a strap wrench.*

6 Slide the puller off the nose of the crankshaft. If the pulley is stuck, use a puller that

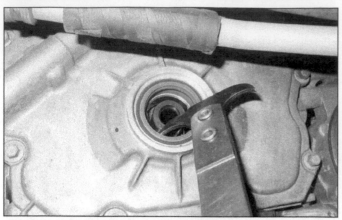

10.7 Use a seal puller to remove the old crankshaft seal, taking care not to damage the crankshaft or the seal bore in the cover

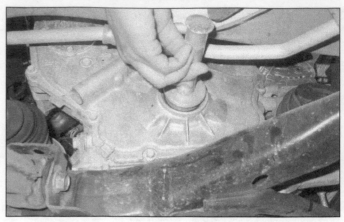

10.9 Driving the new front cover seal in with a seal driver

11.5a Install a camshaft locking tool to hold the sprockets and timing chain in place - make sure the camshaft sprockets are locked properly and the tool is bolted to the cylinder head

11.5b The diamond-shaped hole on the intake camshaft should be in the 12 o'clock position

bolts to the three threaded holes in the pulley hub. Additionally, a spacer, such as a deep socket that just fits into the hole in the pulley and bears on the crankshaft, will be required to avoid damage to the crankshaft.

7   Use a seal puller to remove the crankshaft front oil seal (see illustration). A screwdriver may be used instead, if the tip is wrapped with tape to avoid scratching the crankshaft.

8   Clean the seal bore and check it for nicks or gouges. Also examine the area of the hub that rides in the seal for signs of abnormal wear or scoring. For many popular engines, a repair sleeve is available to restore a smooth finish to the sealing surface. Check with your auto parts store.

Installation

9   Coat the lip of the new seal with clean engine oil and drive it into the bore with a seal driver or a socket slightly smaller in diameter than the seal (see illustration). The open side of the seal faces into the engine.

10   Using clean engine oil, lubricate the sealing surface of the hub. Install the crankshaft

pulley/damper with a special installation tool, available at most auto parts stores. Do not use a hammer to install the pulley/damper. Install a new center bolt and tighten it to the torque listed in this Chapter's Specifications. **Note:** *You must use a new pulley bolt.*

11   The remainder of the installation is the reverse of the removal procedure.

12   Reconnect the battery (see Chapter 5, Section 1).

---

11   Camshafts and hydraulic lash adjusters - removal, inspection and installation

Note: *This is a difficult procedure, involving special tools. Read through the entire Section and obtain the necessary tools before beginning the procedure.*

Removal

1   Disconnect the cable from the negative battery terminal (see Chapter 5).

2   Remove the valve cover (see Section 4).

3   Set the engine to TDC for cylinder num-

ber one (see Section 3), then turn the crankshaft counterclockwise until the engine is set at 60-degrees before TDC. At this point, the diamond-shaped hole on the intake camshaft should be in the 12 o'clock position.

Caution: *Do not remove the camshafts with the engine at TDC number 1 or the valves and pistons will be damaged.*

4   Remove the upper timing chain guide (see Section 8).

5   Install a special tool to secure the camshaft sprockets in position (see illustrations). This camshaft locking tool (jig) can be purchased through a dealership parts department, through specialty automotive tool suppliers, or at some auto parts stores.

Note: *The J44217 tool used in Section 8 can be used instead of the camshaft locking tool shown. The J44217 tool will hold the timing chain tightly in place, then the sprockets can be secured to the chain with plastic cable ties or wire.*

6   Remove the camshaft sprocket bolts and slide the camshaft sprockets forward, then tighten the wingnuts to hold the sprockets securely.

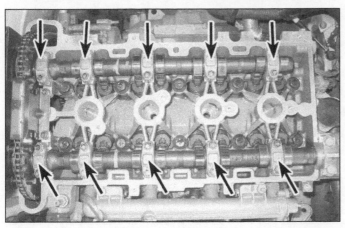

11.7a The camshaft bearing cap designations are stamped onto each cap

11.7b Note that the arrow on the cap faces the timing chain end of the engine

11.8 Remove each rocker arm . . .

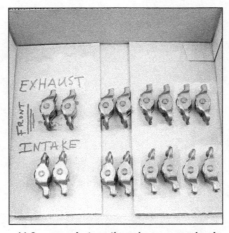

11.9 . . . and store them in an organized manner so they can be returned to their original locations

11.10 Pull the lash adjusters from their bores in the head and store them along with their corresponding rocker arms

Intake camshaft

7   Each camshaft cap is marked with a number indicating its position (see illustrations). A little at a time, loosen each bearing cap bolt slowly and evenly, allowing the camshaft to lift from the cylinder head, parallel to the surface of the cylinder head.

Caution: *The caps must be installed in their original locations. Keep all parts from each camshaft together; never mix parts from one camshaft with those for another.*

8   Remove the rocker arms (see illustration).

9   Place the rocker arms in a suitable container, in order, so they can be reinstalled in their original positions (see illustration).

10   Remove the hydraulic lash adjusters from their bores in the cylinder head (see illustration). Store these with their corresponding rocker arms so they can be reinstalled in their original locations.

Exhaust camshaft

11   Mark the exhaust bearing caps in the original positions and remove them from the cylinder head. Each camshaft cap is designated with a number (see illustrations 11.7a

and 11.7b). Loosen each bearing cap nut slowly and evenly, allowing the camshaft to lift from the cylinder head, parallel to the surface of the cylinder head.

Caution: *The camshaft bearing caps are numbered to identify the locations of the caps. The caps must be installed in their original locations. Keep all parts from each camshaft together; never mix parts from one camshaft with those for another.*

12   Mark the positions of the rocker arms so they can be reinstalled in their original locations, then remove the rocker arms.

13   Place the rocker arms in a suitable container so they can be separated and identified (see illustration 11.9).

14   Lift the hydraulic lash adjusters from their bores in the cylinder head. Identify and separate the adjusters so they can be reinstalled in their original locations (see illustration 11.10).

Inspection

15   Check each hydraulic lash adjuster for excessive wear, scoring, pitting, or an out-of-round condition (see illustration). Replace as necessary.

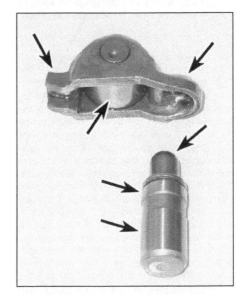

11.15 Check the rocker arms and lash adjusters for wear at the indicated points

11.18 Check the cam lobes for pitting, excessive wear, and scoring. If scoring is excessive, as shown here, replace the camshaft

11.19 Measure each camshaft lobe height with a micrometer

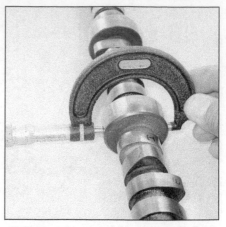

11.20 Measure each journal diameter with a micrometer. If any journal is less than the specified minimum, replace the camshaft

11.21 Lay a strip of Plastigage on each camshaft journal, in line with the camshaft

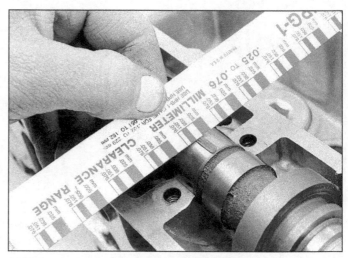

11.22 Compare the width of the crushed Plastigage to the scale on the package to determine the journal oil clearance

16   Measure the outside diameter of each adjuster at the top and bottom of the adjuster. Then take a second set of measurements at a right angle to the first. If any measurement is significantly different from the others, the adjuster is tapered or out of round and must be replaced. If the necessary equipment is available, measure the diameter of the lash adjuster and the inside diameter of the corresponding cylinder head bore. Subtract the diameter of the lash adjuster from the bore diameter to obtain the oil clearance. Compare the measurements obtained to those given in this Chapter's Specifications. If the adjusters or the cylinder head bores are excessively worn, new adjusters or a new cylinder head, or both, may be required. If the valve train is noisy, particularly if the noise persists after a cold start, you can suspect a faulty lash adjuster.

17   Inspect the rocker arms for signs of wear or damage. The areas of wear are the tip that contacts the valve stem, the socket that contacts the lash adjuster and the

roller that contacts the camshaft (see illustration 11.15).

18   Examine the camshaft lobes for scoring, pitting, galling (wear due to rubbing), and evidence of overheating (blue, discolored areas). Look for flaking of the hardened surface layer of each lobe (see illustration). If any such wear is evident, replace the camshaft.

19   Measure the lobe height of each cam lobe on the intake camshaft, and record your measurements (see illustration). Compare the measurements for excessive variation; if the lobe heights vary more than 0.005 inch (0.125 mm), replace the camshaft. Compare the lobe height measurements on the exhaust camshaft and follow the same procedure. Do not compare intake camshaft lobe heights with exhaust camshaft lobe heights, as they are different. Only compare intake lobes with intake lobes and exhaust lobes with exhaust lobes.

20   Inspect the camshaft bearing journals and the cylinder head bearing surfaces for pitting or excessive wear. If any such wear is

evident, replace the component concerned. Using a micrometer, measure the diameter of each camshaft bearing journal at several points (see illustration). If the diameter of any journal is less than specified, replace the camshaft.

21   To check the bearing journal oil clearance, remove the rocker arms and hydraulic lash adjusters (if not already done), use a suitable solvent and a clean lint-free rag to clean all bearing surfaces, then install the camshafts and bearing caps with a piece of Plastigage across each journal (see illustration). Tighten the bearing cap bolts to the specified torque. Don't rotate the camshafts.

22   Remove the bearing caps and measure the width of the flattened Plastigage with the Plastigage scale (see illustration). Scrape off the Plastigage with your fingernail or the edge of a credit card. Don't scratch or nick the journals or bearing caps.

23   If the oil clearance of any bearing is worn beyond the service limit, install a new camshaft and repeat the check. If the clearance is

11.29 Camshaft and timing sprocket alignment details

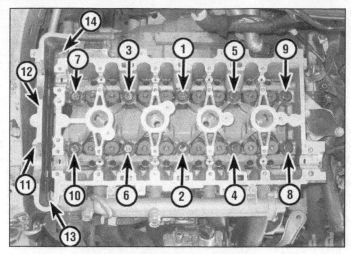

12.16 Cylinder head bolt tightening sequence

still excessive, replace the cylinder head.

24   To check camshaft endplay, remove the hydraulic lash adjusters, clean the bearing surfaces carefully, and install the camshafts and bearing caps. Tighten the bearing cap bolts to the specified torque, then measure the endplay using a dial indicator mounted on the cylinder head so that its tip bears on the camshaft end.

25   Lightly but firmly tap the camshaft fully toward the gauge, zero the gauge, then tap the camshaft fully away from the gauge and note the gauge reading. If the measured endplay is at or beyond the specified service limit, install a new camshaft thrust cap and repeat the check. If the clearance is still excessive, the camshaft or the cylinder head must be replaced.

Installation

26   Lubricate the rocker arms and hydraulic lash adjusters with engine assembly lubricant or fresh engine oil. Install the adjusters into their original bores, then install the rocker arms in their correct locations.

27   Lubricate the camshafts with camshaft installation lubricant and install them in their correct locations. Position the camshafts with the slots in the end of the camshafts positioned as shown in illustration 11.29, aligning them with the slots in the camshaft sprockets.

28   Install the camshaft bearing caps in their correct locations, except for the front end and rear end bearing caps on each camshaft. Install the cap bolts and tighten by hand until snug. Tighten the bolts in four to five steps, starting with the center cap and working to the outside caps, to the torque listed in this Chapter's Specifications.

29   Slide the camshaft sprockets and timing chain along the guide pins toward the camshafts. Rotate the camshafts with an open-end wrench on the hex drive on each camshaft until the slots are aligned with the projections on the sprockets (see illustration). Install new bolts and tighten the camshaft sprockets to the torque listed in this Chapter's

Specifications (see Section 8).

30   Remove the camshaft locking tool from the cylinder head. Then install the front and rear camshaft caps and tighten them to the torque listed in this Chapter's Specifications. Note that the rear cap on the intake camshaft is equipped with larger bolts and requires a different torque.

31   Install the upper timing chain guide (see Section 8). Rotate the engine by hand two revolutions - if you feel any resistance, stop and find out why.

32   The remainder of installation is the reverse of removal.

33   Reconnect the battery (see Chapter 5, Section 1).

---

12   Cylinder head - removal and installation

Caution: *The engine must be completely cool when the head is removed. Failure to allow the engine to cool off could result in head warpage.*

Removal

1   Disconnect the cable from the negative battery (see Chapter 5).

2   Wait until the engine is completely cool, then drain the cooling system (see Chapter 1).

3   Remove the drivebelt (see Chapter 1) and the drivebelt tensioner.

4   Remove the exhaust manifold (see Section 6).

5   Remove the intake manifold (see Section 5).

6   Remove the timing chain (see Section 8).

7   Label and disconnect the electrical connectors from the cylinder head that will interfere with removal. Use tape and mark each connector to insure correct reassembly.

8   Remove the cylinder head bolts and discard them, following the reverse of the tightening sequence (see illustration 12.16). Loosen the bolts in sequence 1/4-turn at a time. If the

head is to be completely overhauled, refer to Section 11 for removal of the camshafts, rocker arms and hydraulic lash adjusters.

9   Use a prybar at the corners of the head-to-block mating surface to break the gasket seal. Do not pry between the cylinder head and engine block in the gasket sealing area.

10   Lift the cylinder head off the engine. If resistance is felt, place a wood block against the end and strike the wood block with a hammer. Store the cylinder head on wood blocks to prevent damage to the gasket sealing surfaces.

11   Remove the old cylinder head gasket. Before removing, note the correct orientation of the gasket for correct installation.

Installation

12   The mating surfaces of the cylinder head and block must be perfectly clean when the head is installed. Use a gasket scraper to remove all traces of carbon and old gasket material, then clean the mating surfaces with lacquer thinner or acetone. If there's oil on the mating surfaces when the cylinder head is installed, the gasket may not seal correctly and leaks may develop. When working on the engine block, cover the open areas of the engine with shop rags to keep debris out during repair and reassembly. Use a vacuum cleaner to remove any debris that falls into the cylinders.

13   Check the engine block and cylinder head mating surfaces for nicks, deep scratches and other damage.

14   Use a tap of the correct size to chase the threads in the cylinder head bolt holes. Dirt, corrosion, sealant and damaged threads will affect torque readings.

15   Make sure the new gasket is located on the dowels in the block.

16   Carefully position the cylinder head on the engine block without disturbing the gasket. Install new cylinder head bolts and, following the recommended sequence (see illustration), tighten the bolts to the torque listed in this Chapter's Specifications. All the main cyl-

13.6a The forward oil pan mounting bolts can be accessed near the subframe

inder head bolts (numbers 1 through 10) are tightened in the first Step and second Step. The four smaller bolts located on the front of the cylinder head are the only ones tightened in the third Step. Mark a stripe on each of the main cylinder head bolts to help keep track of the bolts that have been tightened the additional 155-degrees.

Note: *The method used for the head bolt tightening procedure is referred to as a "torque-angle" method. A special torque angle gauge (available at most auto parts stores) is available to attach to a breaker bar and socket for better accuracy during the tightening procedure.*

17   Install the timing chain (see Section 8).
18   Install the exhaust manifold (see Section 5).
19   Install the intake manifold (see Section 5).
20   The remaining installation steps are the reverse of removal.
21   Reconnect the battery (see Chapter 5,

Section 1).
22   Change the engine oil and filter and refill the cooling system (Chapter 1), then start the engine and check carefully for oil and coolant leaks.

---

13   Oil pan - removal and installation

Removal

1   Drain the engine oil and remove the drivebelt (see Chapter 1).
2   Loosen the right-front wheel lug nuts, raise the front of the vehicle and support it securely on jackstands. Remove the right front wheel.
3   Remove the splash shield from below the right side of the engine compartment.
4   Remove the lower air conditioning compressor mounting bolt (see Chapter 3). Loosen, but don't remove, the other compres-

sor mounting bolts.
5   Remove the dipstick an the dipstick tube (the tube is bolted to the intake manifold) then disconnect the electrical connector and harness to the electric coolant and remove the pump-to-oil pan mounting bolts, if equipped.
6   Remove the oil pan bolts. Follow the reverse of the tightening sequence (see illustrations and illustration 13.10).
7   Carefully remove the oil pan from the lower crankcase.

Caution: *If the oil pan is difficult to separate from the lower crankcase, use a rubber mallet or a block of wood and a hammer to jar it loose. If it's stubborn and still won't come off, pry carefully on casting protrusions (not the mating surfaces!).*

Installation

8   Using a gasket scraper, thoroughly clean all old gasket material from the lower crankcase and oil pan. Remove residue and oil film with a solvent such as acetone or lacquer thinner.
9   Apply a 2 mm bead of RTV sealant to the perimeter of the oil pan, inboard of the bolt holes, and around the oil suction port. Allow the sealant to set-up before installing the oil pan to the engine (but be sure to install the pan in the time given by the sealant manufacturer).
10   Install the oil pan and bolts (see illustration). Follow the correct torque sequence and tighten the bolts to the torque listed in this Chapter's Specifications.
11   The remaining installation is the reverse of removal. Be sure to tighten the wheel lug nuts to the torque listed in the Chapter 1 Specifications.
12   Refill the engine with oil and install a new oil filter (see Chapter 1), then run the engine and check for leaks.

13.6b Oil pan-to-transaxle bolts

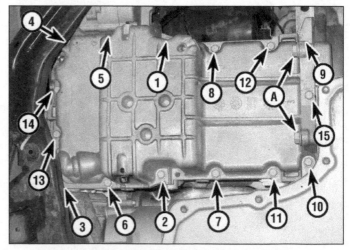

13.10 Oil pan bolt tightening sequence. Tighten the pan-to-transaxle bolts (A, other two bolts not visible in photo) until they're snug (but not too tight), then tighten the pan-to-block bolts in numerical order to the torque listed in this Chapter's Specifications. Finally, tighten bolts (A) securely

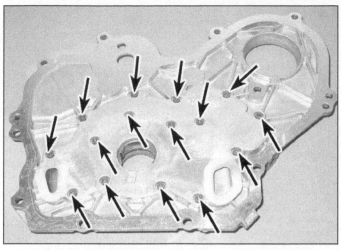

14.5a Location of the oil pump cover mounting bolts

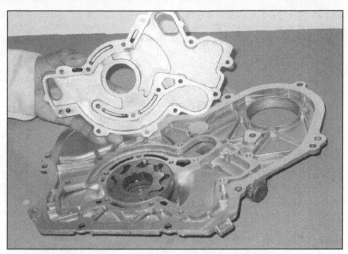

14.5b Lift the oil pump cover from the oil pump assembly

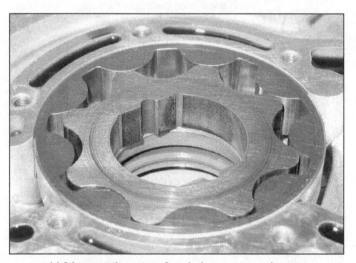

14.8 Inspect the rotors for obvious wear or damage

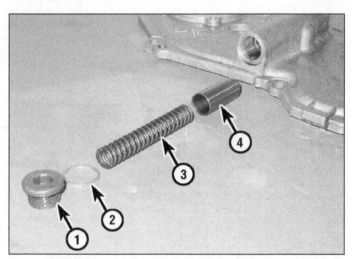

14.10 Oil pressure relief valve component details

1   Oil pressure relief valve access plug
2   Washer
3   Spring
4   Piston

14   Oil pump - removal, inspection and installation

Removal

1   Drain the engine oil (see Chapter 1).
2   Remove the drivebelt (see Chapter 1).
3   Loosen the right-front wheel lug nuts, raise the front of the vehicle and support it securely on jackstands. Remove the right front wheel.
4   Remove the engine front cover (see Section 7).
5   Working on the backside of the engine cover, loosen the oil pump cover screws, a little at a time, until they're all loose (see illustration). When all of the screws are loose, remove the cover (see illustration).

Inspection

6   Note any identification marks on the rotors and withdraw the rotors from the pump body. If no marks can be seen, use a permanent marker and make your own to ensure that they will be installed correctly.
7   Thoroughly clean and dry the components.
8   Inspect the rotors for obvious wear or damage (see illustration). If either rotor, the pump body or the cover is scored or damaged, the complete oil pump assembly must be replaced.
9   If the oil pump components are in acceptable condition, dip the rotors in clean engine oil and install them into the pump body with any identification marks positioned as noted during disassembly.
10   Remove the oil pressure relief valve

components from the oil pump body. Thoroughly clean and dry the components. Inspect the components for obvious wear or damage. Install them in the correct order (see illustration).

Installation

11   Install the rotors into the housing with the hub of the inner rotor facing the engine front cover. The inner rotor hub must be installed correctly or the engine front cover will not fasten properly.
12   Install the oil pump cover and screws and tighten by hand until snug. Then tighten the screws gradually and evenly to the torque listed in this Chapter's Specifications.
13   Install the engine front cover (see Section 7).
14   Refer to Chapter 1 and fill the engine

with fresh engine oil. Install a new oil filter.
15   Start the engine and check for leaks.
16   Run the engine and make sure oil pressure comes up to normal quickly. If it doesn't, stop the engine and find out the cause. Severe engine damage can result from running an engine with insufficient oil pressure!

15   Driveplate - removal and installation

Removal

1   Raise the vehicle and support it securely on jackstands, then refer to Chapter 7B and remove the transaxle. If it's leaking, now would be a very good time to replace the front pump seal/O-ring.
2   Use a center punch or paint to make alignment marks on the driveplate and crankshaft to ensure correct alignment during reinstallation.
3   Remove the bolts that secure the driveplate to the crankshaft. If the crankshaft turns, wedge a screwdriver in the ring gear teeth to jam the driveplate.
4   Remove the driveplate from the crankshaft and, if equipped, retrieve the spacer between the crankshaft and the driveplate.

Installation

5   Clean the driveplate to remove grease and oil. Inspect the surface for cracks or other damage. Check for cracked and broken ring gear teeth. Lay the driveplate on a flat surface and use a straightedge to check for warpage.
6   Clean and inspect the mating surfaces of the driveplate and the crankshaft. If the crankshaft rear seal is leaking, replace it before reinstalling the driveplate (see Section 16).
7   Position the driveplate against the crankshaft. Be sure to align the marks made during removal. Note that some engines have an alignment dowel or staggered bolt holes to ensure correct installation. Before installing the bolts, apply thread locking compound to the threads.
8   Wedge a screwdriver in the ring gear teeth to keep the driveplate from turning and

tighten the bolts to the torque listed in this Chapter's Specifications.
9   The remainder of installation is the reverse of the removal procedure.

16   Rear main oil seal - replacement

1   The one-piece rear main oil seal is pressed into engine block and the crankcase reinforcement section. Remove the transaxle (see Chapter 7B) and the driveplate (see Section 15).
2   Pry out the old seal with a seal removal tool or a flat blade screwdriver. Caution: To prevent an oil leak after the new seal is installed, be very careful not to scratch or otherwise damage the crankshaft sealing surface or the bore in the engine block.
3   Clean the crankshaft and seal bore in the block thoroughly and de-grease these areas by wiping them with a rag soaked in lacquer thinner or acetone. Lubricate the lip of the new seal and the outer diameter of the crankshaft with engine oil.
4   Position the new seal onto the crankshaft, being careful not to roll-over the lip of the seal.
Note: *When installing the new seal, if so marked, the words THIS SIDE OUT on the seal must face out, toward the rear of the engine. Use a special rear main oil seal installation tool or a socket with the exact diameter of the seal to drive the seal in place. Make sure the seal is not off-set; it must be flush along the entire circumference of the engine block and the crankcase reinforcement section.*
5   The remainder of installation is the reverse of removal.

17   Engine mounts - check and replacement

Check

1   Engine mounts seldom require attention, but broken or deteriorated mounts should be replaced immediately or the added strain

placed on the driveline components may cause damage or wear.
2   During the check, the engine must be raised slightly to remove the weight from the mounts.
3   Raise the vehicle and support it securely on jackstands, then position a jack under the engine oil pan. Place a large block of wood between the jack head and the oil pan, then carefully raise the engine just enough to take the weight off the mounts.
Warning: *DO NOT place any part of your body under the engine when it's supported only by a jack!*
4   Check the mounts to see if the rubber is cracked, hardened or separated from the bushing in the center of the mount.
5   Check for relative movement between the mounts and the engine or frame (use a large screwdriver or pry bar to attempt to move the mounts).
6   If movement is noted, lower the engine and tighten the mount fasteners.

Replacement

7   Disconnect the cable from the negative terminal on the battery (see Chapter 5), then raise the vehicle and support it securely on jackstands.
8   Place a large block of wood between the jack head and the oil pan, then carefully raise the engine just enough to take the weight off the mounts.
Caution: *Do not disconnect more than one mount at a time unless the engine will be removed from the vehicle.*
Note: *On 2010 models, the right front wheel and splash shield must be removed to gain access to some of the mount attaching bolts.*
9   Remove the engine mount through-bolt/nuts and detach the mount from the chassis bracket.
10   Remove the nuts holding the mount to the engine bracket.
11   Installation is the reverse of removal. Use thread-locking compound on the mount bolts and be sure to tighten them securely.
12   Reconnect the battery (see Chapter 5, Section 1).

Chapter 2 Part B
3.5L and 3.9L V6 engines

Contents

Specifications

General

Displacement	
3.5L V6	214 cubic inches
3.9L V6	238 cubic inches
Bore and stroke	
3.5L V6	
2004 through 2006	3.70 x 3.31 inches
2007	3.90 x 2.99 inches
3.9L V6	3.90 x 3.31 inches
Cylinder numbers (drivebelt end-to-transaxle end)	
Front bank (radiator side)	2-4-6
Rear bank	1-3-5
Firing order	1-2-3-4-5-6
Compression pressure	See Chapter 2D
Oil pressure	See Chapter 2D

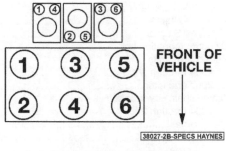

FRONT OF VEHICLE

38027-2B-SPECS HAYNES

Cylinder location and coil terminal identification diagram

Torque specifications

Ft-lbs (unless otherwise indicated)

Note: *One foot-pound (ft-lb) of torque is equivalent to 12 inch-pounds (in-lbs) of torque. Torque values below approximately 15 ft-lbs are expressed in inch-pounds, since most foot-pound torque wrenches are not accurate at these smaller values.*

Camshaft sprocket/variable camshaft timing sprocket bolt(s)	
2004 through 2006 3.5L V6	103
3.9L and 2007 and later 3.5L V6 engines	144 in-lbs
Catalytic converter-to-exhaust manifold flange nuts	23
Crankshaft balancer bolt	
3.5L V6	
2004 and 2005	118
2006 and later	
Step 1	92
Step 2	Tighten an additional 130 degrees
3.9L V6	
Step 1	92
Step 2	Tighten an additional 130 degrees

Torque specifications (continued)   Ft-lbs (unless otherwise indicated)

Note: *One foot-pound (ft-lb) of torque is equivalent to 12 inch-pounds (in-lbs) of torque. Torque values below approximately 15 ft-lbs are expressed in inch-pounds, since most foot-pound torque wrenches are not accurate at these smaller values.*

Cylinder head bolts (in sequence - see illustration 8.16)	
2007 and earlier models	
Step 1	44
Step 2	Tighten an additional 95-degrees
2008 and later models	
Step 1	44
Step 2	Tighten an additional 140-degrees
Coolant crossover pipe (3.9L and 2007 3.5L V6)	37
Driveplate-to-crankshaft bolts	52
Exhaust manifold retaining nuts/bolts	
2004 through 2006	144 in-lbs
2007 and later	15
Exhaust heat shield bolts	89 in-lbs
Idler pulley (right and left side) bolts	22
Idler pulley (center) bolt	37
Intake manifold bolts (lower)	
Step 1	
Bolts 1 through 8	
2004	62 in-lbs
2005 and later	115 in-lbs
Step 2	
Bolts 1 through 4	
3.5L V6 (exclude 2007 and later)	
2004	115 in-lbs
2005 and 2006	15
3.9L V6 and 2007 and later 3.5L V6	144 in-lbs
Step 3	
Bolts 5 through 8	18
Intake manifold bolts (upper)	18
Oil pan bolts/nuts	
To block	18
Side bolts	
2007 and earlier models	37
2008 and later models	
Step 1	37
Step 2	Tighten an additional 50-degrees
Oil pump mounting bolt	30
Rocker arm bolts	
2004 through 2006	24
2007 and later	25
Timing chain cover bolts	
2004 through 2006	
Small	20
Medium	41
Large	41
2007 and later	18
Timing chain tensioner plate bolts	15
Transaxle brace	37
Transaxle-to-oil pan brace bolts	
3.5L V6	37
3.9L V6	46
Valve cover-to-cylinder head bolts	89 in-lbs

1 General Information

1   This Part of Chapter 2A is devoted to in-vehicle repair procedures for the 3.5L and 3.9L V6 engines. These engines utilize cast-iron blocks with six cylinders arranged in a "V" shape at a 60-degree angle between the two banks. The overhead valve aluminum cylinder heads are equipped with replaceable valve guides and seats. Hydraulic lifters actuate the valves through tubular pushrods.
2   The engines are easily identified by looking for the designations printed directly on top of the upper intake plenum.
3   Information concerning engine removal and installation and overhaul can be found in Part D of this Chapter. The following repair procedures are based on the assumption that the engine is installed in the vehicle. If the engine has been removed from the vehicle and mounted on a stand, many of the Steps outlined in this Part of Chapter will not apply.

2 Repair operations possible with the engine in the vehicle

1   Many major repair operations can be accomplished without removing the engine from the vehicle.
2   Clean the engine compartment and the exterior of the engine with some type of degreaser before any work is done. It'll make the job easier and help keep dirt out of the internal areas of the engine.
3   Depending on the components involved, it may be helpful to remove the hood to improve access to the engine as repairs are performed (refer to Chapter 11 if necessary). Cover the fenders to prevent damage to the paint. Special pads are available, but an old bedspread or blanket will also work.
4   If vacuum, exhaust, oil or coolant leaks develop, indicating a need for gasket or seal replacement, the repairs can generally be done with the engine in the vehicle. The intake and exhaust manifold gaskets, timing chain cover gasket, oil pan gasket, crankshaft oil seals and cylinder head gaskets are all accessible with the engine in place.
5   Exterior engine components, such as the intake and exhaust manifolds, the oil pan (and the oil pump), the water pump, the starter motor, the alternator and the fuel system components can be removed for repair with the engine in place.
6   Since the cylinder heads can be removed without pulling the engine, valve component servicing can also be accomplished with the engine in the vehicle. Replacement of the timing chain and sprockets is also possible with the engine in the vehicle, although camshaft removal cannot be performed with the engine in the chassis (see Chapter 2D).
7   In extreme cases caused by a lack of necessary equipment, repair or replacement of piston rings, pistons, connecting rods and

4.12 Loosen the valve cover mounting bolts (arrows indicate three) - the bolts will stay with the cover

rod bearings is possible with the engine in the vehicle. However, this practice is not recommended because of the cleaning and preparation work that must be done to the components involved.

3 Top Dead Center (TDC) - locating

1   Refer to Chapter 2C, Section 3 for this procedure.

4 Valve covers - removal and installation

Removal

1   Disconnect the cable from the negative terminal of the battery (see Chapter 5, Section 1)

Front valve cover
Warning: *Wait until the engine is completely cool before beginning this procedure.*
2   Drain the coolant (see Chapter 1).
3   Remove the intake manifold cover (see Section 6).
4   Remove the air intake duct from the air filter housing (see Chapter 4).
5   Remove the spark plug wires from the front cylinder bank spark plugs (see Chapter 1). Be sure each wire is labeled before removal to ensure correct reinstallation.

2006 and earlier 3.5L V6 models
6   Remove the upper radiator hose (see Chapter 3).
7   Remove the throttle body (see Chapter 4).
8   Disconnect the heater outlet hose from the thermostat bypass pipe.
9   Remove the thermostat bypass pipe.

3.5L V6 models
10   Disconnect the heater inlet and outlet pipes (see Chapter 3).

All models
11   Remove the PCV hose from the valve cover (see Chapter 6).

12   Loosen the valve cover mounting bolts (see illustration).
13   Detach the valve cover.
Note: *If the cover sticks to the cylinder head, use a block of wood and a hammer to dislodge it. If the cover still won't come loose, pry on it carefully, but don't distort the sealing flange.*
14   Trim the gasket material from the lower intake manifold gasket at the cylinder head. The surface must be cleaned and prepared to prevent damage to the intake manifold gasket when the new valve cover and gasket are installed.

Rear valve cover
15   Remove the spark plug wires from the rear cylinder bank spark plugs (see Chapter 1). Be sure each wire is labeled before removal to ensure correct reinstallation.
16   Remove the ignition coil assembly (see Chapter 5).
17   Remove the ignition coil bracket and the bracket studs.
18   Remove the serpentine drivebelt (see Chapter 1).
19   Remove the alternator (see Chapter 5) and the alternator bracket.

2006 and earlier 3.5L V6 models
20   Remove the EVAP purge valve (see Chapter 6).
21   Detach the brake booster vacuum hose from the upper intake manifold.

3.5L V6 models
Warning: *Wait until the engine is completely cool before beginning this procedure.*
22   Drain the coolant (see Chapter 1).
23   Remove the coolant crossover pipe (see Section 6 , Steps 22 through 25).
24   Disconnect the MAP sensor (see Chapter 6) and the fuel injectors (see Chapter 4).

All models
25   Loosen the valve cover mounting bolts.
26   Detach the valve cover.
Note: *If the cover sticks to the cylinder head, use a block of wood and a hammer to dislodge it. If the cover still won't come loose, pry on it carefully, but don't distort the sealing flange.*

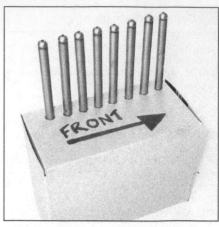

5.3 Rocker arm details - the rocker arms are kept as an assembly by a small sleeve between the bolt and the pedestal - note the projections on the pedestal; they fit into grooves in the head

A   Rocker arm bolt
B   Rocker arm
C   Rocker arm pedestal
D   Pedestal projections
E   Grooves in the head

5.4 A perforated cardboard box can be used to store the pushrods to ensure they are reinstalled in their original locations - note the label indicating the front end of the engine

27   Trim the gasket material from the lower intake manifold gasket at the cylinder head. The surface must be cleaned and prepared to prevent damage to the intake manifold gasket when the new valve cover and gasket are installed.

Installation

28   The mating surfaces of each cylinder head and valve cover must be perfectly clean when the covers are installed. Use a gasket scraper to remove all traces of sealant or old gasket material, then clean the mating surfaces with lacquer thinner or acetone (if there's sealant or oil on the mating surfaces when the cover is installed, oil leaks may develop). The valve covers are made of aluminum, so be extra careful not to nick or gouge the mating surfaces with the scraper.

29   Clean the mounting bolt threads with a die if necessary to remove any corrosion and restore damaged threads. Use a tap to clean the threaded holes in the heads.

30   Apply a dab of RTV sealant to the two joints where the intake manifold and cylinder head meet.

31   Place the valve cover and new gasket in position, then install the bolts. Tighten the bolts in several steps to the torque listed in this Chapter's Specifications.

32   Complete the installation by reversing the removal procedure. Start the engine and check carefully for oil leaks at the valve cover-to-head joints.

5   Rocker arms and pushrods - removal, inspection and installation

Removal

1   Disconnect the cable from the negative terminal of the battery (see Chapter 5, Section 1).

2   Remove the valve cover(s) (see Section 4).

3   Beginning at the drivebelt end of one cylinder head, remove the rocker arm mounting bolts one at a time and detach the rocker arms, pivot balls and pedestals (see illustration). Store each set of rocker arm components separately in a marked plastic bag to ensure they're reinstalled in their original locations.

Note: *The rocker arms have the pedestal mount "captured" on the rocker arm bolt by a metal sleeve inside. The components can be separated if necessary by tapping the bolt out of the pedestal, but normally all components for a particular valve will stay as an assembly.*

4   Remove the pushrods and store them separately to make sure they don't get mixed up during installation (see illustration).

Note: *Intake and exhaust pushrods are different lengths. Intake pushrods are approximately 5-3/4 inches long, while exhausts are 6.0 inches long. They may also have color codes to easily tell them apart.*

Inspection

5   Inspect each rocker arm for wear, cracks and other damage, especially where the pushrods and valve stems make contact.

6   Make sure the rollers operate freely as well.

7   Make sure the hole at the pushrod end of each rocker arm is open.

8   Inspect the pushrods for cracks and excessive wear at the ends. Roll each pushrod across a piece of plate glass to see if it's bent (if it wobbles, it's bent).

Installation

9   Lubricate the lower end of each pushrod with clean engine oil or moly-base grease and install them in their original locations. Make sure each pushrod seats completely in the lifter socket.

10   Apply moly-base grease to the ends of the valve stems and the upper ends of the pushrods.

11   Apply clean engine oil to the pivot balls and to the bearing surfaces of each rocker arm to prevent damage to the mating surfaces before engine oil pressure builds up. Install the rocker arms, pivot balls, pedestals and bolts and tighten them to the torque listed in this Chapter's Specifications. As the bolts are tightened, make sure the pushrods engage properly in the rocker arms and that the projections on the bottom of the pedestals fit into the grooves on the head before tightening the bolts (see illustration 5.3).

12   Install the valve covers. Start and run the engine, then check for oil leaks and unusual sounds coming from the valve cover area.

6   Intake manifold - removal and installation

Warning: *The engine must be completely cool before starting this procedure.*

1   Relieve the fuel system pressure (see Chapter 4).

2   Disconnect the cable from the negative terminal of the battery (see Chapter 5, Section 1).

3   Drain the cooling system (see Chapter 1).

Upper intake manifold

4   Remove the engine cover (see illustration).

5   Remove the spark plug wires (see Chapter 1) and the ignition coils and bracket assembly (see Chapter 5).

6   Disconnect the fuel lines from the fuel rail (see Chapter 4). On 2007 and later 3.5L V6 engines, disconnect the fuel line bracket at the MAP sensor.

7   On 2004 through 2006 3.5L V6 engines, remove the EGR valve and the EGR pipe (see Chapter 6).

8   Disconnect the EVAP purge solenoid line and remove the EVAP purge solenoid valve (see Chapter 6).

9   Label and disconnect the hoses and electrical connectors attached to the upper intake manifold (EGR valve, MAP sensor, IAT sensor, ETC, etc) (see Chapter 6) and throttle

6.4 Remove the oil filler cap, lift up on the upper left and right corners of the engine cover and separate the cover from the engine

6.28 Pry the manifold loose at a casting boss - don't pry between the gasket surfaces!

6.32 Install the intake gaskets against each cylinder head . . .

body (see Chapter 4).

Note: *On 3.9L engines, disconnect the Manifold Tuning Valve (see Chapter 6).*

10   Remove the PCV hose quick connect fitting from the valve cover (see Chapter 6). Remove the PCV hose from the intake manifold and lift it from the engine.

11   On 3.9L and 2007 and later 3.5L V6 engines, remove the coolant expansion tank (see Chapter 3).

12   Remove the heater hoses and the bracket assembly (see Chapter 3). If the throttle body has coolant hose connections at the brackets, remove the mounting nuts before removing the heater lines from the intake manifold

13   Remove the alternator and the mounting bracket (see Chapter 5).

14   Loosen the upper intake manifold bolts a little at a time, starting with the outer bolts and working towards the inner bolts, then remove the upper intake manifold with the throttle body attached.

15   Clean the mounting surfaces of the lower intake manifold and the upper intake manifold with brake system cleaner, removing all traces of the old gasket material or sealant.

16   Install the new gasket over the lower intake manifold. Install the upper intake manifold onto the lower intake manifold and tighten the bolts a little at a time, starting with the inner bolts and working toward the outer bolts, to the torque listed in this Chapter's Specifications. The remainder of the installation is the reverse of removal.

Lower intake manifold

17   Remove the upper intake manifold (see Steps 1 through 14).

18   Label and disconnect any remaining wires, fuel and vacuum lines from the lower intake manifold.

19   Remove the fuel rail and injectors from the lower intake manifold (see Chapter 4).

20   Remove the valve covers (see Section 4).

2006 and earlier 3.5L V6 models

21   Detach the heater pipe from the lower intake manifold and position it off to the side.

3.9L and 2007 3.5L V6 models

22   Remove the drivebelt idler pulleys.

23   Remove the power steering pump and position the pump off to the side without disconnecting the power steering fluid lines (see Chapter 10).

24   Remove the radiator expansion tank hose, the radiator inlet hose and the thermal bypass hose from the crossover pipe.

25   Remove the coolant crossover bolts and separate the coolant crossover pipe from the cylinder heads.

All models

26   Remove the thermostat housing and the thermostat (see Chapter 3).

27   Loosen the manifold mounting bolts/nuts in 1/4-turn increments until they can be removed by hand.

28   The manifold will probably be stuck to the cylinder heads and force may be required to break the gasket seal (see illustration).

Caution: *Don't pry between the manifold and the heads or damage to the gasket sealing surfaces may occur, leading to vacuum leaks.*

29   Loosen the rocker arm bolts, rotate the rocker arms out of the way and remove the pushrods that go through the manifold gaskets (see Section 5).

30   Lift the old gaskets off. Use a gasket scraper to remove all traces of sealant and old gasket material, then clean the mating surfaces with lacquer thinner or acetone.

Note: *The mating surfaces of the cylinder heads, block, coolant crossover housing and manifold must be perfectly clean when the manifold is installed. Gasket removal solvents are available at most auto parts stores and may be helpful when removing old gasket material that's stuck to the heads and manifold (since the manifold and the coolant crossover housing is made of aluminum, aggressive scraping can cause damage). Be sure to follow the di-*

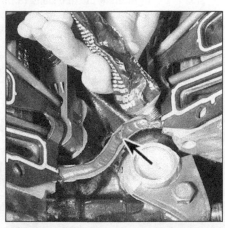

6.33 . . . then apply a bead of sealant to the end ridges between the cylinder heads

rections printed on the container. If there's old sealant or oil on the mating surfaces when the manifold is installed, oil or vacuum leaks may develop. Use a vacuum cleaner to remove any gasket material that falls into the intake ports or the lifter valley.

31   Use a tap of the correct size to chase the threads in the bolt holes, if necessary, then use compressed air (if available) to remove the debris from the holes.

Caution: *Wear safety glasses or a face shield to protect your eyes when using compressed air!*

32   Place the intake manifold gaskets in position on the heads (see illustration). Then install the pushrods and rocker arms (see Section 4).

33   Apply a 3/16-inch (5 mm) bead of RTV sealant to the front and rear ridges of the engine block between the heads (see illustration). Allow the RTV sealer to "set-up" (slightly harden) before installing the intake manifold.

34   Carefully lower the manifold into place and install the mounting bolts/nuts finger-tight.

Note: *Coat the bolt threads with pipe sealant before installing them.*

6.35 Intake manifold TIGHTENING sequence - make sure the bolts in the center (1 through 4) are completely tightened before tightening the end bolts (5 through 8)

7.6 Unbolt the exhaust pipe where it joins the front manifold

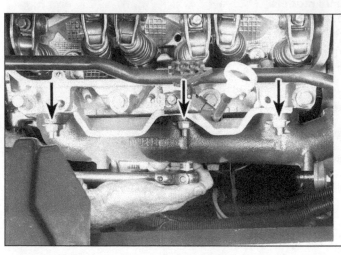

7.7 Remove the six bolts (arrows indicate the upper three) from the exhaust manifold

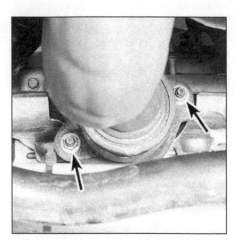

7.13 Remove the nuts holding the exhaust pipe to the rear manifold

35   Tighten the four vertical bolts (1 to 4) at the center of the manifold in the recommended tightening sequence (see illustration) to the torque listed in this Chapter's Specifications.

36   Tighten the four angled bolts (5 to 8) at the ends of the manifold in the recommended tightening sequence to the torque listed in this Chapter's Specifications.

37   Install the coolant crossover housing and tighten the bolts to the torque listed in this Chapter's Specifications.

38   Install the remaining components in the reverse order of removal.

39   Change the oil and filter and refill the cooling system (see Chapter 1). Start the engine and check for leaks.

7   Exhaust manifolds - removal and installation

Removal

1   Disconnect the cable from the negative terminal of the battery (see Chapter 5, Section 1).

2   Remove the air filter housing and air intake duct (see Chapter 4).

3   Remove the upstream oxygen sensors (see Chapter 6).
Note: *To help prevent possible damage to the oxygen sensor(s), it is recommended that the sensor(s) be removed from the exhaust manifold before the manifold is removed from the engine (see Chapter 6).*

4   Remove the spark plugs (see Chapter 1).

Front manifold

5   Remove the exhaust manifold heat shield.

6   Unbolt the catalyst/exhaust pipe assembly from the front exhaust manifold (see illustration). You should first apply penetrating oil to the fastener threads - they're usually rusted.

7   Remove the bolts and detach the manifold from the cylinder head (see illustration).

Rear manifold

8   On 2004 through 2006 3.5L V6 engines, remove the EGR valve mounting bolt and the EGR valve (see Chapter 6).

9   On 2004 through 2006 3.5L V6 engines, remove the EGR pipe from the exhaust manifold (see Chapter 6).

10   Remove the exhaust manifold heat shield.

11   Unbolt the catalyst/exhaust pipe assembly from the rear manifold. You should first apply penetrating oil to the fastener threads - they're usually rusted.

12   Remove the MAP sensor (see Chapter 6).

13   Set the parking brake, block the rear wheels and raise the front of the vehicle, supporting it securely on jackstands. Working under the vehicle, remove the exhaust pipe-to-manifold bolts and position the exhaust pipe aside (see illustration).
Note: *You may have to apply penetrating oil to the fastener threads - they're usually corroded.*

14   Remove the bracket bolt on the transaxle dipstick tube and move the tube aside.

15   Unbolt and remove the rear exhaust manifold.

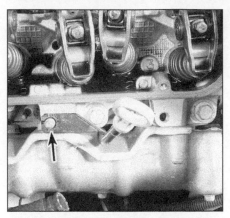

8.3 Remove the bolt holding the oil dipstick tube to the front cylinder head

8.10 Remove the old gasket and carefully scrape off all old gasket material and sealant

8.13a Position the new gasket over the dowel pins . . .

Installation (front or rear)

16   Clean the mating surfaces to remove all traces of old gasket material, then inspect the manifold for distortion and cracks. Warpage can be checked with a precision straightedge held against the mating flange. If a feeler gauge thicker than 0.030-inch can be inserted between the straightedge and flange surface, take the manifold to an automotive machine shop for resurfacing.

17   Remove the exhaust manifold inner heat shield/gasket.

18   Using a new heat shield/gasket, place the manifold against the head and install the bolts.

19   Starting in the middle and working out toward the ends, tighten the mounting bolts a little at a time until all of them are at the torque listed in this Chapter's Specifications.

20   Install the remaining components in the reverse order of removal.

21   Start the engine and check for exhaust leaks between the manifold and cylinder head and between the manifold and exhaust pipe.

8   Cylinder heads - removal and installation

Warning: *The engine must be completely cool before beginning this procedure.*

Removal

1   Disconnect the cable from the negative terminal of the battery (see Chapter 5, Section 1).

2   Remove the air filter housing (see Chapter 4) and then remove the upper and lower intake manifolds as described in Section 6.

3   If you're removing the front cylinder head, remove the oil dipstick tube mounting bolt and dipstick tube (see illustration).

Note: *On 3.9L V6 engines, remove the oxygen sensor harness clip from the oil dipstick tube and position the O2 harness off to the side.*

4   Disconnect all wires and vacuum hoses from the cylinder head(s). Be sure to label them to simplify reinstallation.

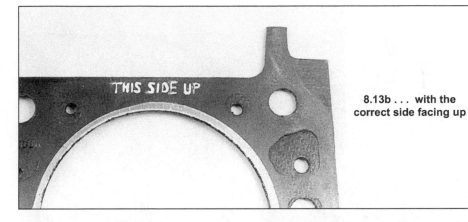

8.13b . . . with the correct side facing up

5   Detach the exhaust manifold from the cylinder head being removed (see Section 7).

6   Remove the lower intake manifold (see Section 6). Remove the rocker arms and pushrods (see Section 5).

7   Loosen each of the cylinder head bolts 1/4-turn at a time until they can be removed by hand - work from bolt-to-bolt in a pattern that's the reverse of the tightening sequence (see illustrations 8.16). Discard the bolts - new ones must be used during installation, but note which ones are studs and their locations. **Caution:** The engine must be completely cool before loosening the cylinder head bolts.

8   Lift the head(s) off the engine. If resistance is felt, don't pry between the head and block, as damage to the mating surfaces will result. Recheck for head bolts that may have been overlooked, then use a hammer and block of wood to tap up on the head and break the gasket seal. Be careful because there are locating dowels in the block which position each head. As a last resort, pry each head up at the rear corner only and be careful not to damage anything. After removal, place the head on blocks of wood to prevent damage to the gasket surfaces.

Installation

9   The mating surfaces of each cylinder head and block must be perfectly clean when the head is installed.

10   Use a gasket scraper to remove all traces of carbon and old gasket material (see illustration), then clean the mating surfaces with lacquer thinner or acetone. If there's oil on the mating surfaces when the head is installed, the gasket may not seal correctly and leaks may develop. When working on the block, it's a good idea to cover the lifter valley with shop rags to keep debris out of the engine. Use a shop rag or vacuum cleaner to remove any debris that falls into the cylinders.

11   Check the block and head mating surfaces for nicks, deep scratches and other damage. If damage is slight, it can be removed with a file; if it's excessive, machining may be the only alternative.

12   Use a tap of the correct size to chase the threads in the head bolt holes. Dirt, corrosion, sealant and damaged threads will affect torque readings.

13   Position the new gasket over the dowel pins in the block. Some gaskets are marked TOP or THIS SIDE UP to ensure correct installation (see illustrations).

14   Carefully position the head on the block without disturbing the gasket.

15   Install the new cylinder head bolts.

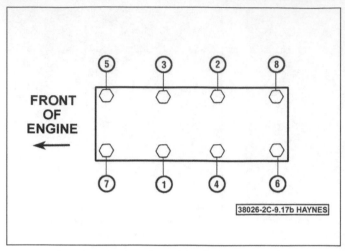

8.15 Cylinder head bolt TIGHTENING sequence

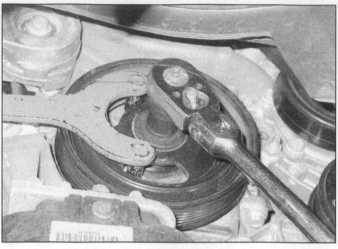

9.6 A pin spanner can be used to prevent the crankshaft from turning while loosening the pulley bolt

9.7 Use a puller that bolts to the crankshaft pulley hub; jaw-type pullers will damage the crankshaft pulley

9.9 The pulley keyway must be aligned with the Woodruff key in the crankshaft nose

Tighten the bolts, using the recommended sequence (see illustration), to the torque listed in this Chapter's Specifications. Then, using the same sequence, turn each bolt the amount of angle listed in this Chapter's Specifications.

16   The remaining installation steps are the reverse of removal.

17   Change the engine oil and filter and refill the cooling system (see Chapter 1).

---

9   Crankshaft pulley - removal and installation

---

1   Disconnect the cable from the negative terminal of the battery (see Chapter 5, Section 1).

2   With the parking brake applied and the shift lever in Park, loosen the lug nuts from the right front wheel, then raise the front of the vehicle and support it securely on jackstands.

3   Remove the right front wheel and the right splash shield from the wheelwell (see Chapter 11).

4   Remove the drivebelt (see Chapter 1).

5   Support the right side of the subframe with a floor jack. Remove the right-side subframe bolts and loosen the left-side subframe bolts, then partially lower the right side of the subframe.

6   Remove the bolt from the front of the crankshaft (see illustration). The bolt is normally very tight, so use a large breaker bar and a six-point socket to remove it. Obtain a new bolt, but save the old one for the initial installation of the pulley.

Note: *Use a pin spanner to prevent the crankshaft from turning. If you don't have a pin spanner, remove the torque converter cover and position a large screwdriver in the ring gear teeth to keep the crankshaft from turning while an assistant removes the crankshaft pulley bolt.*

7   Using a puller that bolts to the crankshaft hub, remove the crankshaft pulley/balancer from the crankshaft (see illustration).

Caution: *On these engines, a rubber sleeve connects the inertia weight to the balancer hub. Be careful when working on the crankshaft pulley/balancer that you do not accidentally shift the inertia weight's position relative to the sleeve or balancer hub, as this will upset the tuning of the balancer. Additionally, a spacer, such as a deep socket that just fits into the hole in the pulley and bears on the nose of the crankshaft will be required to avoid damage to the crankshaft.*

8   Apply a small amount of RTV sealant onto the crankshaft keyway and allow the sealant to "set-up" (slightly harden).

9   Position the crankshaft pulley/balancer on the crankshaft and slide it on as far as it will go. Note that the slot (keyway) in the hub must be aligned with the Woodruff key in the end of the crankshaft (see illustration).

10   Using a crankshaft balancer installation tool, available at most auto parts stores, press the crankshaft pulley/balancer onto the crankshaft. Note that the crankshaft bolt can also be used to press the crankshaft balancer into position, but when doing so, use a liberal amount of clean engine oil on the bolt threads to prevent galling.

11   If you're working on a 2004 or 2005 model, install the new bolt and washer and tighten it to the torque listed in this Chapter's Specifications. If you're working on a 2006 or later model, install the old bolt and washer and tighten the bolt to the Step 1 torque listed in this Chapter's Specifications, then remove the bolt. Install the new bolt and washer, tightening it to the Step 1 torque listed in this Chapter's Specifications, followed by the Step 2 angle torque.

12   The remaining installation Steps are the reverse of removal. When tightening the subframe bolts, tighten them to the torque listed in the Chapter 10 Specifications,.

10 Crankshaft front oil seal - removal and installation

1   Remove the crankshaft pulley (see Section 9).
2   Note how the seal is installed - the new one must be installed to the same depth and facing the same way. Carefully pry the oil seal out of the cover with a seal puller or a large screwdriver (see illustration). Be very careful not to distort the cover or scratch the crankshaft! Wrap electrician's tape around the tip of the screwdriver to avoid damage to the crankshaft.
3   Apply clean engine oil or multi-purpose grease to the outer edge of the new seal, then install it in the cover with the lip (spring side) facing IN. Drive the seal into place (see illustration) with a seal driver or a large socket and a hammer. Make sure the seal enters the bore squarely and stop when the front face is at the proper depth.
4   Check the surface on the pulley hub that the oil seal rides on. If the surface has been grooved from long-time contact with the seal, a press-on sleeve may be available to renew the sealing surface (see illustration). This sleeve is pressed into place with a hammer and a block of wood and is commonly available at auto parts stores for various applications.
5   Lubricate the pulley hub with clean engine oil and reinstall the crankshaft pulley. Use a vibration damper installation tool to press the pulley onto the crankshaft.
6   Install the crankshaft pulley retaining bolt and tighten it to the torque listed in this Chapter's Specifications.
7   The remainder of installation is the reverse of the removal.

11 Timing chain and sprockets - removal, inspection and installation

Removal

1   Disconnect the cable from the negative terminal of the battery (see Chapter 5, Section 1).
2   Remove the hood (see Chapter 11).
3   Drain the coolant and engine oil (see Chapter 1).
4   Remove the drivebelt (see Chapter 1).
5   Unbolt the drivebelt tensioner (see Chapter 1).
6   Remove the thermostat and the water pump pulley (see Chapter 3).
7   Remove the crankshaft pulley (see Section 9).
8   Remove the alternator and loosen the mounting bracket (see Chapter 5).
9   Unbolt the power steering pump and tie it aside (see Chapter 10). Leave the power steering fluid hoses connected.
10   Unbolt the driveplate cover below the transaxle.
11   Remove the starter (see Chapter 5).

10.2 Carefully pry the old seal out of the timing chain cover - don't damage the crankshaft in the process

10.3 Drive the new seal into place with a seal driver or a large socket and hammer

12   Attach an engine support fixture to the engine lifting eyes (see illustration), then remove the passenger side engine mount and the mount support bracket from the front of the timing chain cover (see Section 17).
Note: *Engine support fixtures can usually be obtained at equipment rental yards, as well as some auto parts stores.*
13   Remove the front exhaust manifold (see Section 8).
14   Remove the oil pan (see Section 13).
15   On 3.9L and 2007 and later 3.5L V6 engines, remove the radiator expansion tank hose, the radiator inlet hose and the thermal bypass hose from the crossover pipe (see Chapter 3).
16   On 2004 through 2006 3.5L V6 engines, disconnect the coolant hoses from the coolant bypass pipe from the front cover (see Chapter 3).
17   Remove the crankshaft position sensor and the camshaft actuator solenoid (see Chapter 6).

10.4 If the sealing surface of the pulley hub has a wear groove from contact with the seal, repair sleeves are available at most auto parts stores

11.12 Support the engine with an engine support fixture and chains to the front and rear engine lifting eyes

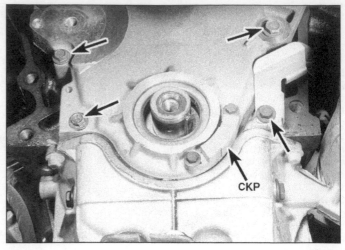

11.18a Timing chain cover bolt locations, upper . . .

11.18b . . . and lower

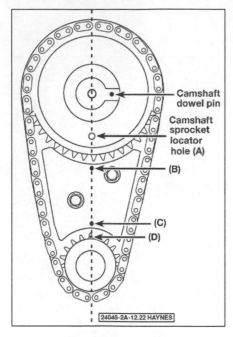

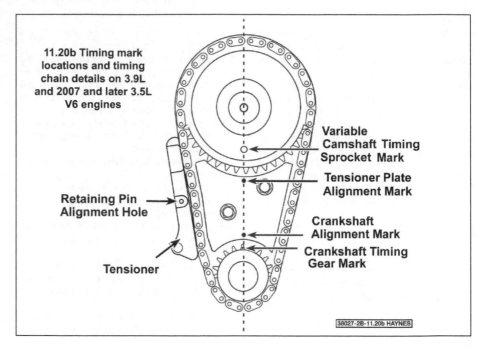

11.20b Timing mark locations and timing chain details on 3.9L and 2007 and later 3.5L V6 engines

Camshaft dowel pin

Camshaft sprocket locator hole (A)

(B)

(C)

(D)

Variable Camshaft Timing Sprocket Mark

Tensioner Plate Alignment Mark

Retaining Pin Alignment Hole

Crankshaft Alignment Mark

Crankshaft Timing Gear Mark

Tensioner

11.20a The timing marks on the sprockets should align as shown - a straight line should pass through the center of the camshaft, camshaft sprocket timing hole (A), the upper mark on the tensioner (B), the lower mark on the tensioner (C), the crankshaft sprocket timing mark (D) and the center of the crankshaft - 2004 through 2006 3.5L V6 engines

18   Remove the timing chain cover-to-engine block bolts (see illustrations).
19   Separate the cover from the engine. If it's stuck, tap it with a soft-face hammer, but don't try to pry it off.
20   Temporarily install the crankshaft pulley bolt and turn the crankshaft with the bolt to align the timing marks on the crankshaft and camshaft sprockets. When aligned at TDC for number 1 piston, the crankshaft sprocket timing mark should align with the mark on the

bottom of the chain tensioner plate, and the small hole in the camshaft sprocket should be at the 6 o'clock position, aligned with the timing mark in the top of the chain tensioner plate (see illustration) or tensioner (see illustration).
21   Remove the camshaft sprocket bolt(s). Do not turn the camshaft in the process (if you do, realign the timing marks before the bolt(s) are removed).
22   Use two large screwdrivers to carefully pry the camshaft sprocket (2004 through 2006 3.5L V6 engines) or variable camshaft timing sprocket (3.9L and 2007 and later 3.5L V6 engines) off the camshaft dowel pin. Slip the timing chain and camshaft sprocket off the engine.
Caution: *On all 3.9L and 2007 and later 3.5L V6 engines, use only a TORX® PLUS socket to remove the variable camshaft timing sprocket bolts. A regular TORX socket will NOT be a substitute and may damage the TORX® PLUS bolts.*

Inspection

23   The timing chain should be replaced with a new one if the engine has high mileage, the chain has visible damage, or total freeplay midway between the sprockets exceeds one-inch. Failure to replace a worn timing chain may result in erratic engine performance, loss of power and decreased fuel mileage. Loose chains can "jump" timing. In the worst case, chain "jumping" or breakage will result in severe engine damage. Always replace the timing chain and sprockets in sets. If you intend to install a new timing chain, remove the crankshaft sprocket with a puller and install a new one. Be sure to align the key in the crankshaft with the keyway in the sprocket during installation.
24   On 2004 through 2006 3.5L V6 engines, inspect the timing chain tensioner plate for cracks and wear and replace it if necessary. The plate is held to the engine block by two

11.24 The timing chain damper (guide) or tensioner plate is retained by two bolts

12.5 Remove the bolts (A) and pull up the roller lifter guides (B)

bolts (see illustration). The plate should be reinstalled before installing the new timing chain and sprockets.

25   On all 3.9L and 2007 and later 3.5L V6 engines, inspect the timing chain tensioner assembly for wear, damage and correct operation. The chain tensioner is held onto the engine block by two bolts (see illustration 11.20b). The tensioner must be reinstalled before installing the new timing chain. Also, use a special tool to retract the tensioner and install a pin into the alignment hole to lock the tensioner in this position.

26   Clean the timing chain and sprockets with solvent and dry them with compressed air (if available).

Warning: *Wear eye protection when using compressed air.*

27   Inspect the components for wear and damage. Look for teeth that are deformed, chipped, pitted and cracked.

Installation

28   Use a gasket scraper to remove all traces of old gasket material and sealant from the cover and engine block. The cover is made of aluminum, so be careful not to nick or gouge it. Clean the gasket sealing surfaces with lacquer thinner or acetone.

29   Install the tensioner plate, tightening the bolts to the torque listed in this Chapter's Specifications.

30   If the camshaft has turned at all since removal of the sprocket, turn the camshaft to position the dowel pin at 3 o'clock. Mesh the timing chain with the camshaft sprocket, then engage it with the crankshaft sprocket. The timing marks should be aligned as shown in illustrations 11.20a or 11.20b.

Note: *If the crankshaft has been disturbed, turn it until the "O" stamped on the crankshaft sprocket is exactly at the top.*

31   Install the camshaft sprocket bolt (2004 through 2006 3.5L V6 engines) or variable camshaft timing sprocket bolts (3.9L and 2007 and later 3.5L V6 engines) (make sure the dowel hole in the sprocket is aligned with the

dowel pin in the camshaft) and tighten to the torque listed in this Chapter's Specifications.

Note: *On all 3.9L and 2007 and later 3.5L V6 engines, be sure to install a new camshaft sprocket filter when the sprocket has been removed from the camshaft.*

32   Lubricate the chain and sprocket with clean engine oil.

33   On all 3.9L and 2007 and later 3.5L V6 engines, remove the tensioner locking tool to release the tensioner.

34   Apply a thin layer of anaerobic sealant to both sides of the new gasket, then position the gasket on the engine block (the dowel pins should keep it in place). Apply sealant to the bottom of the gasket, where it meets the oil pan.

35   Attach the cover to the engine and install the bolts. Follow a criss-cross pattern when tightening the fasteners and work up to the torque listed in this Chapter's Specifications in three steps.

36   The remainder of installation is the reverse of removal.

37   Add oil and coolant, start the engine and check for leaks.

12   Valve lifters - removal, inspection and installation

1   A noisy valve lifter can be isolated when the engine is idling. Hold a mechanic's stethoscope or a length of hose near the location of each valve while listening at the other end. Another method is to remove the valve cover and, with the engine idling, touch each of the valve spring retainers, one at a time. If a valve lifter is defective, it'll be evident from the shock felt at the retainer each time the valve seats.

2   The most likely causes of noisy valve lifters are dirt trapped inside the lifter and lack of oil flow, viscosity or pressure. Before condemning the lifters, check the oil for fuel contamination, correct level, cleanliness and correct viscosity.

12.6a A magnetic pick-up tool . . .

Removal

3   Remove the valve cover (see Section 4) and intake manifold (see Section 6).

4   Remove the rocker arms and pushrods (see Section 5).

5   Remove the bolts holding the roller lifter guide to the block, and remove the two roller lifter guides (see illustration). Mark the guides as to which side they came from.

6   There are several ways to extract the lifters from the bores. A special tool designed to grip and remove lifters is manufactured by many tool companies and is widely available, but it may not be required in every case. On newer engines without a lot of varnish buildup, the lifters can often be removed with a small magnet or even with your fingers. A machinist's scribe with a bent end can be used to pull the lifters out by positioning the point under the retainer ring in the top of each lifter (see illustrations).

Caution: *Don't use pliers to remove the lifters unless you intend to replace them with new ones (along with the camshaft). The pliers may damage the precision machined and hardened lifters, rendering them useless.*

12.6b . . . or a scribe can be used to remove the lifters

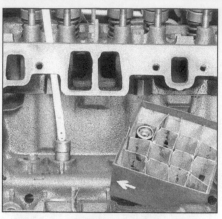

12.7 Store the lifters in order to ensure installation in their original locations

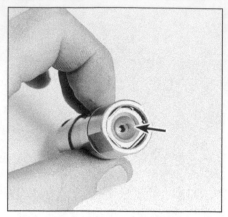

12.10a Check the pushrod seat in the top of each lifter for wear

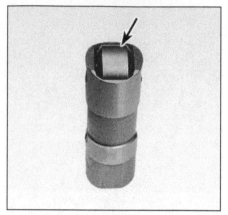

12.10b The roller on the roller lifters must turn freely - check for wear and excessive play as well

13.7 Disconnect the oil level sensor connector, if equipped

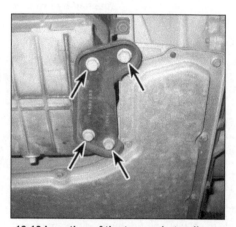

13.10 Location of the transaxle-to-oil pan mount bolts on a 3.5L V6 engine

7   Before removing the lifters, arrange to store them in a clearly labeled box to ensure they're reinstalled in their original locations. Remove the lifters and store them where they won't get dirty (see illustration).

Inspection and installation

8   Parts for valve lifters are not available separately. The work required to remove them from the engine again if cleaning is unsuccessful outweighs any potential savings from repairing them.
9   Clean the lifters thoroughly with solvent and dry them thoroughly, without mixing them up.
10   Check each lifter wall and plunger seat for scuffing, score marks or uneven wear (see illustration). Check the rollers carefully for wear or damage and make sure they turn freely without excessive play (see illustration). If the lifters walls are worn (not very likely), inspect the lifter bores in the block. If the pushrod seats are worn, inspect the pushrods also.
11   When reinstalling used lifters, make sure they're replaced in their original bores. Soak new lifters in oil to remove trapped air. Coat all lifters with moly-base grease or engine assembly lube prior to installation.

12   Install the push rods and the rocker arms (see Section 5).
13   The remaining installation Steps are the reverse of removal.
14   Run the engine and check for oil leaks.

---

13 Oil pan - removal and installation

Removal

1   Disconnect the cable from the negative terminal of the battery (see Chapter 5, Section 1).
2   Install an engine support fixture (see illustration 11.12) and install an engine support fixture if you're working on a 3.5L engine.
3   Remove the serpentine drivebelt (see Chapter 1).
4   Unbolt the air conditioning compressor and set it aside without disconnecting the refrigerant lines (see Chapter 3).
5   Raise the front of the vehicle and place it securely on jackstands. Apply the parking brake and block the rear wheels to keep it from rolling off the stands.
6   Drain the engine oil (see Chapter 1). On 2007 and later 3.5L V6 engines, remove the

oil filter adapter.
7   Disconnect the oil level sensor connector from the sensor, if equipped (see illustration).
8   Remove the right front wheel and the right splash shield from the wheelwell.
9   Remove the starter (see Chapter 5).
10   Refer to Section 17 and remove the passenger side upper engine mount, then remove the passenger side lower mount and bracket, if equipped. Remove the bolt holding the driver's side transaxle mount to the oil pan (see illustration), then remove the retaining bolts securing the mount to transaxle.
11   On 3.9L V6 engines, place floor jacks under the subframe, remove the bolts and lower the subframe from the vehicle (see Chapter 10).
12   Remove the transaxle brace mounting bolts and the transaxle brace.

3.5L V6 engines

13   Remove the ground cable from the transaxle.
14   Remove the transaxle wiring harness clip from the stud on the transaxle. Position the harness away from the transaxle.
15   Remove the bellhousing cover, then remove the driveplate-to-torque converter

13.21 Apply a bead of RTV sealant on either side of the rear main cap, where the pan gasket will meet it

14.2 Oil pump mounting bolt location

15.2a Most driveplates have locating dowels - if the one you're working on doesn't have one, make some marks to ensure proper alignment on reassembly

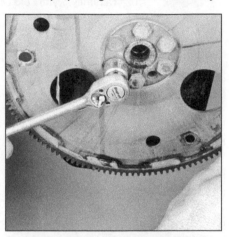

15.2b A large screwdriver wedged in one of the holes in the driveplate can be used to keep the driveplate from turning as the mounting bolts are removed

bolts (see Chapter 7B).
16   Raise the engine slightly and secure it in position using the engine support fixture.
17   Loosen the transaxle-to-engine bolts a few turns and separate the engine from the transaxle approximately 1/2 inch (see Chapter 7B).

All engines
18   Remove the side bolts (connecting the sides of the cast oil pan to the main cap supports) on each side of the oil pan.
19   Remove the remaining oil pan-to-block bolts, then carefully separate the oil pan from the block. Don't pry between the block and the pan or damage to the sealing surfaces could occur and oil leaks may develop. Instead, tap the pan with a soft-face hammer to break the gasket seal.

Installation
20   Clean the pan with solvent and remove all old sealant and gasket material from the block and pan mating surfaces. Clean the mating surfaces with lacquer thinner or acetone and make sure the bolt holes in the block are clear.
21   Apply a bead of RTV sealant to the front of the gasket, where it contacts the front cover, and a short bead (9/32-inch wide) to either side of the rear main cap where it meets the block (see illustration), then install the new one-piece oil pan gasket.
22   Place the oil pan in position on the block and install the nuts/bolts.
23   After the pan-to-block fasteners are installed, tighten them to the torque listed in this Chapter's Specifications. Starting at the center, follow a criss-cross pattern and work up to the final torque in three steps.
24   After all the pan-to-block bolts have been tightened, install the oil pan side bolts and tighten them to torque listed in this Chapter's Specifications.
25   On 2007 and later 3.5L models, re-join the engine and transaxle, tightening the bolts to the torque listed in the. Install the driveplate-to-torque converter bolts, tightening them to the torque listed in the.

26   Install the oil pan-to-transaxle brace and torque the bolts to the Specifications listed in this Chapter.
27   On 3.9L V6 engines, install the transaxle brace and tighten the bolts to the torque listed in this Chapter's Specifications.
28   The remaining steps are the reverse of the removal procedure. Tighten the subframe bolts to the torque listed in the Chapter 10 Specifications.
29   Refill the engine with oil, run it until normal operating temperature is reached and check for leaks.

14   Oil pump - removal and installation

1   Remove the oil pan (see Section 13).
2   Unbolt the oil pump and lower it from the engine (see illustration).
Note: *The oil pump driveshaft will come out with the pump as you lower it. It's a rod with a flat-sided portion at each end.*
3   If the pump is defective, replace it with a new one - don't reuse the original or attempt to rebuild it. Inspect the ends of the oil pump driveshaft and the plastic collar that retains the driveshaft to the oil pump. If there are signs of wear on the shaft or if the plastic collar is cracked or missing, replace the shaft with a new one.
Note: *The plastic collar centers the oil pump driveshaft over the oil pump shaft. If the collar is not used or is missing, damage to the oil pump driveshaft and the oil pump will occur. A new plastic collar is usually included with a new oil pump or driveshaft.*
4   Prime the pump by pouring clean engine oil into the pick-up screen while turning the pump driveshaft.
5   To install the pump, turn the flat on the driveshaft so it mates with the slot in the oil pump shaft. Make sure the plastic collar is fitted over the oil pump-to-oil pump driveshaft joint, then install the oil pump and driveshaft assembly into the block while engaging the upper end of the oil pump driveshaft into the

oil pump drive.
6   Install the pump mounting bolt and tighten it to the torque listed in this Chapter's Specifications.
7   The remainder of assembly is the reverse of the removal procedure.

15   Driveplate - removal and installation

Removal
1   Raise the vehicle and support it securely on jackstands, then refer to Chapter 7B and remove the transaxle.
2   Remove the bolts that secure the driveplate to the crankshaft (see illustration). If the crankshaft turns, wedge a screwdriver in the ring gear teeth to jam the driveplate (see illustration).
Note: *If there is a retaining ring between the bolts and the driveplate, note which side faces the driveplate when removing it.*

16.4 Carefully pry the old seal out

17.1a The engine mount strut is removed in two steps; first, remove the bracket-to-engine bolts, then lift the strut to access the strut-to-body bolts

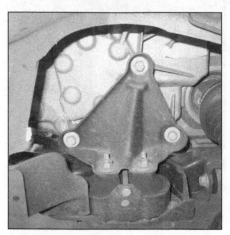

17.1b Location of the transaxle mount bolts on a 3.5L V6 model

3   Remove the driveplate from the crankshaft.

Caution: *When removing a driveplate, wear gloves to protect your fingers - the edges of the ring gear teeth may be sharp.*

4   Clean the driveplate to remove grease and oil. Inspect the surface for cracks, and check for cracked and broken ring gear teeth. Lay the driveplate on a flat surface to check for warpage.

5   Clean and inspect the mating surfaces of the driveplate and the crankshaft. If the crank-

shaft rear seal is leaking, replace it before reinstalling the driveplate (see Section 17).

Installation

6   Position the driveplate against the crankshaft. Be sure to align the marks made during removal. Note that some engines have an alignment dowel or staggered bolt holes to ensure correct installation. Before installing the bolts, apply thread locking compound to the threads and place the retaining ring in position on the driveplate.

7   Wedge a screwdriver through the ring gear teeth to keep the driveplate from turning as you tighten the bolts to the torque listed in this Chapter's Specifications. If the front pump seal/O-ring is leaking, now would be a very good time to replace it.

8   The remainder of installation is the reverse of the removal procedure.

---

16   Rear main oil seal - replacement

1   Remove the transaxle (see Chapter 7B).
2   Remove the driveplate (see Section 15).
3   Inspect the oil seal, as well as the oil pan and engine block surface for signs of leakage. Sometimes an oil pan gasket leak can appear to be a rear oil seal leak.
4   Pry the oil seal from the block with a screwdriver (see illustration). Be careful not to nick or scratch the crankshaft or the seal

bore. Thoroughly clean the seal bore in the block with a shop towel. Remove all traces of oil and dirt.
5   Lubricate the lips of the new seal with engine oil or multi-purpose grease. Install the seal over the end of the crankshaft (make sure the lips of the seal point toward the engine) and carefully tap it into place. A special aftermarket tool may be available at your local auto parts store. The tool just fits the diameter of the seal and, used with a hammer, drives the seal in.

Note: *Do not drive it in any further than the original seal was installed.*

6   Install the driveplate (see Section 15).
7   Install the transaxle (see Chapter).

---

17   Powertrain mounts - check and replacement

1   This procedure is essentially the same for all engines except for the engine mount strut and bracket on V6 engines (see illustration). 3.5L engines are also equipped with a passenger's side lower engine mount and bracket assembly. The transaxle mount is attached to the subframe and transaxle (see illustration). Refer to Chapter 2A and follow the procedures outlined there for all other mounts except the engine mount strut.

Chapter 2 Part C
3.6L V6 engine

Contents

Specifications

General

Displacement	217 cubic inches
Bore and stroke	3.70 x 3.37 inches
Cylinder numbers (front to rear)	
Left (front) side	2-4-6
Right (rear) side	1-3-5
Firing order	1-2-3-4-5-6
Compression pressure	See Chapter 2D
Oil pressure	See Chapter 2D

Camshaft

Journal diameters	
Front journal	1.376 inches
Other journals	1.061 inches
Camshaft endplay	0.0018 to 0.0085 inch
Valve lift	0.425 inch
Lobe height	
Intake	1.668 to 1.680 inches
Exhaust	1.670 to 1.682 inches

Torque specifications   Ft-lbs (unless otherwise indicated)

Note: *One foot-pound (ft-lb) of torque is equivalent to 12 inch-pounds (in-lbs) of torque. Torque values below approximately 15 foot-pounds are expressed in inch-pounds, because most foot-pound torque wrenches are not accurate at these smaller values.*

Camshaft bearing cap bolts	89 in-lbs
Camshaft sprocket bolts (use NEW bolts)	43
Camshaft actuator oil valve bolts	89 in-lbs
Crankshaft pulley bolt (use NEW bolt)	
Step 1	74
Step 2	Tighten an additional 150 degrees
Cylinder head bolts (in sequence; see illustrations 9.15a and 9.15b)	
8 mm bolts	
Step 1	132 in-lbs
Step 2	Tighten an additional 75 degrees
11 mm bolts (use NEW bolts)	
Step 1	22
Step 2	Tighten an additional 150 degrees
Driveplate bolts (use NEW bolts)	
Step 1	22
Step 2	Tighten an additional 45 degrees
Engine mount nuts	59
Engine mount bracket bolts	
8 mm	28
10 mm	45
Exhaust manifold bolts	15
Exhaust heat shield bolts	89 in-lbs
Exhaust manifold-to-catalytic converter nuts	33
Intake manifold (upper and lower) bolts	17
Oil pan baffle bolts	89 in-lbs
Oil pan-to-block bolts	17
Oil pan-to-rear seal housing bolts	89 in-lbs
Timing chain cover bolts	17
Timing chain guide bolts	17
Timing chain tensioner bolts	17
Timing chain idler sprocket bolts	43
Valve cover bolts	89 in-lbs

1　General Information

Warning: *The models covered by this manual are equipped with airbags. Always disable the airbag system before working in the vicinity of the impact sensors, steering column or instrument panel to avoid the possibility of accidental deployment of the airbag(s), which could cause personal injury (see Chapter 12).*

1　This Part of Chapter 2A is devoted to in-vehicle repair procedures for the 3.6L V6 engine. Information concerning engine removal and installation and overhaul on this engine can be found in Chapter 2D.

2　Since the repair procedures included in this Part are based on the assumption the engine is still in the vehicle, if they are being used during a complete engine overhaul (with the engine already removed from the vehicle and on a stand) many of the Steps included here will not apply.

3　The engine utilizes an aluminum block with cast-iron sleeves. The six cylinders are arranged in a "V" shape at a 60-degree angle between the two banks. The cylinder heads utilize a twin overhead camshaft arrangement with four valves per cylinder. Variable cam timing is used via an adjustable camshaft drive sprocket controlled by the engine control system. The engine uses aluminum cylinder heads with powdered metal guides and valve seats. Hydraulic lash adjusters negate the need for valve adjustments and roller rocker arms actuate the valves. The oil pump is mounted at the front of the engine behind the timing chain cover and is driven by the crankshaft.

2　Repair operations possible with the engine in the vehicle

1　Many major repair operations can be accomplished without removing the engine from the vehicle.

2　Clean the engine compartment and the exterior of the engine with some type of pressure washer before any work is done. A clean engine will make the job easier and will help keep dirt out of the internal areas of the engine.

3　Depending on the components involved, it may be a good idea to remove the hood to improve access to the engine as repairs are performed (refer to Chapter 11 if necessary).

4　If oil or coolant leaks develop, indicating a need for gasket or seal replacement, the repairs can generally be made with the engine in the vehicle. The oil pan gasket, the cylinder head gaskets, intake and exhaust manifold gaskets, timing chain cover gaskets and the crankshaft oil seals are all accessible with the engine in place.

5　Exterior engine components, such as the water pump, the starter motor, the alternator, the power steering pump and the fuel injection components, as well as the intake and exhaust manifolds, can be removed for repair with the engine in place.

6　Since the cylinder heads can be removed without removing the engine, valve component servicing can also be accomplished with the engine in the vehicle.

7　Replacement of, repairs to or inspection of the timing chain and sprockets and the oil pump are all possible with the engine in place.

8　In extreme cases caused by a lack of necessary equipment, repair or replacement of piston rings, pistons, connecting rods and rod bearings is possible with the engine in the vehicle. However, this practice is not recommended because of the cleaning and preparation work that must be done to the components involved.

3　Top Dead Center (TDC) for number one piston - locating

1　Top Dead Center (TDC) is the highest point in the cylinder that each piston reaches as it travels up the cylinder bore. Each piston reaches TDC on the compression stroke and again on the exhaust stroke, but TDC generally refers to piston position on the compression stroke.

2　Positioning the piston(s) at TDC is an essential part of many procedures such as timing chain/sprocket removal.

3　Before beginning this procedure, be sure to place the transmission in Neutral and apply the parking brake or block the rear wheels. Disable the ignition system by disconnecting the primary electrical connectors at the ignition coils, then remove the spark plugs (see Chapter 1). Also disable the fuel system (see Chapter 4, Section 2).

4　Install a compression gauge in the number one spark plug hole and rotate the crankshaft clockwise using a socket and rachet on the crankshaft pulley bolt until pressure registers on the gauge, which indicates the cylinder has started the compression stroke. Once the compression stroke has begun, TDC for the number one cylinder is obtained when the piston reaches the top of the cylinder on the compression stroke.

5　To bring the piston to the top of the cylinder, insert a long wooden dowel into the number one spark plug hole until it touches the top of the piston. Use the dowel (as a feeler gauge) to tell where the top of the piston is located in the cylinder while slowly rotating the crankshaft. As the piston rises, the dowel will be pushed out. The point at which the dowel stops moving outward is TDC.

7　If you go past TDC, rotate the crankshaft counterclockwise until the piston is approximately 20 mm below TDC, then slowly rotate the crankshaft clockwise again until TDC is reached.

8　After the number one piston has been positioned at TDC on the compression stroke, TDC for any of the remaining pistons can be rotating the engine clockwise, 120-degrees at a time, and following the firing order.

4　Valve covers - removal and installation

Removal

1　Disconnect the cable from the negative battery terminal (see Chapter 5).

2　Refer to Chapter 5 and remove the ignition coils.

3　Disconnect the engine wiring harness from the valve cover and position it aside.

4　Remove the upper intake manifold (see Section 7).

5　Remove the valve cover bolts, then detach the cover from the cylinder head.

Note: *If the cover is stuck to the engine, bump one end with a block of wood and a hammer. If that doesn't work, try to slip a putty knife under it to break the seal. Don't try to pry the cover off as damage to the sealing surface could occur, leading to oil leaks in the future.*

Installation

6　The mating surfaces of each cylinder head and valve cover must be perfectly clean when the covers are reinstalled. Use a gasket scraper to remove all traces of sealant and old gasket material, then clean the mating surfaces with brake system cleaner. If there's sealant or oil on the mating surfaces when the cover is reinstalled, oil leaks may develop.

7　Clean the mounting bolt threads with a die to remove any corrosion and restore damaged threads. Make sure the threaded holes in the cylinder head are clean - run a tap into them to remove corrosion and restore damaged threads.

8　The manufacturer recommends installing tool EN-46101 to the spark plug tube holes to allow the spark plug tube seals to slide over the tubes without tearing. However, after applying a small amount of engine oil to the spark plug tube seal inner faces and with care, the valve cover with the seals attached can slide over the spark plug tubes without the special tools.

9　Position the gasket inside the cover

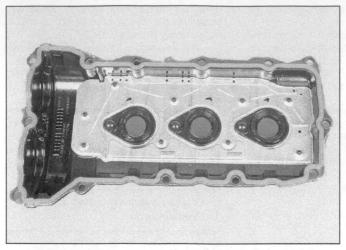

4.9 Make sure the gasket is seated in the valve cover groove

4.10 Place some RTV sealant at the joint where the timing chain cover and cylinder head meet

groove (see illustration). If the gasket will not stay in place, apply a thin coat of RTV sealant to the cover flange so the gasket adheres to the cover.

Note: *If the gasket wasn't leaking, is not damaged and is still pliable, it can be re-used.*

10   Apply an 8 mm wide by 4 mm high bead of RTV sealant to the joints between the timing chain cover and cylinder head mating faces where the valve cover will sit (see illustration).

11   Inspect the valve cover bolt grommets for damage. If the grommets aren't damaged they can be reused. Carefully position the valve cover(s) on the cylinder head and install the grommets and bolts.

12   Tighten the bolts to the torque listed in this Chapter's Specifications and in the proper sequence (see illustrations).

13   The remaining installation steps are the reverse of removal.

14   Start the engine and check carefully for oil leaks as the engine warms up.

---

5   **Rocker arms and hydraulic lash adjusters - removal, inspection and installation**

Removal

1   Refer to Section 4 and remove the valve covers.

2   Refer to Section 13 and remove the camshafts (see illustration).

3   Place markings on each rocker arm to ensure they are returned to the same position, then lift each one from the engine. Once the rocker arms are removed, the hydraulic lash adjusters can be lifted from the cylinder head.

Note: *It is important that any rocker arms or*

lash adjusters being used again must be returned to their original position on the engine.

Inspection

4   Check each rocker arm for wear, cracks and other damage, especially where the valve stems contact the rocker arm (see illustration).

5   Check the rollers for binding and roughness. If the bearings are worn or damaged, replacement of the entire rocker arm will be necessary.

Note: *Keep in mind that there is no valve adjustment on these engines, so excessive wear or damage in the valve train can easily result in excessive valve clearance, which in turn will cause valve noise when the engine is running.*

6   Make sure the oil passage hole in each hydraulic lash adjuster is not blocked.

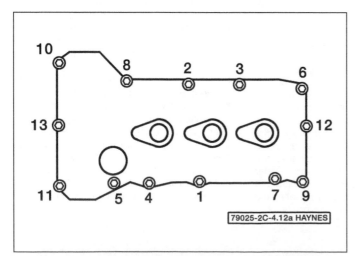

4.12a Valve cover bolt tightening sequence - left side (front)

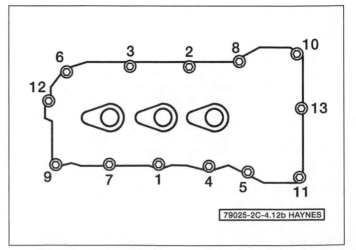

4.12b Valve cover bolt tightening sequence - right side (rear)

5.2 Remove the camshafts (1) to remove the rockers (2). (3) is the lash adjuster socket on the rocker

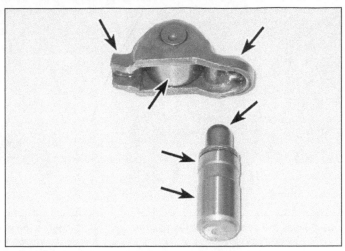

5.4 Check the rocker arms and lash adjusters for wear at the indicated points

Installation

7   Apply some camshaft/lifter pre-lube to the pivot pockets, roller and the slot that contacts the valve on each rocker arm. Apply clean engine oil into the bores for the hydraulic lash adjusters and then install them into the cylinder head. Position the rocker arms onto the lash adjusters.

8   The remaining installation steps are the reverse of removal.

9   Before starting and running the engine, change the oil and install a new oil filter (see Chapter 1).

---

6   Valve springs, retainers and seals - replacement

Note: *Broken valve springs and defective valve stem seals can be replaced without removing the cylinder head. Two special tools and a compressed air source are normally re-*

quired to perform this operation, so read through this Section carefully and rent or buy the tools before beginning the job.

1   Remove the spark plugs (see Chapter 1).

2   Remove the upper intake manifold (see Section 7), valve covers (see Section 4), camshafts (see Section 13) and rocker arms (see Section 5).

3   Thread an adapter into the spark plug hole and connect an air hose from a compressed air source to it. Most auto parts stores can supply the air hose adapter.

Note: *Many cylinder compression gauges utilize a screw-in fitting that may work with your air hose quick-disconnect fitting. If a cylinder compression gauge fitting is used, it will be necessary to remove the Schrader valve from the end of the fitting before using it in this procedure.*

4   Apply compressed air to the cylinder. The valves should be held in place by the air pressure (see illustration).

5   Using a socket and a hammer, gently tap

on the top of each valve spring retainer several times (this will break the seal between the valve keeper and the spring retainer and allow the keeper to separate from the valve spring retainer as the valve spring is compressed), then use a valve-spring compressor to compress the spring. Remove the keepers with small needle-nose pliers or a magnet (see illustration).

Note: *Several different types of tools are available for compressing the valve springs with the head in place. One type grips the lower spring coils and presses on the retainer as the knob is turned, while the lever-type shown here utilizes the rocker arm bolt for leverage. Both types work very well, although the lever type is usually less expensive.*

6   Remove the valve spring and retainer.

Note: *If air pressure fails to retain the valve in the closed position during this operation, the valve face or seat may be damaged. If so, the cylinder head will have to be removed for repair.*

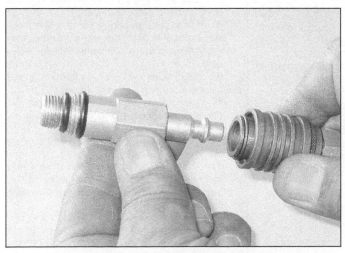

6.4 Thread the air hose adapter into the spark plug hole - adapters are commonly available at auto parts stores

6.5 After compressing the valve spring, remove the keepers with a magnet or needle-nose pliers

6.7 Ensure the seals are installed on the correct valve stems

1   *Intake valve seal*
2   *Exhaust valve seal*

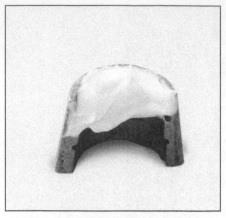

6.16 Apply a small dab of grease to each keeper as shown here before installation - it'll hold them in place on the valve stem as the spring is released

7   Remove the old valve stem seals, noting differences between the intake and exhaust seals (see illustration).

8   Wrap a rubber band or tape around the top of the valve stem so the valve won't fall into the combustion chamber, then release the air pressure.

9   Inspect the valve stem for damage. Rotate the valve in the guide and check the end for eccentric movement, which would indicate that the valve is bent.

10   Move the valve up-and-down in the guide and make sure it does not bind. If the valve stem binds, either the valve is bent or the guide is damaged. In either case, the head will have to be removed for repair.

11   Reapply air pressure to the cylinder to retain the valve in the closed position, then remove the tape or rubber band from the valve stem.

12   If you're working on an exhaust valve, fit the new exhaust valve seal on the valve stem and press it down over the valve guide. Don't force the seal against the top of the guide.

13   If you're working on an intake valve, install a new intake valve stem seal over the valve stem and press it down over the valve guide. Don't force the intake valve seal against the top of the guide.

Caution: *Do not fit an exhaust valve seal on an intake valve, as high oil consumption will result.*

14   Install the spring and retainer in position over the valve.

15   Compress the valve spring assembly only enough to install the keepers in the valve stem.

16   Position the keepers in the valve stem groove. Apply a small dab of grease to the inside of each keeper to hold it in place if necessary (see illustration). Remove the pressure from the spring tool and make sure the keepers are seated.

17   Disconnect the air hose and remove the adapter from the spark plug hole.

18   Repeat the above procedure on the remaining cylinders.

19   Install the rocker arm assemblies, camshafts, timing chains and the valve covers (see appropriate Sections).

20   Start the engine, then check for oil leaks and unusual sounds coming from the valve cover area. Allow the engine to idle for at least five minutes before revving the engine.

---

7   Intake manifold - removal and installation

Warning: *Wait until the engine is completely cool before starting this procedure.*

Removal

Upper intake manifold

1   Refer to Chapter 4 and relieve the fuel system pressure.

2   Disconnect the cable from the negative battery terminal (see Chapter 5).

3   Remove the engine top cover.

4   Remove the outlet tube from the air filter housing (see Chapter 4).

5   Disconnect the fuel supply tube (see Chapter 4) and move it out of the way.

6   Disconnect the small coolant air bleed hose from the intake manifold and the coolant outlet and move it aside also. Plug the hose and fitting to prevent spillage.

7   Disconnect the brake booster vacuum hose from the intake manifold.

8   Disconnect the wiring from the MAP sensor, the throttle body, the intake manifold tuning valve and the EVAP canister purge solenoid.

9   Disconnect the PCV tube from the intake manifold and move it out of the way.

10   Disconnect the quick-connect fitting at the tube from the EVAP canister purge solenoid. Move the tube aside.

11   Unbolt the fuel rail from its end support bracket.

12   Remove the bolt from the fuel rail wiring harness connector. Detach the fuel rail wiring harness from the intake manifold and move it out of the way.

13   Remove the six mounting bolts from the top of the upper intake manifold, then lift the manifold off of the lower manifold (see illustration).

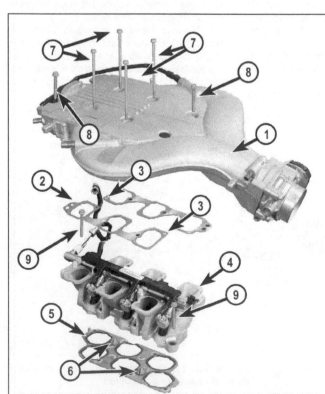

7.13 Typical intake manifold components

1   *Upper intake manifold*
2   *Gasket*
3   *Locating tabs*
4   *Lower intake manifold*
5   *Gasket*
6   *Locating tabs*
7   *Upper intake manifold-to-cylinder head bolts*
8   *Upper intake manifold-to-lower intake manifold bolts*
9   *Lower intake manifold-to-cylinder head bolts*

Lower intake manifold

14   Refer to Chapter 4 and remove the fuel rail and the fuel injectors.
15   Remove the lower intake manifold mounting bolts, then lift off the manifold.

Installation

Note: *The sealing surfaces of the intake manifold and cylinder heads must be perfectly clean when the manifold is installed.*

16   Carefully remove all traces of old gasket material. After the gasket surfaces are cleaned and free of any gasket material, wipe the mating surfaces with a cloth saturated with solvent. If there is old sealant or oil on the mating surfaces when the manifold is reinstalled, oil or vacuum leaks may develop. Use a vacuum cleaner to remove any gasket material that falls into the intake ports in the heads.
17   Use a tap of the correct size to chase the threads in the bolt holes, then use compressed air (if available) to remove the debris from the holes.
Warning: *Wear safety glasses or a face shield to protect your eyes when using compressed air.*
18   Install new gaskets on the cylinder heads, making sure that they're correctly aligned.
19   Install the lower intake manifold. Evenly tighten all the mounting bolts to the torque listed in this Chapter's Specifications.
20   Refer to Chapter 4 and replace the fuel injectors and the fuel rail.
21   Install new gaskets to the top of the lower intake manifold, then replace the upper intake manifold.
22   Install the bolts, then evenly tighten them to the torque listed in this Chapter's Specifications.
23   The remainder of the installation procedure is the reverse of removal. Check the coolant level, adding as necessary (see Chapter 1).

8   Exhaust manifolds - removal and installation

Removal

Warning: *Use caution when working around the exhaust manifolds - the sheetmetal heat shields can be sharp on the edges. Also, the engine should be cold when this procedure is followed.*

1   Disconnect the cable from the negative battery terminal (see Chapter 5).
2   Raise the vehicle and support it securely on jackstands.
3   Apply penetrating oil to the studs and nuts that attach the catalytic converters to the exhaust manifolds (they're usually rusty). Allow it to soak for as long as possible.
4   Remove the catalytic converter bolts and disconnect them from the exhaust manifolds.
5   If you're working on a left (front) manifold, remove the oil dipstick tube and the manifold heat shield.

8.6 Exhaust manifold upper fastener locations (left side shown, right side similar)

6   Remove the exhaust manifold mounting bolts (see illustration). The lower bolts for the right (rear) manifold can be removed from under the vehicle.
7   Lift off the exhaust manifolds and discard the gaskets.

Installation

8   Check the manifold for cracks and make sure the bolt threads are clean and undamaged. The manifold and cylinder head mating surfaces must be clean before the manifolds are reinstalled - use a gasket scraper to remove all carbon deposits and gasket material.
Note: *The cylinder heads are made of aluminum, therefore aggressive scraping is not suggested and will damage the sealing surfaces. Also, if any of the manifold bolts are broken it indicates a warped manifold. Have the manifold machined at a cylinder head reconditioning workshop prior to installing the manifold to the vehicle. If this is not done, the manifold may not seal properly and the new manifold bolts will probably break.*
9   Place the manifold on the cylinder head and install the mounting bolts finger-tight.
10   When tightening the manifold bolts, work from the center to the ends and be sure to use a torque wrench. Tighten the bolts to the torque listed in this Chapter's Specifications. If required, bend the exposed end of the exhaust manifold gasket back against the cylinder head.
11   Apply an anti-seize compound to the threads of the outer heat shield bolts, then install the bolts and tighten them to the torque listed in this Chapter's Specifications.
12   The remaining installation steps are the reverse of removal.
13   Start the engine and check for exhaust leaks.

9   Cylinder heads - removal and installation

Warning: *The engine must be completely cool before beginning this procedure.*
Note: *It will be necessary to purchase a new set of 11 mm cylinder head bolts.*

Removal

1   Relieve the fuel system pressure (see Chapter 4), then disconnect the cable from the negative battery terminal (see Chapter 5). Drain the cooling system (see Chapter 1).
2   Remove the lower intake manifold (see Section 7), the valve covers (see Section 4) and the upper (secondary) timing chains (see Section 12).
3   Disconnect the wiring from the engine coolant temperature sensor and remove the dipstick tube if you're removing the left (front) cylinder head. If you're removing the right (rear) head, refer to Chapter 11 and remove the hood.
4   Disconnect the catalytic converters from the exhaust manifolds (see Section 8). The heads will be removed with the exhaust manifolds attached to them.
5   Loosen the head bolts in 1/4-turn increments in the reverse order of the tightening sequence (see illustrations 9.15a and 9.15b) until they can be removed by hand.
Note: *There will be different lengths and sizes of bolts in different locations. Make notes of where each bolt belongs to aid in assembly.*
6   Carefully lift the heads off the engine block. If you feel resistance, don't pry the head off. To dislodge the head, put a pry bar into an intake port and pry with caution.
7   Store the heads on wooden blocks to prevent damage.

Installation

8   The mating surfaces of the cylinder heads and block must be perfectly clean when the heads are reinstalled. Gasket removal solvents are available at auto parts stores and may prove helpful.
9   Use a gasket scraper to remove all traces of carbon and old gasket material, then wipe the mating surfaces with a cloth saturated with brake system cleaner. If there is oil on the mating surfaces when the heads are reinstalled, the gaskets may not seal correctly and leaks may develop. When working on the block, use a vacuum cleaner to remove any debris that falls into the cylinders.
Note: *The cylinder heads are made of aluminum, therefore aggressive scraping is not suggested and will damage the sealing surfaces.*

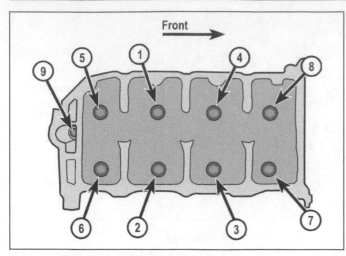

9.15a RH (rear) cylinder head tightening sequence. Bolts 1 to 8 are M11 bolts. Bolt 9 is an M8 bolt

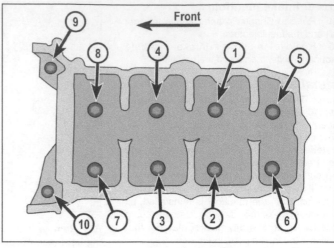

9.15b LH (front) cylinder head bolt tightening sequence. Bolts 1 to 8 are M11 bolts. Bolts 9 and 10 are M8 bolts

10   Check the block and head mating surfaces for nicks, deep scratches and other damage. If damage is slight, it can be removed with emery cloth. If it is excessive, machining may be the only alternative.

11   Use a tap of the correct size to chase the threads in the head bolt holes in the block. If a tap is not available, spray a liberal amount of brake cleaner into each hole. Use compressed air (if available) to remove the debris from the holes.

Warning: *Wear safety glasses or a face shield to protect your eyes when using compressed air. Use a wire brush to remove the thread locking compound from the 8 mm head bolts. Corrosion, sealant and damaged threads will affect torque readings, so be sure the threads are clean.*

12   Position the new gaskets over the dowels in the block.

13   Carefully position the heads on the block

without disturbing the gaskets.

14   Before installing the 8 mm head bolt(s), coat the threads with a medium-strength thread locking compound. Then install the 8 mm head bolt(s) (9 [and 10 on LH cylinder head]) finger-tight.

15   Install new 11 mm head bolts (bolts 1 through 8) and tighten them finger-tight. Following the recommended sequence (see illustrations), tighten the bolts to the Step 1 torque listed in this Chapter's Specifications. Then, in sequence, tighten them to the Step 2 angle of rotation listed in this Chapter's Specifications.

Warning: *DO NOT reuse 11 mm head bolts - always replace them.*

16   Tighten the 8 mm bolt(s) to the torque and angle of rotation listed in this Chapter's Specifications.

17   The remaining installation steps are the reverse of removal.

18   Add coolant and change the engine oil and filter (see Chapter 1). Start the engine and check for proper operation and coolant or oil leaks.

10   Crankshaft pulley - removal and installation

Note: *This procedure requires a special pulley installation tool that is only available through specialized tool manufacturers and a new crankshaft pulley bolt. Read through the entire procedure and obtain the tools and materials before proceeding.*

1   Disconnect the cable from the negative battery terminal (see Chapter 5).

2   Raise the vehicle and support it securely on jackstands.

3   Refer to Chapter 1 and remove the drivebelt.

4   Install an engine support fixture or an engine hoist to the engine lifting bracket on the right side.

Note: *As an alternative, the engine may be supported from below the oil pan with a floor jack and a block of wood. If this is done, be sure to position the wood and the jack to avoid damaging the oil pan.*

5   Tighten the engine support chain to remove the weight from the engine mounts.

6   Remove the right engine mount (see Section 18).

7   Carefully lower the engine about two inches for clearance.

8   Remove the crankshaft pulley bolt. Prevent the crankshaft from turning by holding the hub of the tool with a large pin spanner, or by removing the starter motor (see Chapter 5) and wedging a large screwdriver or prybar in the driveplate ring gear teeth.

Note: *Make note of how far the pulley is pressed onto the crankshaft. It must be installed in the same position.*

9   Install a large three-jaw puller on the pulley (see illustration).

Caution: *The hooks of the puller must contact only the hub - not the outer ring. Also, be sure to use the proper adapter so as not to damage the end of the crankshaft or the threads in the end of the crankshaft. Use the puller to draw the pulley from the crankshaft.*

10   To install the pulley, first lubricate the inside of the pulley bore with clean engine oil.

Note: *Don't let oil contact the outside of the pulley's snout or the oil seal in the timing cover. The pulley must only be installed into a dry seal.*

11   Obtain a crankshaft pulley installation tool. They are available at automotive tool dealers and most auto parts stores. Use only this tool to press the pulley into place.

Caution: *Don't try to hammer the pulley on or try to press it into place using a bolt screwed into the crankshaft.*

12   Install the pulley so that it's in the same position as before disassembly. The installation tool should be bottomed on the end of the crankshaft. Remove the installation tool.

10.9 The use of a three jaw puller will be necessary to remove the crankshaft pulley - always place the puller jaws around the hub, not the outer ring

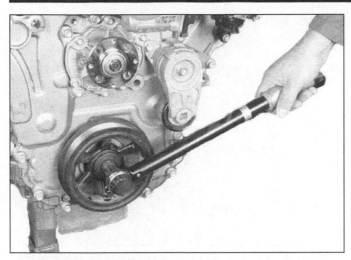

10.13a Use a torque wrench to tighten the new crankshaft pulley bolt to the Step 1 torque . . .

10.13b . . . then use a torque angle gauge to finish the tightening sequence

13   Install the NEW crankshaft pulley bolt. Tighten it to the torque and angle of rotation listed in this Chapter's Specifications (see illustrations).
14   The remainder of installation is the reverse of removal.

11   Crankshaft front oil seal - removal and installation

1   Remove the crankshaft pulley (see Section 10).
2   Note how the seal is installed - the new one must be installed to the same depth and facing the same way. Carefully pry the oil seal out of the cover with a seal puller (see illustration).
3   If the seal is being replaced with the timing chain cover removed, support the cover on top of two blocks of wood and drive the seal out from the backside with a hammer and punch.
Caution: *Be careful not to scratch, gouge or distort the area that the seal fits into or a leak will develop.*
4   Apply clean engine oil or multi-purpose grease to the outer edge of the new seal, then install it in the cover with the lip (spring side) facing IN. Drive the seal into place with a large socket and a hammer (if a large socket isn't available, a piece of pipe will also work). Make sure the seal enters the bore squarely and stop when the front face is at the proper depth.
5   Check the surface on the pulley hub that the oil seal rides on. If the surface has been grooved from long-time contact with the seal, the pulley should be replaced.
6   Lubricate only the inside of the pulley hub with clean engine oil and install the crankshaft pulley as described in Section 10.
7   The remainder of installation is the reverse of the removal.

12   Timing chains and camshaft sprockets - removal, inspection and installation

Warning: *Wait until the engine is completely cool before beginning this procedure.*
Caution: *The timing system is complex, and severe engine damage will occur if you make any mistakes. Do not attempt this procedure unless you are highly experienced with this type of repair. If you are at all unsure of your abilities, be sure to consult an expert. Double-check all your work and be sure everything is correct before you attempt to start the engine.*
Caution: *Before removing the timing chain(s) from their sprockets(s), check to make sure that all of the sprockets are equipped with timing marks like the ones shown in the illustrations. If the crankshaft sprocket is not equipped with a mark, a special tool (EN 46111) will be required.*
Note: *The engine shown in the illustrations has only one variable camshaft sprocket on each cylinder head. Procedures are identical, however.*

Removal

Note: *This procedure requires two special tools to lock the camshafts on each cylinder head in position. The part numbers for the tools are EN 46105-1 and EN 46105-2, available through SPX Tools. Read through the entire procedure and obtain the tools before proceeding.*
1   Disconnect the cable from the negative battery terminal (see Chapter 5)
2   Relieve the fuel system pressure (see Chapter 4).
3   Drain the engine coolant and remove the drivebelt and its tensioner (see Chapter 1).
4   Remove the lower intake manifold (see Section 7).
5   Remove the valve covers (see Sec-

11.2 Another way of removing an old oil seal is to screw a self tapping screw partially into the seal, then use pliers as a lever to pull it from the engine

tion 4).
6   Remove the water pump (see Chapter 3).
7   Unbolt the power steering pump and move it out of the way.
8   Remove the crankshaft pulley (see Section 10).
9   Remove all of the camshaft position sensors (CMP) from the front cover (see Chapter 6).
10   Disconnect the electrical connectors from the camshaft position actuator valves. Remove the bolts, then remove the actuator solenoid valves.
11   Remove the thermostat housing (see Chapter 3).
12   Remove all of the bolts from the timing chain cover (see illustration 12.41 for locations).
13   Screw a 10 mm x 1.5 mm bolt in the threaded hole near the top of the front cover. Tighten this screw while prying at the correct

12.13a Use a screwdriver to pry the cover from the engine at the indicated points

12.13b A 10 x 1.5 mm bolt can be threaded into the hole near the water pump to press the center portion of the cover away from the engine

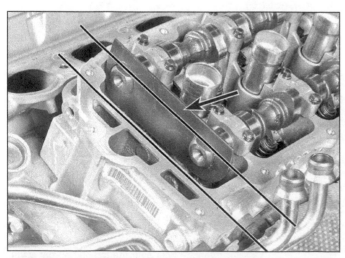

12.14 Crankshaft sprocket in correct position

locations to break the cover loose from the engine block (see illustrations).

14   Reinstall the crankshaft pulley bolt and use it to rotate the crankshaft clockwise until the timing mark on the timing chain sprocket aligns with the mark on the oil pump housing in the 9 o'clock position (see illustration).

15   In this position, the flat segment on the rear of each camshaft on the RH (rear) cylinder head should be facing up and parallel with the valve cover mating surface, and the flat segment on the rear of each camshaft on the LH (front) cylinder head should be at approximately a 45-degree angle with the valve cover mating surface. If this is not the case, rotate the crankshaft clockwise one full turn and re-align the crankshaft timing marks. Install the special tools to the camshafts of each cylinder head (see illustrations).

16   Remove the timing chain tensioner bolts for the RH cylinder head and carefully remove the tensioner and discard the gasket.

Note: *The tensioner is under spring pressure. Use care, as the tensioner can fly apart once the bolts are removed.*

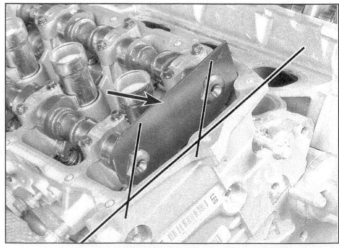

12.15a RH (rear) cylinder head with camshaft locking tool installed

12.15b LH (front) cylinder head with camshaft locking tool installed

17   Remove the RH cylinder head timing chain guides, then remove the chain (see illustration). Remove the special holding tools from the rear of the camshafts.

18   Repeat the process to remove the lower, or primary timing chain and the timing chain for the LH cylinder head.

Note: *Again use care as the tensioners are under spring pressure. Also, do not attempt to remove the timing chain guide from the oil pump housing. At the time of publication, this was only supplied as a complete assembly with the oil pump.*

19   If necessary, after noting their installed position, remove the bolts retaining the idler sprockets and remove the sprockets.

20   If necessary, mark the relationship between the camshaft sprockets and their respective camshaft. Hold the camshaft steady with an open ended wrench placed on the hex in the camshaft and remove the camshaft sprocket bolt. Remove the sprocket from the camshaft.

Inspection

21   Inspect the timing chain cover for damage. Remove all old sealant from the cover and the engine block.

22   Remove and discard the seal for the coolant passage from the rear of the housing.

23   Clean all components with suitable solvent and inspect the guides and timing chain sprockets for wear or damage. Any gouges or deformities on the guides or tensioners will require replacement parts. Ensure the rear face of the tensioner is clean and there are no marks that would affect the sealing quality once reassembled. Remove any old gasket material from the tensioner mating faces with the engine.

24   Inspect the timing chains for wear, stiff or loose rollers and binding.

25   Inspect the tensioners for damage or wear. Reset the tensioner by removing the plunger from the tensioner body and screw

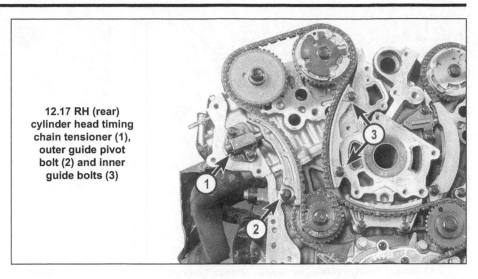

12.17 RH (rear) cylinder head timing chain tensioner (1), outer guide pivot bolt (2) and inner guide bolts (3)

the rear of the plunger into the plunger shaft using a flat bladed screwdriver. Install the plunger to the body of the tensioner and while maintaining pressure on the plunger, insert a paper clip into the hole in the tensioner body to maintain the tensioner in the retracted position (see illustrations). Repeat this on the other two timing chain tensioners.

Note: *The wire must be inserted into the tensioner to maintain it in the compressed position. If this is not done, the spring pressure will not tension the chain once it is installed to the engine.*

Installation

Caution: *Before starting the engine, carefully rotate the crankshaft by hand through at least two full revolutions (use a socket and breaker bar on the crankshaft pulley center bolt). If you feel any resistance, STOP! There is something wrong - most likely valves are contacting the pistons. You must find the problem before proceeding.*

Note: *Timing chains must be replaced as a*

set with the camshaft and crankshaft sprockets. Never put a new chain on old sprockets.

26   Align the crankshaft sprocket with the Woodruff key and slide the sprocket onto the crankshaft.

Caution: *If resistance is encountered, do not hammer the sprocket onto the crankshaft. It may eventually move onto the shaft, but it may be cracked in the process and fail later, causing extensive engine damage. Ensure the timing mark on the sprocket is visible and the timing marks are aligned. Rotate the crankshaft clockwise and align the crankshaft sprocket mark with the mark on the oil pump housing that is in the 5 o'clock position.*

27   Rotate the camshafts for the LH cylinder head counterclockwise using an open ended wrench on the hexagonal portion of the camshafts until the flats on the rear of the camshaft are parallel to the valve cover gasket face on the rear of the cylinder head.

28   Install the special locking tool EN46105-1 onto the LH (front) cylinder head camshafts. Ensure the tool is correctly seated against the camshafts. Rotate the camshafts for the

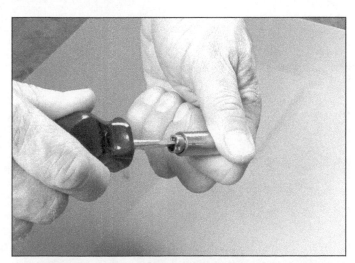

12.25a Remove the plunger from the tensioner and reset it by screwing the bottom of the plunger shaft into the top half of the plunger shaft with a screwdriver

12.25b Slowly push the plunger back into the tensioner body and secure it in place with a paper clip

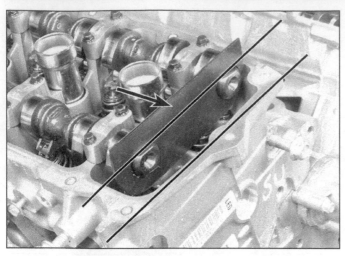

12.28a LH (front) cylinder head with camshaft locking tool installed

12.28b RH (rear) cylinder head with camshaft locking tool installed

RH (rear) cylinder head and install the other special locking tool (EN 46105-2) to the RH cylinder head camshafts (see illustrations).

29   Install the LH (front) cylinder head timing chain first, aligning the bright (or marked) chain links with the letter L stamped on the camshaft sprockets. Then install the LH lower idler sprocket onto the chain ensuring the large sprocket and the markings "LB Front" are facing outward. The lower bright (or marked) link on the chain must align with the hole in the face of the outer sprocket (see illustrations). Slide the idler sprocket onto the engine and install the bolt. Tighten the idler sprocket bolt to the torque listed in this Chapter's Specifications and check that all of the timing marks are still aligned correctly.

30   Install the outer LH (front) timing chain guide that runs between the idler sprocket and the exhaust camshaft, then install the inner LH (front) timing chain guide to the engine.

Tighten the bolts to the torque listed in this Chapter's Specifications.

Note: *Ensure the inner timing chain guide is not fouling on the mounting pad for the tensioner before tightening the bolts.*

31   Using a new gasket, install the tensioner for the LH (front) timing chain. Ensure the tensioner is aligned with the inner guide and tighten the tensioner bolts to the torque listed in this Chapter's Specifications. Release the pin from the tensioner. Grasp the inner guide and force it into the tensioner. This will allow the tensioner to tension the chain and check that the timing chain alignment marks are all correct.

32   Install the RH (rear) timing chain idler sprocket to the engine ensuring that the small sprocket is facing outward along with the markings "RB Front." Tighten the idler sprocket bolt to the torque listed in this Chapter's Specifications.

33   Ensure the crankshaft sprocket is still aligned with the timing mark on the oil pump housing that is in the 5 o'clock position and install the lower, or primary timing chain. Ensure the shiny chain links align with the arrows on both large idler sprockets and also with the timing mark on the crankshaft sprocket. Install the timing chain guide to the lower chain running between the two idler sprockets, tightening the bolts to the torque listed in this Chapter's Specifications. Then using a new gasket, install the tensioner and tighten the tensioner bolts. Remove the pin from the tensioner, force the tensioner guide and pushrod into the tensioner body to unlock it. Allow the tensioner to tension the timing chain. Check that the timing chain alignment marks are all correct. Remove the camshaft special locking tools from the camshafts on both cylinder heads (see illustration).

34   Rotate the crankshaft until the timing

12.29a LH (front) cylinder head timing chain-to-camshaft sprocket alignment marks

12.29b Alignment mark between LH timing chain and idler sprocket

12.33 Lower timing chain sprocket alignment marks (A). The tensioner (1), upper timing chain guide (2) and lower timing chain guide (3)

12.35 RH (rear) cylinder head timing chain alignment marks: (1) exhaust camshaft, (2) intake camshaft, (3) idler sprocket

mark on the crankshaft sprocket is aligned with the mark on the oil pump housing that is in the 9 o'clock position. Install the special locking tool EN-46105-2 onto the LH cylinder head camshafts. Rotate the camshafts for the RH cylinder head and install the other special locking tool to the RH (rear) cylinder head (EN-46105-1). Ensure the tools are correctly seated against the camshafts (see illustrations 12.11a and 12.11b).

Note: *The camshaft flats for the RH cylinder head must be parallel to the valve cover gasket face on the rear of the cylinder head.*

35   Install the RH cylinder head timing chain, aligning the bright (or marked) chain links with the letter R stamped on the camshaft sprockets. The lower bright link on the chain must align with the hole in the face of the inner idler sprocket. Check that all of the timing marks are still aligned correctly (see illustration).

36   Install the inner RH timing chain guide that runs between the idler sprocket and the intake camshaft, then install the outer RH timing chain guide to the engine. Tighten the bolts to the torque listed in this Chapter's Specifications.

37   Using a new gasket, install the tensioner for the RH timing chain. Ensure the tensioner is aligned with the outer guide and tighten the tensioner bolts to the torque listed in this Chapter's Specifications. Remove the pin from the tensioner, force the tensioner guide and pushrod into the tensioner body to unlock it. Allow the tensioner to tension the timing chain (see illustration). Check that the timing chain alignment marks are all correct, then rotate the engine slowly by hand until oil is forced into the three timing chain tensioners.

Caution: *If you encounter any resistance while rotating the engine, STOP! There is something wrong - most likely valves are contacting the pistons. You must find the problem before proceeding.*

Note: *While doing this, ensure that the timing chains don't jump teeth on any of the timing gears.*

12.37 Force the tensioner guide and chain into the tensioner body to unlock the tensioner

38   Carefully clean all old sealant from the block and the timing chain cover, being careful to avoid scratching the sealing surfaces. Remove all residue with brake system cleaner. Remove the special locking tools from the camshafts.

39   Install a new seal to the inside of the timing cover and apply a bead of RTV sealant to

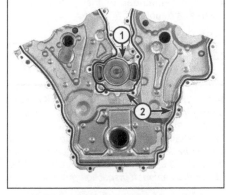

12.39 Seal location (1) and sealant location (2)

the faces of the cover that will mate with the engine (see illustration). Slide the timing chain cover over the crankshaft and the two studs, then install the bolts hand tight into the cover.

40   Install two 8 mm studs into the engine block to guide the timing chain cover into the correct position (see illustration).

12.40 Fabricate two timing chain cover aligning studs from M8 x 1.25 mm bolts with the heads removed and use the plastic sleeves that were between the original timing cover bolts and the timing cover to ensure the cover is in the correct position

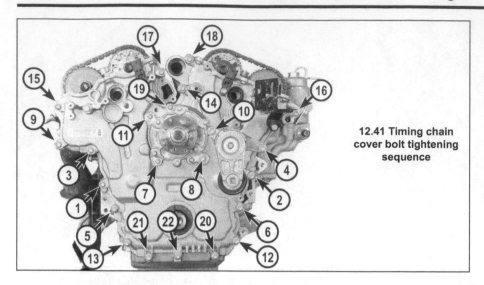

12.41 Timing chain cover bolt tightening sequence

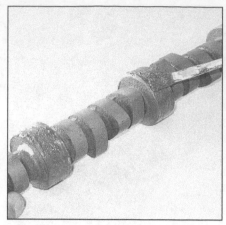

13.14 Be sure to apply camshaft assembly lube to the cam lobes and bearing journals before installing the camshaft

41   Remove the two 8 mm studs before tightening the timing chain cover bolts in the illustrated order (see illustration).

42   The remainder of installation is the reverse of removal.

13 Camshafts - removal and installation

Note: *The camshaft should always be thoroughly inspected before installation and camshaft endplay should always be checked prior to camshaft removal. The camshaft inspection procedure for this engine is identical to that of the 3.5L V6 engine, therefore refer to Chapter 2D for the camshaft inspection procedure.*

13.16 The camshafts are in the correct position if the special tool EN-46105-1 (1) can be installed to the rear of the camshafts. The camshaft bearing caps should have the tang (2) facing toward the center of the cylinder head. The number (3) signifies the cylinder number for that cap and the letter next to the number indicates whether it is for the intake camshaft (I) or exhaust camshaft (E)

Note: *The camshaft bearing caps are marked with a number and letter and a raised arrow. The number signifies the bearing journal (cylinder number) position from the front of the engine. The letter indicates whether it originates from the intake camshaft (I), or the exhaust (E) camshaft and the arrow always points toward the front of the engine.*

Note: *It is necessary to obtain a special timing chain holding tools EN 46108 or EN 48313, available from SPX Tools, to perform this procedure. If you don't use this tool set or a similar fixture, you'll still be able to do the job, however you'll have to remove the timing chain cover to ensure that the chains stay properly attached to their lower sprockets. If you decide to fabricate timing chain holding tools yourself, make sure that they hold the timing chains and upper sprockets securely and tightly in place.*

Removal

1   Disconnect the cable from the negative terminal of the battery (see Chapter 5).

2   Remove the valve cover(s) (see Section 4).

3   Remove the lower intake manifold (see Section 7). There may be other interfering components that must be removed for access; these components vary by model.

4   Remove the camshaft position sensors (see Chapter 6).

5   Remove the camshaft position actuator solenoid(s).

6   Use a ratchet and socket on the crankshaft pulley bolt to rotate the engine until the flats at the rear of the camshafts you're working on are facing up and are parallel to the surface of the top of the cylinder head. This puts the camshafts in a low-tension position.

7   Check the camshaft endplay by mounting a dial indicator to the front of the engine with the plunger against the end of the camshaft. Lever the camshaft toward the rear of the engine and zero the indicator. Lever it toward the front of the engine and note the reading. Compare this with the endplay listed in the Specifications in this Chapter. Repeat

the test on the other camshafts. A camshaft with excessive endplay should be replaced.

Note: *Use care to avoid damaging the camshaft when levering it. It can easily be chipped.*

8   Hold the camshafts with an open-end wrench, then loosen (but don't remove) the camshaft sprocket bolts.

9   Install the EN 46108 tools, or the EN 48313 (or similar tool) to hold the timing chain and sprockets in position. Make sure the tools are engaged securely. Make absolutely sure there's no way a sprocket can come loose before proceeding. Make matchmarks on the sprockets and the chain in case they get moved.

10   Remove the camshaft sprocket bolts, then detach the sprockets from the camshafts.

11   Before removing the bearing caps, arrange to store them in a clearly labeled box to ensure that they're installed in their original locations.

12   Remove the camshaft bearing cap bolts in gradual steps in the reverse of the tightening sequence (see illustrations 13.17 and 13.18).

13   Inspect the camshafts for wear and damage (see Chapter 2A).

Installation

14   Ensure the sealing rings on the front of the camshafts are in place and lubricate the camshaft bearing journals and cam lobes with camshaft assembly lube (see illustration).

15   Working on the RH cylinder head, lubricate the bearing journals and place the camshafts onto the bearing journals. Position the camshafts so the flat portions on the rear of the camshaft are parallel to the valve cover gasket face on the top of the cylinder head.

Note: *It should be possible to install the camshaft lock tool no. EN 46105-1 used when removing the timing chains and sprockets.*

16   Apply engine oil to the inside of the bearing caps, then install the front bearing cap first, ensuring the thrust face is correctly posi-

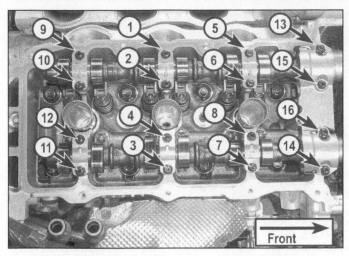

13.17 RH (rear) cylinder head bearing cap TIGHTENING sequence

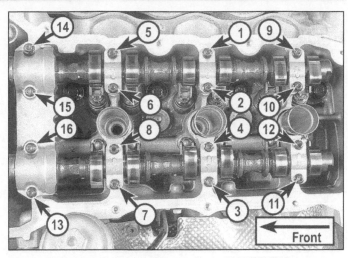

13.18 LH (front) cylinder head bearing cap TIGHTENING sequence

tioned. Install the remaining bearing caps and hand tighten the bolts (see illustration).

Note: *Ensure the bearing caps are returned to their original positions and the arrows face the center of the cylinder head.*

17   Tighten the bearing cap bolts for both camshafts in the order illustrated (see illustration) to the torque listed in this Chapter's Specifications. Once the torque procedure is complete, loosen the bolts labeled 1, 2 and 3, 4 and re-torque them.

18   Repeat Steps 14 through 17 to install the LH cylinder head camshafts (see illustration). Caution: If there is any question about the timing chain holding tool retaining the sprockets and chain during the previous Steps, remove the timing chain cover to verify that all chains and sprockets are correctly positioned before starting the engine.

19   The remainder of installation is the reverse of removal.

20   Before starting and running the engine, change the oil and install a new oil filter (see Chapter 1).

5   Remove the right engine mount (see Section 18).

6   Drain the engine oil and remove the oil filter (see Chapter 1).

7   Remove the front portion of the exhaust system (see Chapter 4).

8   Remove the air conditioning compressor (see Chapter 3) or, if possible, simply unbolt it and position it out of the way without disconnecting the refrigerant lines.

9   Remove the oil pan mounting bolts, then remove the oil pan by prying only at the proper points (see illustration).

10   The oil pickup can be removed after the bolts in the bottom of the oil pan are removed. The pickup seal must be replaced if removed.

Installation

11   Thoroughly clean the mounting surfaces of the oil pan and engine block of old gasket material and sealer. Wipe the gasket surfaces clean with a rag soaked in brake system cleaner.

12   Check the oil pump pickup for cracks or signs of leakage. Ensure the screen in the bottom of the pickup is not blocked or damaged.

13   Ensure a new oil seal is installed to the oil pump pickup. Install the oil pickup and tighten the bolts in the bottom of the oil pan. Install the baffle to the inside of the oil pan, tightening the retaining bolts.

14   Fabricate two aligning studs from M8 x 1.25 mm bolts with the heads removed and screw them into the oil pan bolt holes labeled 1 and 2 to allow for correct oil pan alignment (see illustration 14.16).

15   Apply a 3 mm wide bead of oxygen sensor safe RTV sealant to the oil pan mating face with the engine block. Also apply some sealant to the corners of the block where the front cover and the rear cover meet the engine block. Install the oil pan and tighten the bolts finger-tight.

16   Remove the aligning studs before tightening the oil pan bolts in the order shown to

---

14   Oil pan - removal and installation

Removal

1   Have the air conditioning system discharged at an authorized shop.

Note: *It may be possible on some models to avoid discharging the system by moving the air conditioning compressor with the lines still connected. Check your vehicle before having the system discharged.*

2   Disconnect the cable from the negative battery terminal (see Chapter 5).

3   Install an engine support fixture, with the chain connected to a solid engine lifting bracket and tighten it to take the weight off of the engine mounts.

4   Raise the vehicle and support it securely on jackstands.

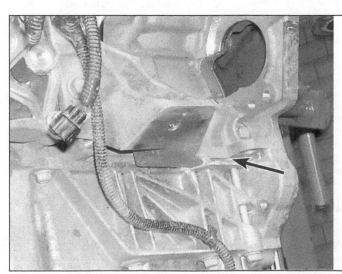

14.9 Arrow indicates a lever point to break the seal between the oil pan and cylinder block. There is another one at the front on the opposite side of the engine

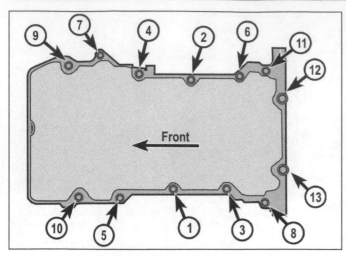

14.16 Oil pan bolt tightening sequence

15.2 Oil pump retaining bolts

the torque listed in this Chapter's Specifications (see illustration).

Note: *Check the Specifications for the correct torque. The two bolts holding the oil pan to the rear main oil seal housing are a different torque to the remaining bolts.*

17   The remainder of installation is the reverse of removal.

18   Add the proper type and quantity of oil (see Chapter 1), start the engine and check for leaks before placing the vehicle back in service.

15   Oil pump - removal, inspection and installation

Removal

1   Remove the timing chains (see Section 12). After noting the installed position, remove the timing chain sprocket from the crankshaft.

2   Remove the oil pump retaining bolts and slide the pump off the end of the crankshaft (see illustration).

Inspection

3   Remove the timing chain guide from the oil pump housing then remove the oil pump cover and withdraw the rotors from the pump body. Remove the circlip from the side of the pump housing and withdraw the cap, spring and plunger from the oil pump housing. These components are the oil pressure relief valve assembly. Clean the components with solvent, dry them thoroughly and inspect for any obvious damage. Also check the bolt holes for damaged threads and the splined surfaces on the crankshaft sprocket for any apparent damage. If any of the components are scored, scratched or worn, replace the entire oil pump assembly. There are no serviceable parts currently available.

Installation

4   If re-using the oil pump, assemble the pressure relief valve components and the rotors to the oil pump housing. Prior to installing the cover, prime the pump by pouring clean motor oil between the rotors. Install the cover and tighten the retaining bolts.

5   Position the oil pump over the end of the crankshaft and align the teeth on the crankshaft sprocket with the teeth on the oil pump drive gear. Make sure the pump is fully seated against the block.

6   Install the oil pump mounting bolts and tighten them to the torque listed in this Chapter's Specifications.

7   The remainder of installation is the reverse of removal.

8   Add oil and coolant as necessary. Run the engine and check for oil and coolant leaks. Also check the oil pressure as described in Chapter 2D.

16   Driveplate - removal and installation

1   The driveplate replacement for the 3.6L V6 engine is identical to the driveplate replacement procedure for the 3.5L V6 engine. Refer to Chapter 2B for the procedure and use the torque figures in this Chapter's Specifications.

Note: *New bolts must be used when re-installing the driveplate.*

17   Rear main oil seal - replacement

Note: *At the time of publication, the rear main oil seal is supplied as an assembly consisting of the oil seal and housing. The manufacturer recommends removing the oil pan and using special studs EN 46109, seal alignment tool EN 47839 and handle J 42183 to ensure cor-*

rect alignment of the seal. These tools are available through GM dealers and specialty automotive tool suppliers.

1   Disconnect the cable from the negative battery terminal (see Chapter 5).

2   Refer to Section 16 and remove the driveplate.

3   Refer to Section 14 and remove the oil pan.

Note: *Although it is possible to replace the rear seal housing without first removing the oil pan, the manufacturer recommends this method to ensure correct alignment of the oil seal housing.*

4   Remove the retaining bolts and remove the seal housing from the engine (see illustration). It may be necessary to use a knife to cut the sealant away from the joint between the rear main oil seal housing and the cylinder block.

Caution: *Use care not to scratch the machined faces of the cylinder block, rear main seal housing or oil pan when scraping old gasket material from components.*

5   Thoroughly clean the mounting surfaces of the oil pan, rear main oil seal housing and engine block of old gasket material and sealer. Wipe the gasket surfaces clean with a rag soaked in brake system cleaner.

6   Install the two dowels into the oil seal housing bolt holes (see illustration).

7   Apply a 1/8-inch wide bead of RTV sealant around the sealing area of the oil seal housing, passing on the inside of the bolt holes.

8   Install the special alignment tool EN 47839 on the rear of the crankshaft, attaching it with the two supplied screws.

9   Install the oil seal housing over the alignment tool and studs onto the rear of the engine block. Remove the guide studs but leave the alignment tool in place.

10   Install the housing bolts and tighten them to the torque listed in this Chapter's Specifications.

17.4 Rear main oil seal housing bolts

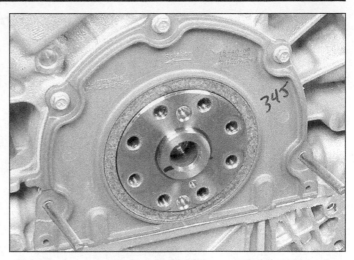

17.6 Alignment dowels installed in the rear main oil seal housing bolt holes

11   Remove the alignment tool.

12   The remainder of installation is the reverse of removal.

13   Add oil as necessary, run the engine and check for leaks.

18   Engine mount - check and replacement

1   Engine mounts seldom require attention, but broken or deteriorated mounts should be replaced immediately or the added strain placed on the driveline components may cause damage.

Check

2   During the check, the engine must be raised slightly to remove the weight from the mount.

3   Raise the vehicle and support it securely on jackstands, then position the jack under the engine oil pan. Place a large block of wood between the jack head and the oil pan, then carefully raise the engine just enough to take the weight off the mount. Do not use the jack to support the entire weight of the engine.

4   Check the mount to see if the rubber is cracked, hardened or separated from the metal plates. Sometimes the rubber will split right down the center.

5   Check for relative movement between the mount plates and the engine or frame (use a large screwdriver or pry bar to attempt to move the mount). If movement is noted, check the tightness of the mount fasteners first before condemning the mount. Usually when engine mounts are broken, they are very obvious as the engine will easily move away from the mount when pried or under load.

Replacement

6   Disconnect the cable from the negative battery terminal (see Chapter 5).

7   Refer to Chapter 1 and remove the drivebelt.

8   Remove the engine top cover and the air filter assembly (see Chapter 4).

9   Raise the vehicle and support it securely on jackstands.

10   Support the engine. This can be done by using an engine support fixture that attaches between the fenders and holds the engine from above (the preferred method) or by using a floor jack with a block of wood placed under the oil pan.

11   Loosen the topmost bolt in the upper engine mount bracket. Remove the other three bracket bolts to allow the bracket to pivot on the top bolt.

12   Remove the two upper nuts from the engine mount that attach to the upper mount bracket.

13   Unbolt the lower mount bracket from the frame of the vehicle, then remove the mount and bracket. The mount can now be separated from the lower bracket.

14   Installation is the reverse of removal.

Notes

Chapter 2 Part D
General engine overhaul procedures

Contents

Specifications

General
Displacement
    2.2L four-cylinder .... 134 cubic inches
    2.4L four-cylinder .... 146 cubic inches
    3.5L V6.. 214 cubic inches
    3.6L V6.. 217 cubic inches
    3.9L V6.. 238 cubic inches
Bore and Stroke
    2.2L four-cylinder .... 3.39 x 3.73 inches
    2.4L four-cylinder .... 3.47 x 3.86 inches
    3.5L V6
        2004 through 2006... 3.70 x 3.31 inches
        2007... 3.90 x 2.99 inches
    3.6L V6.. 3.70 x 3.37 inches
    3.9L V6.. 3.90 x 3.31 inches
Cylinder compression.. Lowest cylinder must be within 75 percent of highest cylinder
Oil pressure (engine at operating temperature)
    Four-cylinder models.. 50 to 80 psi @ 1,000 rpm
    V6 models
        2004... 60 psi @ 1,850 rpm
        2005 and later.. 30 to 45 psi @ 1,850 rpm

Camshaft (V6 models)
Camshaft bearing journal diameter
    2004 through 2006 3.5L V6...................................... 1.868 to 1.869 inches
    3.9L V6 and 2007 and later 3.5L V6......................... 2.024 to 2.025 inches
    3.6L V6
        Front No. 1... 1.375 to 1.376 inches
        No. 2, 3, 4.. 1.060 to 1.061 inches
Bearing oil clearance .... 0.002 to 0.004 inch
Lobe lift
    All except 3.6L V6 .... 0.273 inch
    3.6L V6
        Intake... 1.669 to 1.681 inches
        Exhaust... 1.670 to 1.682 inches

Torque specifications

Ft-lbs (unless otherwise indicated)

Note: *One foot-pound (ft-lb) of torque is equivalent to 12 inch-pounds (in-lbs) of torque. Torque values below approximately 15 foot-pounds are expressed in inch-pounds, because most foot-pound torque wrenches are not accurate at these smaller values.*

Subframe mounting bolts*	See Chapter 10
Driveplate-to-torque converter bolts	
All except 3.6L V6 engines..	46
3.6L V6 engines	
Step 1	22
Step 2	Tighten an additional 45-degrees
Camshaft thrust plate screws (all except 3.6L V6 engines)....................	89 in-lbs
Camshaft drive gear bolt (all except 3.6L V6 engines)............................	27
Connecting rod bearing cap bolts*	
Four-cylinder models	
Step 1	18
Step 2	Tighten an additional 100-degrees
V6 models	
All except 3.6L V6	
Step 1	18
Step 2	Tighten an additional 110-degrees
3.6L V6	
Step 1	22
Step 2	Loosen fully
Step 3	18
Step 4	Tighten an additional 110-degrees
Main bearing cap bolts*	
Four-cylinder models **(see illustration 11.19a)**	
Step 1	15
Step 2	Tighten an additional 70-degrees
3.5L and 3.9L V6 models	
Step 1	37
Step 2	Tighten an additional 77-degrees
3.6L V6 models **(see illustration 11.19b)** **	
Inner bolts	
Step 1..	15
Step 2	Tighten an additional 80-degrees
Outer bolts	
Step 1	10
Step 2	Tighten an additional 110-degrees
Side bolts	
Step 1	22
Step 2	Tighten an additional 60-degrees
Lower crankcase perimeter bolts (four-cylinder engines) **(see illustration 11.30)**...	18
Transaxle-to-engine bolts	66

* Bolts must be replaced with new ones upon final assembly after clearance check
** Tighten inner bolts followed by outer bolts, then side bolts

1.9a An engine block being bored. An engine rebuilder will use special machinery to recondition the cylinder bores

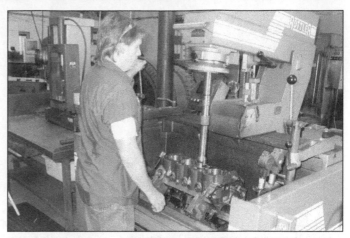

1.9b If the cylinders are bored, the machine shop will normally hone the engine on a machine like this

1 General information - engine overhaul

1 Included in this portion of Chapter 2A are general information and diagnostic testing procedures for determining the overall mechanical condition of your engine.

2 The information ranges from advice concerning preparation for an overhaul and the purchase of replacement parts and/or components to detailed, step-by-step procedures covering removal and installation.

3 The following Sections have been written to help you determine whether your engine needs to be overhauled and how to remove and install it once you've determined it needs to be rebuilt. For information concerning in-vehicle engine repair, see Chapter 2B or Chapter 2C.

4 It's not always easy to determine when, or if, an engine should be completely overhauled, because a number of factors must be considered.

5 High mileage is not necessarily an indication that an overhaul is needed, while low mileage doesn't preclude the need for an overhaul. Frequency of servicing is probably the most important consideration. An engine that's had regular and frequent oil and filter changes, as well as other required maintenance, will most likely give many thousands of miles of reliable service. Conversely, a neglected engine may require an overhaul very early in its service life.

6 Excessive oil consumption is an indication that piston rings, valve seals and/or valve guides are in need of attention. Make sure that oil leaks aren't responsible before deciding that the rings and/or guides are bad. Perform a cylinder compression check to determine the extent of the work required (see Section 3). Also check the vacuum readings under various conditions (see Section 4).

7 Check the oil pressure with a gauge installed in place of the oil pressure sending unit and compare it to this Chapter's Specifications (see Section 2). If it's extremely low, the bearings and/or oil pump are probably worn out.

8 Loss of power, rough running, knocking or metallic engine noises, excessive valve train noise and high fuel consumption rates may also point to the need for an overhaul, especially if they're all present at the same time. If a complete tune-up doesn't remedy the situation, major mechanical work is the only solution.

9 An engine overhaul involves restoring the internal parts to the specifications of a new engine. During an overhaul, the piston rings are replaced and the cylinder walls are reconditioned (rebored and/or honed) (see illustrations 1.9a and 1.9b). If a rebore is done by an automotive machine shop, new oversize pistons will also be installed. The main bearings, connecting rod bearings and camshaft bearings are generally replaced with new ones and, if necessary, the crankshaft may be reground to restore the journals (see illustration 1.9c). Generally, the valves are serviced as well, since they're usually in less-than-perfect condition at this point. While the engine is being overhauled, other components, such as the distributor, starter and alternator, can be rebuilt as well. The end result should be similar to a new engine that will give many trouble free miles.

Note: *Critical cooling system components such as the hoses, drivebelts, thermostat and water pump should be replaced with new parts when an engine is overhauled. The radiator should be checked carefully to ensure that it isn't clogged or leaking (see Chapter 3). If you purchase a rebuilt engine or short block, some rebuilders will not warranty their engines unless the radiator has been professionally flushed. Also, we don't recommend overhauling the oil pump - always install a new one when an engine is rebuilt.*

10 Overhauling the internal components on today's engines is a difficult and time-consuming task which requires a significant amount of specialty tools and is best left to a professional engine rebuilder (see illustrations 1.10a, 1.10b and 1.10c). A competent engine rebuilder will handle the inspection of your old parts and offer advice concern-

1.9c A crankshaft having a main bearing journal ground

1.10a A machinist checks for a bent connecting rod, using specialized equipment

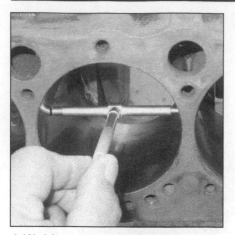

1.10b A bore gauge being used to check a cylinder bore

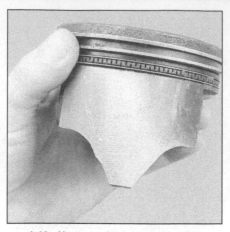

1.10c Uneven piston wear like this indicates a bent connecting rod

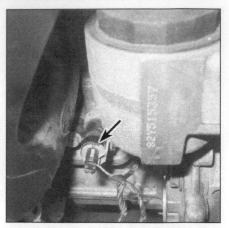

2.2a Location of the oil pressure sending unit on four-cylinder engines

ing the reconditioning or replacement of the original engine, never purchase parts or have machine work done on other components until the block has been thoroughly inspected by a professional machine shop. As a general rule, time is the primary cost of an overhaul, especially since the vehicle may be tied up for a minimum of two weeks or more. Be aware that some engine builders only have the capability to rebuild the engine you bring them while other rebuilders have a large inventory of rebuilt exchange engines in stock. Also be aware that many machine shops could take as much as two weeks time to completely rebuild your engine depending on shop workload. Sometimes it makes more sense to simply exchange your engine for another engine that's already rebuilt to save time.

2   Oil pressure check

1   Low engine oil pressure can be a sign of an engine in need of rebuilding. A "low oil pressure" indicator (often called an "idiot light") is not a test of the oiling system. Such indicators only come on when the oil pressure is dangerously low. Even a factory oil pressure gauge in the instrument panel is only a relative indication, although much better for driver information than a warning light. A better test is with a mechanical (not electrical) oil pressure gauge.
2   Locate the engine oil pressure sending unit on the engine block:

a) On four-cylinder engines, the oil pressure sending unit is located on the front left side of the engine block, below the oil filter cartridge housing (see illustration).
b) On V6 engines, the oil pressure sending unit is located near the oil filter housing on the front side of the engine (see illustration). **Note:** On some models a heat shield will have to be removed for access to the sending unit.

3   Unscrew and remove the oil pressure sending unit and screw in the hose for your oil pressure gauge (see illustration). If necessary, install an adapter fitting. Use Teflon tape or thread sealant on the threads of the adapter and/or the fitting on the end of your gauge's hose.
4   Connect an accurate tachometer to the engine, according to the tachometer manufacturer's instructions.
5   Check the oil pressure with the engine running (normal operating temperature) at the specified engine speed, and compare it to this Chapter's Specifications. If it's extremely low, the bearings and/or oil pump are probably worn out.

3   Cylinder compression check

1   A compression check will tell you what mechanical condition the upper end of your engine (pistons, rings, valves, head gaskets) is in. Specifically, it can tell you if the compression is down due to leakage caused by worn piston rings, defective valves and seats or a blown head gasket. **Note:** The engine must be at normal operating temperature and the battery must be fully charged for this check.
2   Begin by cleaning the area around the spark plugs before you remove them (compressed air should be used, if available). The idea is to prevent dirt from getting into the cylinders as the compression check is being done.
3   Remove all of the spark plugs from the engine (see Chapter 1).
4   Remove the air intake duct (see Chapter 4) and block the throttle plate wide open.
5   On 3.5L and 3.9L V6 models, disable the ignition system by disconnecting the primary (low voltage) wires from the coil pack (see Chapter 5). (The ignition system on four-cylinder and 3.6L V6 models will already have been disabled, since the ignition coils were removed for spark plug removal.) On all models, disable the fuel system by removing the fuel pump fuse (see Chapter 4, Section 2).

2.2b Location of the oil pressure sending unit on V6 engines

2.3 Remove the oil pressure sending unit and install an oil pressure gauge

6   Install a compression gauge in the number one cylinder spark plug hole (see illustration).

7   Crank the engine over at least seven compression strokes and watch the gauge. The compression should build up quickly in a healthy engine. Low compression on the first stroke, followed by gradually increasing pressure on successive strokes, indicates worn piston rings. A low compression reading on the first stroke, which doesn't build up during successive strokes, indicates leaking valves or a blown head gasket (a cracked head could also be the cause). Deposits on the undersides of the valve heads can also cause low compression. Record the highest gauge reading obtained.

8   Repeat the procedure for the remaining cylinders and compare the results to this Chapter's Specifications.

9   Add some engine oil (about three squirts from a plunger-type oil can) to each cylinder, through the spark plug hole, and repeat the test.

10   If the compression increases after the oil is added, the piston rings are definitely worn. If the compression doesn't increase significantly, the leakage is occurring at the valves or head gasket. Leakage past the valves may be caused by burned valve seats and/or faces or warped, cracked or bent valves.

11   If two adjacent cylinders have equally low compression, there's a strong possibility that the head gasket between them is blown. The appearance of coolant in the combustion chambers or the crankcase would verify this condition.

12   If one cylinder is slightly lower than the others, and the engine has a slightly rough idle, a worn lobe on the camshaft could be the cause.

13   If the compression is unusually high, the combustion chambers are probably coated with carbon deposits. If that's the case, the cylinder head(s) should be removed and decarbonized.

14   If compression is way down or varies greatly between cylinders, it would be a good idea to have a leak-down test performed by an automotive repair shop. This test will pinpoint exactly where the leakage is occurring and how severe it is.

---

4   Vacuum gauge diagnostic checks

1   A vacuum gauge provides inexpensive but valuable information about what is going on in the engine. You can check for worn rings or cylinder walls, leaking head or intake manifold gaskets, incorrect carburetor adjustments, restricted exhaust, stuck or burned valves, weak valve springs, improper ignition or valve timing and ignition problems.

2   Unfortunately, vacuum gauge readings are easy to misinterpret, so they should be used in conjunction with other tests to confirm the diagnosis.

3   Both the absolute readings and the rate of needle movement are important for accurate interpretation. Most gauges measure vacuum in inches of mercury (in-Hg). The following references to vacuum assume the diagnosis is being performed at sea level. As elevation increases (or atmospheric pressure decreases), the reading will decrease. For every 1,000 foot increase in elevation above approximately 2,000 feet, the gauge readings will decrease about one inch of mercury.

4   Connect the vacuum gauge directly to the intake manifold vacuum, not to ported (throttle body) vacuum. Be sure no hoses are left disconnected during the test or false readings will result.

5   Before you begin the test, allow the engine to warm up completely. Block the wheels and set the parking brake. With the transaxle in Park, start the engine and allow it to run at normal idle speed.

Warning: *Keep your hands and the vacuum gauge clear of the fans.*

6   Read the vacuum gauge; an average, healthy engine should normally produce about 17 to 22 in-Hg with a fairly steady needle (see illustration). Refer to the following vacuum gauge readings and what they indicate about the engine's condition:

3.6 Use a compression gauge with a threaded fitting for the spark plug hole, not the type that requires hand pressure to maintain the seal

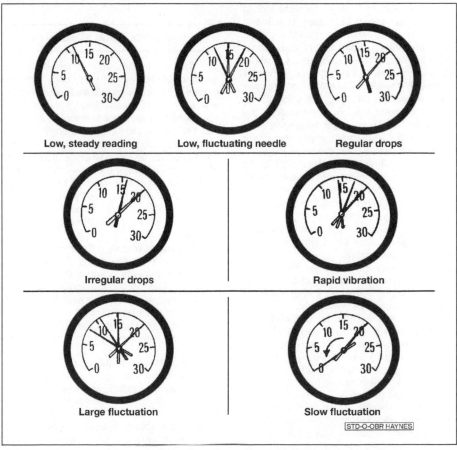

Low, steady reading   Low, fluctuating needle   Regular drops

Irregular drops   Rapid vibration

Large fluctuation   Slow fluctuation

STD-O-OBR HAYNES

4.6 Typical vacuum gauge readings

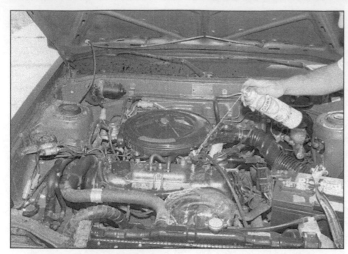

6.3a After tightly wrapping water-vulnerable components, use a spray cleaner on everything, with particular concentration on the greasiest areas, usually around the valve cover and lower edges of the block. If one section dries out, apply more cleaner

6.3b Depending on how dirty the engine is, let the cleaner soak in according to the directions and hose off the grime and cleaner. Get the rinse water down into every area you can get at; then dry important components with a hair dryer or paper towels

7   A low steady reading usually indicates a leaking gasket between the intake manifold and cylinder head(s) or throttle body, a leaky vacuum hose, late ignition timing or incorrect camshaft timing. Check ignition timing with a timing light and eliminate all other possible causes, utilizing the tests provided in this Chapter before you remove the timing chain cover to check the timing marks.

8   If the reading is three to eight inches below normal and it fluctuates at that low reading, suspect an intake manifold gasket leak at an intake port or a faulty fuel injector.

9   If the needle has regular drops of about two-to-four inches at a steady rate, the valves are probably leaking. Perform a compression check or leak-down test to confirm this.

10   An irregular drop or down-flick of the needle can be caused by a sticking valve or an ignition misfire. Perform a compression check or leak-down test and read the spark plugs.

11   A rapid vibration of about four in-Hg vibration at idle combined with exhaust smoke indicates worn valve guides. Perform a leak-down test to confirm this. If the rapid vibration occurs with an increase in engine speed, check for a leaking intake manifold gasket or head gasket, weak valve springs, burned valves or ignition misfire.

12   A slight fluctuation, say one inch up and down, may mean ignition problems. Check all the usual tune-up items and, if necessary, run the engine on an ignition analyzer.

13   If there is a large fluctuation, perform a compression or leak-down test to look for a weak or dead cylinder or a blown head gasket.

14   If the needle moves slowly through a wide range, check for a clogged PCV system, incorrect idle fuel mixture, throttle body or intake manifold gasket leaks.

15   Check for a slow return after revving the engine by quickly snapping the throttle open until the engine reaches about 2,500 rpm and let it shut. Normally the reading should drop to near zero, rise above normal idle reading

(about 5 in-Hg over) and return to the previous idle reading. If the vacuum returns slowly and doesn't peak when the throttle is snapped shut, the rings may be worn. If there is a long delay, look for a restricted exhaust system (often the muffler or catalytic converter). An easy way to check this is to temporarily disconnect the exhaust ahead of the suspected part and redo the test.

---

5   Engine rebuilding alternatives

1   The do-it-yourselfer is faced with a number of options when purchasing a rebuilt engine. The major considerations are cost, warranty, parts availability and the time required for the rebuilder to complete the project. The decision to replace the engine block, piston/connecting rod assemblies and crankshaft depends on the final inspection results of your engine. Only then can you make a cost effective decision whether to have your engine overhauled or simply purchase an exchange engine for your vehicle.

2   Some of the rebuilding alternatives include:

3   **Individual parts** - If the inspection procedures reveal that the engine block and most engine components are in reusable condition, purchasing individual parts and having a rebuilder rebuild your engine may be the most economical alternative. The block, crankshaft and piston/connecting rod assemblies should all be inspected carefully by a machine shop first.

4   **Short block** - A short block consists of an engine block with a crankshaft and piston/connecting rod assemblies already installed. All new bearings are incorporated and all clearances will be correct. The existing camshafts, valve train components, cylinder head and external parts can be bolted to the short block with little or no machine shop work necessary.

5   **Long block** - A long block consists of a short block plus an oil pump, oil pan, cylinder head, valve cover, camshaft and valve train components, timing sprockets and chain or gears and timing cover. All components are installed with new bearings, seals and gaskets incorporated throughout. The installation of manifolds and external parts is all that's necessary.

6   **Low mileage used engines** - Some companies now offer low mileage used engines which is a very cost effective way to get your vehicle up and running again. These engines often come from vehicles which have been in totaled in accidents or come from other countries which have a higher vehicle turn over rate. A low mileage used engine also usually has a similar warranty like the newly remanufactured engines.

7   Give careful thought to which alternative is best for you and discuss the situation with local automotive machine shops, auto parts dealers and experienced rebuilders before ordering or purchasing replacement parts.

---

6   Engine removal - methods and precautions

1   If you've decided that an engine must be removed for overhaul or major repair work, several preliminary steps should be taken. Read all removal and installation procedures carefully prior to committing to this job.

2   Locating a suitable place to work is extremely important. Adequate work space, along with storage space for the vehicle, will be needed. If a shop or garage isn't available, at the very least a flat, level, clean work surface made of concrete or asphalt is required.

3   Cleaning the engine compartment and engine before beginning the removal procedure will help keep tools clean and organized (see illustrations 6.3a and 6.3b).

6.7a Get an engine stand sturdy enough to firmly support the engine while you're working on it. Stay away from three-wheeled models; they have a tendency to tip over more easily, so get a four-wheeled unit

6.7b Since many of the fasteners on these engines are tightened using the "angle torque" method, a torque angle gauge is essential for proper assembly

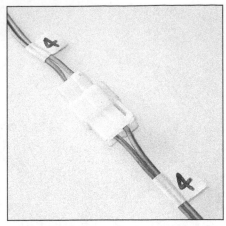

7.11 Label both ends of each wire and hose before disconnecting it

4   An engine hoist will also be necessary. Make sure the hoist is rated in excess of the combined weight of the engine and transaxle. Safety is of primary importance, considering the potential hazards involved in removing the engine from the vehicle.

5   A vehicle hoist will be necessary for engine removal on four-cylinder engines, since the engine and transaxle are removed as an assembly out the bottom of the vehicle. On V6 models, the engine is removed out the top.

6   If you're a novice at engine removal, get at least one helper. One person cannot easily do all the things you need to do to remove a big heavy engine and transaxle assembly from the engine compartment. Also helpful is to seek advice and assistance from someone who's experienced in engine removal.

7   Plan the operation ahead of time. Arrange for or obtain all of the tools and equipment you'll need prior to beginning the job (see illustrations 6.7a and 6.7b). Some of the equipment necessary to perform engine removal and installation safely and with relative ease are (in addition to a vehicle hoist and an engine hoist) a heavy duty floor jack (preferably fitted with a transaxle jack head adapter), complete sets of wrenches and sockets as described in the front of this manual, wooden blocks, plenty of rags and cleaning solvent for mopping up spilled oil, coolant and gasoline.

8   Plan for the vehicle to be out of use for quite a while. A machine shop can do the work that is beyond the scope of the home mechanic. Machine shops often have a busy schedule, so before removing the engine, consult the shop for an estimate of how long it will take to rebuild or repair the components that may need work.

7   Engine - removal and installation

Warning: *Gasoline is extremely flammable, so take extra precautions when you work on any part of the fuel system. Don't smoke or allow open flames or bare light bulbs near the work area, and don't work in a garage where a gas-type appliance (such as a water heater or clothes dryer) is present. Since gasoline is carcinogenic, wear fuel-resistant gloves when there's a possibility of being exposed to fuel, and, if you spill any fuel on your skin, rinse it off immediately with soap and water. Mop up any spills immediately and do not store fuel-soaked rags where they could ignite. The fuel system is under constant pressure, so, if any fuel lines are to be disconnected, the fuel pressure in the system must be relieved first (see Chapter 4 for more information). When you perform any kind of work on the fuel system, wear safety glasses and have a Class B type fire extinguisher on hand.*
Warning: *The engine must be completely cool before beginning this procedure.*

V6 models

Removal

Note: *Keep in mind that during this procedure you'll have to adjust the height of the vehicle to perform certain operations.*

1   Relieve the fuel system pressure (see Chapter 4).

2   Disconnect the cable from the negative terminal of the battery (see Chapter 5, Section 1)

3   Remove the air filter housing (see Chapter 4).

4   Remove the hood (see Chapter 11). Cover the fenders and cowl using special pads. An old bedspread or blanket will also work.

5   Remove the engine cover (see Chapter 2B, Section 6).

6   Remove the passenger's side engine mount strut (see Chapter 2B, Section 17).

7   Remove the drivebelt (see Chapter 1).

8   Drain the cooling system and the engine oil (see Chapter 1).

9   Remove the starter, alternator and ignition coil pack (see Chapter 5).

10   Unplug the electrical connectors from the:

a) *Oil pressure sending unit (see Section 2).*

b) *Knock sensors, Crankshaft Position (CKP) sensor, oxygen sensors, canister purge solenoid, Manifold Absolute Pressure (MAP) sensor, Camshaft Position (CMP) sensor, Powertrain Control Module (PCM), transmission speed sensors and Transmission Range (TR) sensor (see Chapter 6).*

c) *Air conditioning compressor (see Chapter 3).*

d) *Throttle body and fuel injector harness (see Chapter 4).*

e) *Transmission Control Module (TCM) (see Chapter 7B). Also disconnect the large electrical connector from the front of the transaxle.*

f) *In-line electrical connectors between the engine harness and the chassis harness.*

11   Clearly label the connectors, where necessary, to avoid confusion on reassembly. Masking tape and/or a touch up paint applicator work well for marking items (see illustration).

12   Detach all wiring harness clips and ground cables from the engine and transaxle, then position the wiring harness out of the way.

13   On models equipped with hydraulic

7.24 Support the engine from with an engine hoist securely attached by heavy-duty chains to the engine lifting brackets

power steering, remove the power steering pump and reposition it out of the way without disconnecting the lines (see Chapter 10).

14   Remove the catalytic converters (see Chapter 6).

15   Unbolt the air conditioning compressor and position it out of the way without disconnecting the refrigerant lines (see Chapter 3). Use wire to tie the compressor to the subframe.

16   Remove the driveplate-to-torque converter bolts (see Chapter 7B, Section 8).

17   Remove the brace between the transaxle and the engine oil pan (see Chapter 2B).

18   Detach the radiator hoses from the engine (see Chapter 3).

19   Disconnect the coolant expansion tank hoses from the fittings on the engine. Also disconnect the heater hoses from their fittings on the engine.

20   Free the EVAP/fuel line clip from the MAP sensor bracket.

21   Detach the fuel feed line from the fuel rail (see Chapter 4).

22   Detach the purge line from the EVAP canister purge solenoid (see Chapter 6).

23   Detach the brake booster vacuum hose from the intake manifold.

24   Support the transaxle with a floor jack. Support the engine from above with an engine hoist securely attached by heavy-duty chains to the engine lifting brackets (see illustration). Warning: If bolts are used to secure or join the hoist chain, be sure the bolts are tightened securely, using spacers and/or washers on either side of the chain links, if necessary, to ensure this. Raise the hoist just enough to take the weight off the powertrain mounts.

25   Recheck to be sure nothing is still connecting the engine to the vehicle. Disconnect anything still remaining.

26   Remove the engine mount (see Chapter 2B).

27   Remove the transaxle-to-engine bolts, then separate the engine from the transaxle.

28   Raise the engine slightly and inspect it thoroughly once more to make sure that nothing is still attached, then slowly raise the engine out of the engine compartment. Check carefully to make sure nothing is hanging up.

It may be necessary to tilt or turn the engine as it is being raised. Warning: Don't place any part of your body under the engine or between the engine and the vehicle.

29   Remove the driveplate (see Chapter 2A, Chapter 2B or Chapter 2C) and mount the engine on an engine stand.

Installation

30   Installation is the reverse of the removal procedure, noting the following points:

a) *Check the powertrain mounts. If they're worn or damaged, replace them.*

b) *When installing the subframe, tighten the subframe mounting bolts to the torque listed in the Chapter 10 Specifications.*

c) *Refill the cooling system with the proper mixture of antifreeze. Refill the crankcase with the recommended engine oil (see Chapter 1).*

d) *Reconnect the battery (see Chapter 5, Section 1).*

e) *Run the engine and check for proper operation and leaks. Shut off the engine and recheck fluid levels.*

Four-cylinder models

Note: *Engine removal on these models is a difficult job, especially for the do-it-yourself mechanic working at home. Because of the vehicle's design, the manufacturer states that the engine and transaxle have to be removed as a unit from the bottom of the vehicle, not the top. With a floor jack and jackstands, the vehicle can't be raised high enough and supported safely enough for the engine/transaxle assembly to slide out from underneath. The manufacturer recommends that removal of the engine transaxle assembly only be performed on a frame-contact type vehicle hoist.*

Note: *Read through the entire Section before beginning this procedure. The engine and transaxle are removed as a unit from below and then separated outside the vehicle.*

Removal

31   Park the vehicle on a frame-contact type vehicle hoist, then engage the arms of the hoist with the jacking points of the vehicle. Raise the hoist arms until they contact the

vehicle, but not so much that the wheels come off the ground. Position the steering wheel so the front wheels point straight ahead, then disconnect the cable from the negative battery terminal (see Chapter 5).

32   Relieve the fuel system pressure (see Chapter 4).

33   Disconnect the cable from the negative terminal of the battery (see Chapter 5, Section 1).

34   Remove the hood (see Chapter 11). Cover the fenders and cowl using special pads. An old bedspread or blanket will also work.

35   Remove the air filter housing and the air intake duct (see Chapter 4).

36   Disconnect the fuel line from the fuel rail (see Chapter 4).

37   Drain the cooling system (see Chapter 1). Support the radiator/condenser assembly to the body with wire or large plastic tie-wraps, and remove the expansion tank and its hoses.

38   Remove the upper radiator hose.

39   Follow the heater hoses from the firewall and detach them from the pipes on the engine.

40   Clearly label and disconnect all vacuum lines, emissions hoses, wiring harness connectors and fuel lines. Masking tape and/or a touch up paint applicator work well for marking items (see illustration 7.11). Take instant photos or sketch the locations of components and brackets. Move the wiring harness out of the way.

41   Loosen the wheel lug nuts and the driveaxle/hub nuts, then raise the vehicle on the hoist. Remove the wheels.

42   Remove the drivebelt (see Chapter 1).

43   Drain the engine oil (see Chapter 1).

44   Unbolt the air conditioning compressor from its bracket (don't disconnect the refrigerant lines), position it aside and secure it out of the way with wire or rope. **Caution:** *Don't fasten it to the subframe.*

45   Detach the battery cable and the electrical connectors from the starter (see Chapter 5).

46   Detach the battery cable and the electrical connectors from the alternator (see Chapter 5).

47   Remove the lower radiator hose.

48   Unbolt the exhaust pipe from the exhaust manifold.

49   Disconnect the shift cable(s) from the transaxle (see Chapter 7B). Also disconnect any wiring harness connectors from the transaxle and cable brackets from the engine.

50   Refer to Chapter 10 and disconnect the stabilizer links, tie-rod ends, and the steering intermediate shaft. **Caution:** *Don't allow the steering shaft to rotate after the intermediate shaft has been disconnected, as damage to the airbag clockspring could occur. To prevent this, run the seat belt through the steering wheel and click it into its latch.*

51   Disconnect the lower control arms from the steering knuckles (see Chapter 10) and remove the driveaxles (see Chapter 8).

7.52a Attach the chain to the engine lifting bracket on the front of the cylinder head

7.52b Attach the other end of the chain to the engine lifting bracket on the rear of the cylinder head

52   Support the engine/transaxle assembly from above with a engine hoist securely attached by heavy-duty chains to the engine lifting brackets (see illustrations). Place blocks of wood between the engine/transaxle and the subframe, then loosen all the engine and transaxle mount fasteners. With the hoist taking the weight off the mounts, remove the engine/transaxle mounts.
Warning: *DO NOT place any part of your body under the engine when it's supported only by a hoist or other lifting device.*
53   Recheck to be sure nothing is still connecting the engine or vehicle. Disconnect anything still remaining.
54   Scribe or make paint marks where the subframe meets the chassis for installation alignment purposes.
55   Support the subframe with two floor jacks - one positioned under each side of the subframe. Remove the subframe bolts (see Chapter 10).
56   With an assistant to help, carefully and slowly lower each jack until the subframe is down far enough to be slid out from under the vehicle.
57   Inspect the engine/transaxle assembly thoroughly once more to make sure that nothing is still attached, then slowly lower the powertrain down out of the engine compartment and onto the floor. Check carefully to make sure nothing is hanging up as this is done.
58   Once the powertrain is on the floor, disconnect the engine lifting chains and move the engine hoist out of the way, then raise the vehicle hoist until the vehicle clears the powertrain.
59   Reconnect the engine hoist to the engine, raise the engine/transaxle up a little and support the engine with blocks of wood. Support the transaxle with a floor jack; preferably one with a transmission adapter. Secure the transaxle to the jack with safety chains.
60   Remove the starter (see Chapter 5), then mark the torque converter to the driveplate and remove the driveplate-to-torque converter bolts (see illustration).
61   Remove engine-to-transaxle mount-

ing bolts and separate the engine from the transaxle.
62   Remove the driveplate (see Chapter 2A) and mount the engine on an engine stand (see illustration 6.7a).

Installation
63   Installation is the reverse of the removal procedure, noting the following points:

 a) *Check the engine/transaxle mounts. If they're worn or damaged, replace them.*
 b) *Attach the transaxle to the engine following the procedure described in Chapter 7B.*
 c) *When installing the subframe, align the marks made during removal, then tighten the subframe mounting bolts to the torque listed in the Chapter 10 Specifications.*
 d) *Refill the cooling system with the proper mixture of antifreeze. Refill the crankcase with the recommended engine oil (see Chapter 1).*
 e) *Reconnect the battery (see Chapter 5, Section 1).*
 f) *Run the engine and check for proper operation and leaks. Shut off the engine and recheck fluid levels. Check the transaxle fluid level, adding as necessary (see Chapter 1).*

---

8   Engine overhaul - disassembly sequence

1   It's much easier to remove the external components if it's mounted on a portable engine stand. A stand can often be rented quite cheaply from an equipment rental yard. Before the engine is mounted on a stand, the driveplate should be removed from the engine.
2   If a stand isn't available, it's possible to remove the external engine components with it blocked up on the floor. Be extra careful not to tip or drop the engine when working without a stand.

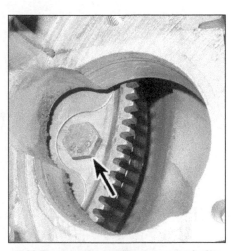

7.60 Remove the torque converter bolts through the starter opening

3   If you're going to obtain a rebuilt engine, all external components must come off first, to be transferred to the replacement engine. These components include:

Driveplate
Ignition system components
Emissions-related components
Engine mounts and mount brackets
Engine rear cover (spacer plate between driveplate and engine block), if equipped
Intake/exhaust manifolds
Fuel injection components
Oil filter
Spark plugs and ignition coil pack (and spark plug wires) or coil-over plug assemblies
Thermostat and housing assembly
Water pump

Note: *When removing the external components from the engine, pay close attention to details that may be helpful or important during installation. Note the installed position of gaskets, seals, spacers, pins, brackets, washers, bolts and other small items.*

9.2 Remove the bolt and pull out the oil pump drive

9.4 Remove the retaining bolts and the camshaft thrust plate

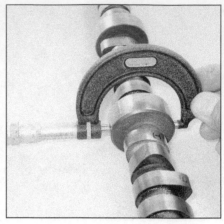

9.7a Measure the camshaft bearing journals with a micrometer

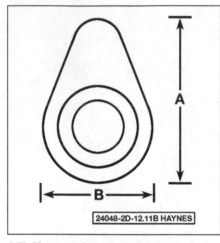

24048-2D-12.11B HAYNES

9.7b Measure the camshaft lobe maximum diameter (A) and the minimum diameter (B) - subtract B from A and the difference is the lobe lift

4   If you're going to obtain a short block (assembled engine block, crankshaft, pistons and connecting rods), then remove the timing chain or belt, cylinder head(s), oil pan, oil pump pick-up tube, oil pump and water pump from your engine so that you can turn in your old short block to the rebuilder as a core. See *Engine rebuilding alternatives* for additional information regarding the different possibilities to be considered.

9   Camshaft (3.5L and 3.9L V6 models) - removal, inspection and installation

Removal

1   Remove the engine (see Section 7).
2   Remove the bolt and clamp holding the oil pump drive and pull the oil pump drive straight up and out of the block (see illustration).

3   Refer to Chapter 2B and remove the timing chain and sprockets and the valve lifters.
4   Remove the bolts holding the camshaft thrust plate to the block (see illustration) and remove the thrust plate.
5   Slide the camshaft straight out of the engine, using a long bolt (with the same thread as the camshaft sprocket bolt) screwed into the front of the camshaft as a "handle." Support the shaft near the block and be careful not to scrape or nick the bearings.

Inspection

6   After the camshaft has been removed, clean it with solvent and dry it, then inspect the bearing journals for uneven wear, pitting and evidence of seizure. If the journals are damaged, the camshaft bearings are probably damaged as well. Both the shaft and bearings will have to be replaced.
7   Measure the bearing journals with a micrometer (see illustration) to determine whether they are excessively worn or out-of-round. Measure the camshaft lobes also to check for wear. Measure the camshaft lobes at their highest point, then subtract the measurement of the lobe at its smallest diameter - the difference is the lobe lift (see illustration). Refer to the Specifications listed in this Chapter.
8   Inspect the camshaft lobes for heat discoloration, score marks, chipped areas, pitting and uneven wear. If the lobes are in good condition and if the lobe lift measurements are as specified, you can reuse the camshaft.
9   Check the camshaft bearings in the block for wear and damage. Look for galling, pitting and discolored areas. Inspect the housing journals for damage and replace them if necessary.
10   The inside diameter of each bearing can be determined with a small hole gauge and outside micrometer or an inside micrometer. Subtract the camshaft bearing journal diameter(s) from the corresponding bearing

inside diameter(s) to obtain the bearing oil clearance. If it's excessive, new bearings or housings will be required regardless of the condition of the originals. Refer to the Specifications listed in this Chapter.
11   Camshaft bearing replacement requires special tools and expertise that place it outside the scope of the home mechanic. Take the block to an automotive machine shop to ensure the job is done correctly.

Installation

12   Lubricate the camshaft bearing journals and cam lobes with a special camshaft installation lubricant.
13   Slide the camshaft into the engine, using a long bolt (the same thread as the camshaft sprocket bolt) screwed into the front of the camshaft as a "handle." Support the cam near the block and be careful not to scrape or nick the bearings. Install the camshaft retainer plate and tighten the bolts to the torque listed in this Chapter's Specifications.
14   Dip the gear portion of the oil pump drive in engine oil and insert it into the block. It should be flush with its mounting boss before inserting the retaining bolt.
Note: *Position a new O-ring on the oil pump driveshaft before installation.*
15   Complete the installation of the timing chain and sprockets by referring to Chapter 2B.

10   Pistons and connecting rods - removal and installation

Removal

Note: *Prior to removing the piston/connecting rod assemblies, remove the cylinder head and oil pan.*

1   Use your fingernail to feel if a ridge has formed at the upper limit of ring travel (about 1/4-inch down from the top of each cylinder). If carbon deposits or cylinder wear have

10.1 Before you try to remove the pistons, use a ridge reamer to remove the raised material (ridge) from the top of the cylinders

10.3 Checking the connecting rod endplay (side clearance)

10.4 If the connecting rods and caps are not marked, use paint to mark the caps to the rods by cylinder number (for example, this would be the No. 4 connecting rod)

produced ridges, they must be completely removed with a special tool (see illustration). Follow the manufacturer's instructions provided with the tool. Failure to remove the ridges before attempting to remove the piston/connecting rod assemblies may result in piston breakage.

2   After the cylinder ridges have been removed, turn the engine so the crankshaft is facing up.

3   Before the main bearing cap assembly an d connecting rods are removed, check the connecting rod endplay with feeler gauges. Slide them between the first connecting rod and the crankshaft throw until the play is removed (see illustration). Repeat this procedure for each connecting rod. The endplay is equal to the thickness of the feeler gauge(s). Check with an automotive machine shop for the endplay service limit (a typical endplay limit should measure between 0.005 to 0.015 inch [0.127 to 0.369 mm]). If the play exceeds the service limit, new connecting rods will be required. If new rods (or a new crankshaft) are installed, the endplay may fall under the minimum allowable. If it does, the rods will have to be machined to restore it. If necessary, consult an automotive machine shop for advice.

4   Check the connecting rods and caps for identification marks. If they aren't plainly marked, use paint or marker to clearly identify each rod and cap (1, 2, 3, etc., depending on the cylinder they're associated with) (see illustration).

5   Remove the connecting rod cap bolts.

Note: *New connecting rod cap bolts must be used when reassembling the engine. Save the old bolts for the oil clearance check.*

6   Remove the number one connecting rod cap and bearing insert. Don't drop the bearing insert out of the cap.

7   Remove the bearing insert and push the connecting rod/piston assembly out through the top of the engine. Use a wooden or plastic hammer handle to push on the upper bearing surface in the connecting rod. If resistance is

10.13 Install the piston ring into the cylinder then push it down into position using a piston so the ring will be square in the cylinder

felt, double-check to make sure that all of the ridge was removed from the cylinder.

8   Repeat the procedure for the remaining cylinders.

9   After removal, reassemble the connecting rod caps and bearing inserts in their respective connecting rods and install the cap bolts finger tight. Leaving the old bearing inserts in place until reassembly will help prevent the connecting rod bearing surfaces from being accidentally nicked or gouged.

10   The pistons and connecting rods are now ready for inspection and overhaul at an automotive machine shop.

Piston ring installation

11   Before installing the new piston rings, the ring end gaps must be checked. It's assumed that the piston ring side clearance has been checked and verified correct.

Note: *3.9L and 2007 3.5L V6 engines are equipped with two different brand pistons. The correct piston rings must match the piston. To identify the piston brand, turn the piston over and look between the pin bores for a FM mark*

10.14 With the ring square in the cylinder, measure the ring end gap with a feeler gauge

(Federal Mogul) or a no mark (Mahle) brand type. Use only the correct type of piston rings that match the piston manufacturer.

12   Lay out the piston/connecting rod assemblies and the new ring sets so the ring sets will be matched with the same piston and cylinder during the end gap measurement and engine assembly.

13   Insert the top (number one) ring into the first cylinder and square it up with the cylinder walls by pushing it in with the top of the piston (see illustration). The ring should be near the bottom of the cylinder, at the lower limit of ring travel.

14   To measure the end gap, slip feeler gauges between the ends of the ring until a gauge equal to the gap width is found (see illustration). The feeler gauge should slide between the ring ends with a slight amount of drag. A typical ring gap should fall between 0.010 and 0.020 inch [0.25 to 0.50 mm] for compression rings and up to 0.030 inch [0.76 mm] for the oil ring steel rails. If the gap is larger or smaller than specified, double-check to make sure you have the correct rings

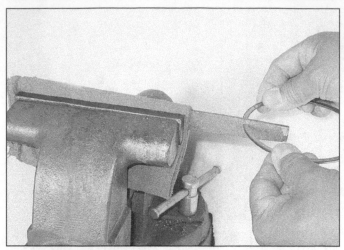

10.15 If the ring end gap is too small, clamp a file in a vise as shown and file the piston ring ends - be sure to remove all raised material

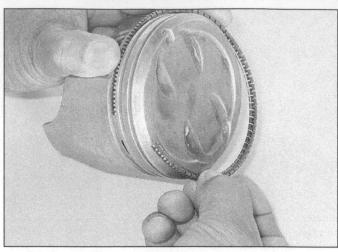

10.19a Installing the spacer/expander in the oil ring groove

before proceeding.

15   If the gap is too small, it must be enlarged or the ring ends may come in contact with each other during engine operation, which can cause serious damage to the engine. If necessary, increase the end gaps by filing the ring ends very carefully with a fine file. Mount the file in a vise equipped with soft jaws, slip the ring over the file with the ends contacting the file face and slowly move the ring to remove material from the ends. When performing this operation, file only by pushing the ring from the outside end of the file towards the vise (see illustration).

16   Excess end gap isn't critical unless it's greater than 0.040 inch (1.01 mm). Again, double-check to make sure you have the correct ring type.

17   Repeat the procedure for each ring that will be installed in the first cylinder and for each ring in the remaining cylinders. Remember to keep rings, pistons and cylinders matched up.

18   Once the ring end gaps have been checked/corrected, the rings can be installed on the pistons.

19   The oil control ring (lowest one on the piston) is usually installed first. It's composed of three separate components. Slip the spacer/expander into the groove (see illustration). If an anti-rotation tang is used, make sure it's inserted into the drilled hole in the ring groove. Next, install the upper side rail in the same manner (see illustration). Don't use a piston ring installation tool on the oil ring side rails, as they may be damaged. Instead, place one end of the side rail into the groove between the spacer/expander and the ring land, hold it firmly in place and slide a finger around the piston while pushing the rail into the groove. Finally, install the lower side rail.

20   After the three oil ring components have been installed, check to make sure that both the upper and lower side rails can be rotated smoothly inside the ring grooves.

21   The number two (middle) ring is installed next. It's usually stamped with a mark which must face up, toward the top of the piston. Do not mix up the top and middle rings, as they have different cross-sections.

Note: *Always follow the instructions printed on the ring package or box - different manufacturers may require different approaches.*

22   Use a piston ring installation tool and make sure the identification mark is facing the top of the piston, then slip the ring into the middle groove on the piston (see illustration). Don't expand the ring any more than necessary to slide it over the piston.

23   Install the number one (top) ring in the same manner. Make sure the mark is facing up. Be careful not to confuse the number one and number two rings.

24   Repeat the procedure for the remaining pistons and rings.

Installation

25   Before installing the piston/connecting rod assemblies, the cylinder walls must be perfectly clean, the top edge of each cylinder bore must be chamfered, and the crankshaft must be in place.

26   Remove the cap from the end of the number one connecting rod (refer to the marks made during removal). Remove the original bearing inserts and wipe the bearing surfaces of the connecting rod and cap with a clean, lint-free cloth. They must be kept spot-

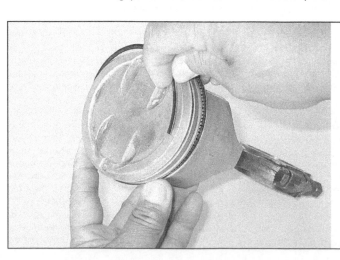

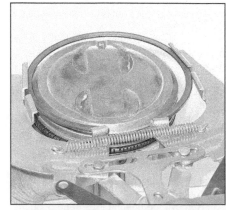

10.19b DO NOT use a piston ring installation tool when installing the oil control side rails

10.22 Use a piston ring installation tool to install the number 2 and the number 1 (top) rings - be sure the directional mark on the piston ring(s) is facing toward the top of the piston

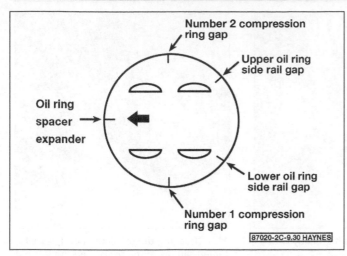

10.30a Piston ring end gap positions on four-cylinder engines

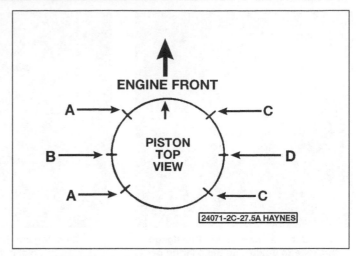

10.30b Piston ring end gap positions on V6 engines

A Oil ring side rail gaps	C Oil ring spacer gap
B Second compression ring gap	D Top compression ring gap

lessly clean.

Connecting rod bearing oil clearance check

27   Clean the back side of the new upper bearing insert, then lay it in place in the connecting rod.

28   Make sure the tab on the bearing fits into the recess in the rod. Don't hammer the bearing insert into place and be very careful not to nick or gouge the bearing face. Don't lubricate the bearing at this time.

29   Clean the back side of the other bearing insert and install it in the rod cap. Again, make sure the tab on the bearing fits into the recess in the cap, and don't apply any lubricant. It's critically important that the mating surfaces of the bearing and connecting rod are perfectly clean and oil free when they're assembled.

30   Position the piston ring gaps at the specified intervals around the piston as shown (see illustrations).

31   Lubricate the piston and rings with clean engine oil and attach a piston ring compressor to the piston. Leave the skirt protruding about 1/4-inch to guide the piston into the cylinder. The rings must be compressed until they're flush with the piston.

32   Rotate the crankshaft until the number one connecting rod journal is at BDC (bottom dead center) and apply a liberal coat of engine oil to the cylinder walls.

33   With the arrow on top of the piston facing the front (timing chain) of the engine, gently insert the piston/connecting rod assembly into the number one cylinder bore and rest the bottom edge of the ring compressor on the engine block.

34   Tap the top edge of the ring compressor to make sure it's contacting the block around its entire circumference.

35   Gently tap on the top of the piston with the end of a wooden or plastic hammer handle (see illustration) while guiding the end of the connecting rod into place on the crankshaft journal. The piston rings may try to pop out of the ring compressor just before entering the cylinder bore, so keep some downward pressure on the ring compressor. Work slowly, and if any resistance is felt as the piston enters the cylinder, stop immediately. Find out what's hanging up and fix it before proceeding. Do not, for any reason, force the piston into the cylinder - you might break a ring and/or the piston.

36   Once the piston/connecting rod assembly is installed, the connecting rod bearing oil clearance must be checked before the rod cap is permanently installed.

37   Cut a piece of the appropriate size Plastigage slightly shorter than the width of the connecting rod bearing and lay it in place on the number one connecting rod journal, parallel with the journal axis (see illustration).

38   Clean the connecting rod cap bearing face and install the rod cap. Make sure the mating mark on the cap is on the same side as the mark on the connecting rod (see illustration 10.4).

39   Install the old rod bolts, at this time, and tighten them to the torque listed in this Chapter's Specifications.

Note: *Use a thin-wall socket to avoid erroneous torque readings that can result if the socket is wedged between the rod cap and the bolt. If the socket tends to wedge itself between the fastener and the cap, lift up on it slightly until it no longer contacts the cap. DO NOT rotate the crankshaft at any time during this operation.*

40   Remove the fasteners and detach the rod cap, being very careful not to disturb the Plastigage. Discard the cap bolts at this time as they cannot be reused.

10.35 Use a plastic or wooden hammer handle to push the piston into the cylinder

10.37 Place Plastigage on each connecting rod bearing journal parallel to the crankshaft centerline

10.41 Use the scale on the Plastigage package to determine the bearing oil clearance - be sure to measure the widest part of the Plastigage and use the correct scale; it comes with both standard and metric scales

11.1 Checking crankshaft endplay with a dial indicator

11.3 Checking the crankshaft endplay with feeler gauges at the thrust bearing journal

Note: *You MUST use new connecting rod bolts.*

41   Compare the width of the crushed Plastigage to the scale printed on the Plastigage envelope to obtain the oil clearance (see illustration). The connecting rod oil clearance is usually about 0.001 to 0.002 inch. Consult an automotive machine shop for the clearance specified for the rod bearings on your engine.

42   If the clearance is not as specified, the bearing inserts may be the wrong size (which means different ones will be required). Before deciding that different inserts are needed, make sure that no dirt or oil was between the bearing inserts and the connecting rod or cap when the clearance was measured. Also, recheck the journal diameter. If the Plastigage was wider at one end than the other, the journal may be tapered. If the clearance still exceeds the limit specified, the bearing will have to be replaced with an undersize bearing.

Caution: *When installing a new crankshaft always use a standard size bearing.*

Final installation

43   Carefully scrape all traces of the Plastigage material off the rod journal and/or bearing face. Be very careful not to scratch the bearing - use your fingernail or the edge of a plastic card.

44   Make sure the bearing faces are perfectly clean, then apply a uniform layer of clean moly-base grease or engine assembly lube to both of them. You'll have to push the piston into the cylinder to expose the face of the bearing insert in the connecting rod.

45   Slide the connecting rod back into place on the journal, install the rod cap, install the new bolts and tighten them to the torque listed in this Chapter's Specifications.

Caution: *Install new connecting rod cap bolts. Do NOT reuse old bolts - they have stretched and cannot be reused.*

46   Repeat the entire procedure for the remaining pistons/connecting rods.

47   The important points to remember are:

a) *Keep the back sides of the bearing inserts and the insides of the connecting rods and caps perfectly clean when assembling them.*

b) *Make sure you have the correct piston/ rod assembly for each cylinder.*

c) *The arrow on the piston must face the front (timing chain) of the engine.*

d) *Lubricate the cylinder walls liberally with clean oil.*

e) *Lubricate the bearing faces when installing the rod caps after the oil clearance has been checked.*

48   After all the piston/connecting rod assemblies have been correctly installed, rotate the crankshaft a number of times by hand to check for any obvious binding.

49   As a final step, check the connecting rod endplay as described in Step 3. If it was correct before disassembly and the original crankshaft and rods were reinstalled, it should still be correct. If new rods or a new crankshaft were installed, the endplay may be inadequate. If so, the rods will have to be removed and taken to an automotive machine shop for resizing.

11   Crankshaft - removal and installation

Removal

Note: *The crankshaft can be removed only after the engine has been removed from the vehicle. It's assumed that the driveplate, crankshaft pulley, timing chain, oil pan, oil pump body, oil filter and piston/connecting rod assemblies have already been removed. The rear main oil seal retainer must be unbolted and separated from the block before proceeding with crankshaft removal.*

1   Before the crankshaft is removed, measure the endplay. Mount a dial indicator with the indicator in line with the crankshaft and just touching the end of the crankshaft as shown (see illustration).

2   Pry the crankshaft all the way to the rear and zero the dial indicator. Next, pry the crankshaft to the front as far as possible and check the reading on the dial indicator. The distance traveled is the endplay. A typical crankshaft endplay will fall between 0.003 to 0.010 inch (0.076 to 0.254 mm). If it is greater than that, check the crankshaft thrust surfaces for wear after it's removed. If no wear is evident, new main bearings should correct the endplay.

3   If a dial indicator isn't available, feeler gauges can be used. Gently pry the crankshaft all the way to the front of the engine. Slip feeler gauges between the crankshaft and the front face of the thrust bearing or washer to determine the clearance (see illustration).

4   If you're working on a V6 engine, loosen the main bearing cap bolts 1/4-turn at a time each, until they can be removed by hand. If you're working on a four-cylinder engine, loosen the lower crankcase bolts, following the reverse of the tightening sequence (see illustration 11.19a).

5   If you're working on a V6 engine, remove the main bearing caps. Pull the main bearing cap straight up and off the cylinder block. Gently tap the main bearing cap with a soft-face hammer, if necessary. If you're working on a four-cylinder engine, remove the lower crankcase, tapping it with a soft-face mallet if necessary. Try not to drop the bearing inserts if they come out with the assembly.

6   Carefully lift the crankshaft out of the engine. It may be a good idea to have an assistant available, since the crankshaft is quite heavy and awkward to handle. With the bearing inserts in place inside the engine block and main bearing caps, reinstall the main bearing cap or lower crankcase assembly onto the engine block and tighten the bolts finger tight. On V6 engines, make sure the caps are in the exact order they were removed with the arrow pointing

11.17 Place the Plastigage onto the crankshaft bearing journal as shown

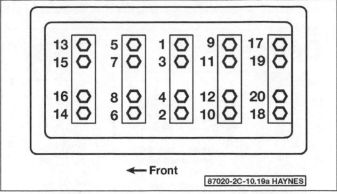

11.19a Lower crankcase bolt tightening sequence on four-cylinder engines

toward the front (timing chain and front cover) of the engine.

Installation

7   Crankshaft installation is the first step in engine reassembly. It's assumed at this point that the engine block and crankshaft have been cleaned, inspected and repaired or reconditioned.

8   Position the engine block with the bottom facing up.

9   Remove the bolts and lift off the main bearing caps (V6 engine) or lower crankcase assembly (four-cylinder engine).

10   If they're still in place, remove the original bearing inserts from the block and from the main bearing caps. Wipe the bearing surfaces of the block and main bearing cap assembly with a clean, lint-free cloth. They must be kept spotlessly clean. This is critical for determining the correct bearing oil clearance.

Main bearing oil clearance check

11   Without mixing them up, clean the back sides of the new upper main bearing inserts (with grooves and oil holes) and lay one in each main bearing saddle in the engine block.

Each upper bearing (engine block) has an oil groove and oil hole in it.

Caution: *The oil holes in the block must line up with the oil holes in the engine block inserts. The thrust washer or thrust bearing insert must be installed in the correct location.*

Note: *The thrust bearing on the four-cylinder engine is located on the engine block number 2 journal (counting from the front of the engine). The thrust bearing on the V6 engine is located on the 3rd journal in the main bearing cap journals. Clean the back sides of the lower main bearing inserts and lay them in the corresponding location in the main bearing caps. Make sure the tab on the bearing insert fits into the recess in the block or main bearing caps.*

Caution: *Do not hammer the bearing insert into place and don't nick or gouge the bearing faces. DO NOT apply any lubrication at this time.*

12   Clean the faces of the bearing inserts in the block and the crankshaft main bearing journals with a clean, lint-free cloth.

13   Check or clean the oil holes in the crankshaft, as any dirt here can go only one way - straight through the new bearings.

14   Once you're certain the crankshaft is clean, carefully lay it in position in the cylinder block.

15   Before the crankshaft can be permanently installed, the main bearing oil clearance must be checked.

16   Cut several strips of the appropriate size of Plastigage. They must be slightly shorter than the width of the main bearing journal.

17   Place one piece on each crankshaft main bearing journal, parallel with the journal axis as shown (see illustration).

18   Clean the faces of the bearing inserts in the main bearing caps or lower crankcase. Install the caps without disturbing the Plastigage.

19   Apply clean engine oil to all bolt threads prior to installation, install all bolts finger-tight, then tighten them to the torque listed in this Chapter's Specifications.

Caution: *On V6 engines, use the old bolts for all checking procedures; the new bolts will be used only for final assembly. The new bolts used on V6 engines can be tightened only once. Once they have been removed, they must be replaced with new ones. DO NOT rotate the engine during this operation. On four cylinder engines, tighten the lower crankcase bolts in the sequence shown (see illustration). On 3.6L V6 engines, tighten the inner cap bolts first, followed by the outer cap bolts and finally the side bolts (see illustration).*

20   Remove the bolts (in the reverse order of the tightening sequence on four-cylinder models) and carefully lift the main bearing caps or

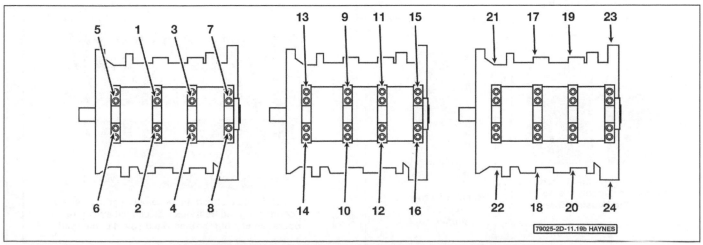

11.19b Main bearing cap bolt tightening sequence for 3.6L V6 engines

ENGINE BEARING ANALYSIS

Debris

Babbitt bearing embedded with debris from machinings

Microscopic detail of debris

Microscopic detail of gouges

Overplated copper alloy bearing gouged by cast iron debris

Aluminum bearing embedded with glass beads

Microscopic detail of glass beads

Damaged lining caused by dirt left on the bearing back

Misassembly

Result of a lower half assembled as an upper - blocking the oil flow

Excessive oil clearance is indicated by a short contact arc

Polished and oil-stained backs are a result of a poor fit in the housing bore

Result of a wrong, reversed, or shifted cap

Overloading

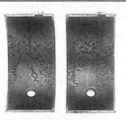

Damage from excessive idling which resulted in an oil film unable to support the load imposed

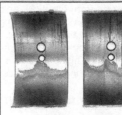

Damaged upper connecting rod bearings caused by engine lugging; the lower main bearings (not shown) were similarly affected

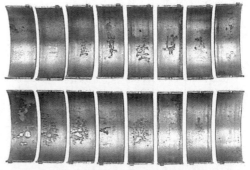

The damage shown in these upper and lower connecting rod bearings was caused by engine operation at a higher-than-rated speed under load

Misalignment

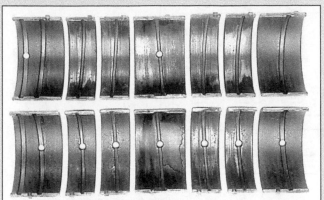

A warped crankshaft caused this pattern of severe wear in the center, diminishing toward the ends

A poorly finished crankshaft caused the equally spaced scoring shown

A bent connecting rod led to the damage in the "V" pattern

A tapered housing bore caused the damage along one edge of this pair

Lubrication

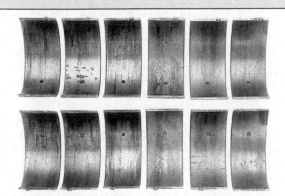

Result of dry start: The bearings on the left, farthest from the oil pump, show more damage

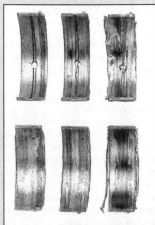

Result of a low oil supply or oil starvation

Severe wear as a result of inadequate oil clearance

Corrosion

Microscopic detail of corrosion

Corrosion is an acid attack on the bearing lining generally caused by inadequate maintenance, extremely hot or cold operation, or inferior oils or fuels

Microscopic detail of cavitation

Example of cavitation - a surface erosion caused by pressure changes in the oil film

Damage from excessive thrust or insufficient axial clearance

Bearing affected by oil dilution caused by excessive blow-by or a rich mixture

11.21 Use the scale on the Plastigage package to determine the bearing oil clearance - be sure to measure the widest part of the Plastigage and use the correct scale; it comes with both standard and metric scales

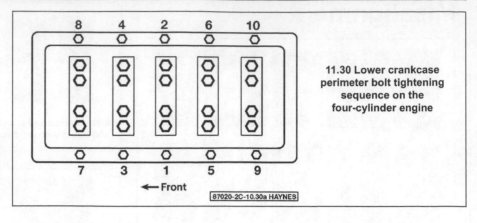

11.30 Lower crankcase perimeter bolt tightening sequence on the four-cylinder engine

lower crankcase straight up and off the block. Do not disturb the Plastigage or rotate the crankshaft.

21   Compare the width of the crushed Plastigage on each journal to the scale printed on the Plastigage envelope to determine the main bearing oil clearance (see illustration). Check with an automotive machine shop for the crankshaft bearing oil clearance for your engine.

22   If the clearance is not as specified, the bearing inserts may be the wrong size (which means different ones will be required). Before deciding if different inserts are needed, make sure that no dirt or oil was between the bearing inserts and the caps or block when the clearance was measured. If the Plastigage was wider at one end than the other, the crankshaft journal may be tapered. If the clearance still exceeds the limit specified, the bearing insert(s) will have to be replaced with an undersize bearing insert(s).

Caution: *When installing a new crankshaft always install a standard bearing insert set.*

23   Carefully scrape all traces of the Plastigage material off the main bearing journals and/or the bearing insert faces. Be sure to remove all residue from the oil holes. Use your fingernail or the edge of a plastic card - don't nick or scratch the bearing faces.

Final installation

24   Carefully lift the crankshaft out of the cylinder block.

25   Clean the bearing insert faces in the cylinder block, then apply a thin, uniform layer of moly-base grease or engine assembly lube to each of the bearing surfaces. Be sure to coat the thrust faces as well as the journal face of the thrust bearing.

26   Make sure the crankshaft journals are clean, then lay the crankshaft back in place in the cylinder block.

27   Clean the bearing insert faces and apply the same lubricant to them. Clean the engine block and the bearing cap mating surfaces (or lower crankcase) thoroughly. The surfaces

must be free of oil residue.

28   Prior to installation, apply clean engine oil to all bolt threads and under their heads. Wipe off excess. Use all NEW bolts if you're working on a V6 engine.

29   On 3.6L V6 engines, install and tighten the old cap bolts one last time, then tap the crankshaft bearing caps with a soft-face hammer. Remove the old bolts and discard them.

30   Install the main bearing cap bolts. Caution: Use NEW bolts on V6 engines. Tighten the bolts to the torque listed in this Chapter's Specifications in the correct sequences (see illustrations 11.19a, 11.19b and the accompanying illustration).

31   Recheck the crankshaft endplay with a feeler gauge or a dial indicator. The endplay should be correct if the crankshaft thrust faces aren't worn or damaged and if new bearings have been installed.

32   Rotate the crankshaft a number of times by hand to check for any obvious binding. It should rotate with a running torque of 50 in-lbs or less. If the running torque is too high, correct the problem at this time.

33   Install the new rear main oil seal (see Chapter 2A, Chapter 2B or Chapter 2C).

---

12   Engine overhaul - reassembly sequence

---

1   Before beginning engine reassembly, make sure you have all the necessary new parts, gaskets and seals as well as the following items on hand:

   Common hand tools
   A 1/2-inch drive torque wrench
   New engine oil
   Gasket sealant
   Thread locking compound

2   If you obtained a short block it will be necessary to install the cylinder head, the oil pump and pick-up tube, the oil pan, the water pump, the timing belt and timing cover, and the valve cover (see Chapter 2A, Chapter 2B or Chapter 2C). In order to save time and avoid problems, the external components must be installed in the following general order:

   Thermostat and housing cover
   Water pump
   Intake and exhaust manifolds

   Fuel injection components
   Emission control components
   Spark plugs
   Ignition coils or coil pack and spark
    plug wires
   Oil filter
   Engine mounts and mount brackets
   Driveplate

---

13   Initial start-up and break-in after overhaul

Warning: *Have a fire extinguisher handy when starting the engine for the first time.*

1   Once the engine has been installed in the vehicle, double-check the engine oil and coolant levels.

2   With the spark plugs out of the engine and the ignition system and fuel pump disabled (see Section 3, Step 5), crank the engine until oil pressure registers on the gauge or the light goes out.

3   Install the spark plugs, hook up the plug wires and restore the ignition system and fuel pump functions.

4   Start the engine. It may take a few moments for the fuel system to build up pressure, but the engine should start without a great deal of effort.

5   After the engine starts, it should be allowed to warm up to normal operating temperature. While the engine is warming up, make a thorough check for fuel, oil and coolant leaks.

6   Shut the engine off and recheck the engine oil and coolant levels.

7   Drive the vehicle to an area with minimum traffic, accelerate from 30 to 50 mph, then allow the vehicle to slow to 30 mph with the throttle closed. Repeat the procedure 10 or 12 times. This will load the piston rings and cause them to seat properly against the cylinder walls. Check again for oil and coolant leaks.

8   Drive the vehicle gently for the first 500 miles (no sustained high speeds) and keep a constant check on the oil level. It is not unusual for an engine to use oil during the break-in period.

9   At approximately 500 to 600 miles, change the oil and filter.

10   For the next few hundred miles, drive the

COMMON ENGINE OVERHAUL TERMS

B

Backlash - The amount of play between two parts. Usually refers to how much one gear can be moved back and forth without moving the gear with which it's meshed.

Bearing Caps - The caps held in place by nuts or bolts which, in turn, hold the bearing surface. This space is for lubricating oil to enter.

Bearing clearance - The amount of space left between shaft and bearing surface. This space is for lubricating oil to enter.

Bearing crush - The additional height which is purposely manufactured into each bearing half to ensure complete contact of the bearing back with the housing bore when the engine is assembled.

Bearing knock - The noise created by movement of a part in a loose or worn bearing.

Blueprinting - Dismantling an engine and reassembling it to EXACT specifications.

Bore - An engine cylinder, or any cylindrical hole; also used to describe the process of enlarging or accurately refinishing a hole with a cutting tool, as to bore an engine cylinder. The bore size is the diameter of the hole.

Boring - Renewing the cylinders by cutting them out to a specified size. A boring bar is used to make the cut.

Bottom end - A term which refers collectively to the engine block, crankshaft, main bearings and the big ends of the connecting rods.

Break-in - The period of operation between installation of new or rebuilt parts and time in which parts are worn to the correct fit. Driving at reduced and varying speed for a specified mileage to permit parts to wear to the correct fit.

Bushing - A one-piece sleeve placed in a bore to serve as a bearing surface for shaft, piston pin, etc. Usually replaceable.

C

Camshaft - The shaft in the engine, on which a series of lobes are located for operating the valve mechanisms. The camshaft is driven by gears or sprockets and a timing chain. Usually referred to simply as the cam.

Carbon - Hard, or soft, black deposits found in combustion chamber, on plugs, under rings, on and under valve heads.

Cast iron - An alloy of iron and more than two percent carbon, used for engine blocks and heads because it's relatively inexpensive and easy to mold into complex shapes.

Chamfer - To bevel across (or a bevel on) the sharp edge of an object.

Chase - To repair damaged threads with a tap or die.

Combustion chamber - The space between the piston and the cylinder head, with the piston at top dead center, in which air-fuel mixture is burned.

Compression ratio - The relationship between cylinder volume (clearance volume) when the piston is at top dead center and cylinder volume when the piston is at bottom dead center.

Connecting rod - The rod that connects the crank on the crankshaft with the piston. Sometimes called a con rod.

Connecting rod cap - The part of the connecting rod assembly that attaches the rod to the crankpin.

Core plug - Soft metal plug used to plug the casting holes for the coolant passages in the block.

Crankcase - The lower part of the engine in which the crankshaft rotates; includes the lower section of the cylinder block and the oil pan.

Crank kit - A reground or reconditioned crankshaft and new main and connecting rod bearings.

Crankpin - The part of a crankshaft to which a connecting rod is attached.

Crankshaft - The main rotating member, or shaft, running the length of the crankcase, with offset throws to which the connecting rods are attached; changes the reciprocating motion of the pistons into rotating motion.

Cylinder sleeve - A replaceable sleeve, or liner, pressed into the cylinder block to form the cylinder bore.

D

Deburring - Removing the burrs (rough edges or areas) from a bearing.

Deglazer - A tool, rotated by an electric motor, used to remove glaze from cylinder walls so a new set of rings will seat.

E

Endplay - The amount of lengthwise movement between two parts. As applied to a crankshaft, the distance that the crankshaft can move forward and back in the cylinder block.

F

Face - A machinist's term that refers to removing metal from the end of a shaft or the face of a larger part, such as a flywheel.

Fatigue - A breakdown of material through a large number of loading and unloading cycles. The first signs are cracks followed shortly by breaks.

Feeler gauge - A thin strip of hardened steel, ground to an exact thickness, used to check clearances between parts.

Free height - The unloaded length or height of a spring.

Freeplay - The looseness in a linkage, or an assembly of parts, between the initial application of force and actual movement. Usually perceived as slop or slight delay.

Freeze plug - See Core plug.

G

Gallery - A large passage in the block that forms a reservoir for engine oil pressure.

Glaze - The very smooth, glassy finish that develops on cylinder walls while an engine is in service.

H

Heli-Coil - A rethreading device used when threads are worn or damaged. The device is installed in a retapped hole to reduce the thread size to the original size.

I

Installed height - The spring's measured length or height, as installed on the cylinder head. Installed height is measured from the spring seat to the underside of the spring retainer.

J

Journal - The surface of a rotating shaft which turns in a bearing.

K

Keeper - The split lock that holds the valve spring retainer in position on the valve stem.

Key - A small piece of metal inserted into matching grooves machined into two parts fitted together - such as a gear pressed onto a shaft - which prevents slippage between the two parts.

Knock - The heavy metallic engine sound, produced in the combustion chamber as a result of abnormal combustion - usually detonation. Knock is usually caused by a loose or worn bearing. Also referred to as detonation, pinging and spark knock. Connecting rod or main bearing knocks are created by too much oil clearance or insufficient lubrication.

L

Lands - The portions of metal between the piston ring grooves.

Lapping the valves - Grinding a valve face and its seat together with lapping compound.

Lash - The amount of free motion in a gear train, between gears, or in a mechanical assembly, that occurs before movement can begin. Usually refers to the lash in a valve train.

Lifter - The part that rides against the cam to transfer motion to the rest of the valve train.

M

Machining - The process of using a machine to remove metal from a metal part.

Main bearings - The plain, or babbit, bearings that support the crankshaft.

Main bearing caps - The cast iron caps, bolted to the bottom of the block, that support the main bearings.

O

O.D. - Outside diameter.

Oil gallery - A pipe or drilled passageway in the engine used to carry engine oil from one area to another.

Oil ring - The lower ring, or rings, of a piston; designed to prevent excessive amounts of oil from working up the cylinder walls and into the combustion chamber. Also called an oil-control ring.

Oil seal - A seal which keeps oil from leaking out of a compartment. Usually refers to a dynamic seal around a rotating shaft or other moving part.

O-ring - A type of sealing ring made of a special rubberlike material; in use, the O-ring is compressed into a groove to provide the sealing action.

Overhaul - To completely disassemble a unit, clean and inspect all parts, reassemble it with the original or new parts and make all adjustments necessary for proper operation.

P

Pilot bearing - A small bearing installed in the center of the flywheel (or the rear end of the crankshaft) to support the front end of the input shaft of the transmission.

Pip mark - A little dot or indentation which indicates the top side of a compression ring.

Piston - The cylindrical part, attached to the connecting rod, that moves up and down in the cylinder as the crankshaft rotates. When the fuel charge is fired, the piston transfers the force of the explosion to the connecting rod, then to the crankshaft.

Piston pin (or wrist pin) - The cylindrical and usually hollow steel pin that passes through the piston. The piston pin fastens the piston to the upper end of the connecting rod.

Piston ring - The split ring fitted to the groove in a piston. The ring contacts the sides of the ring groove and also rubs against the cylinder wall, thus sealing space between piston and wall. There are two types of rings: Compression rings seal the compression pressure in the combustion chamber; oil rings scrape excessive oil off the cylinder wall.

Piston ring groove - The slots or grooves cut in piston heads to hold piston rings in position.

Piston skirt - The portion of the piston below the rings and the piston pin hole.

Plastigage - A thin strip of plastic thread, available in different sizes, used for measuring clearances. For example, a strip of plastigage is laid across a bearing journal and mashed as parts are assembled. Then parts are disassembled and the width of the strip is measured to determine clearance between journal and bearing. Commonly used to measure crankshaft main-bearing and connecting rod bearing clearances.

Press-fit - A tight fit between two parts that requires pressure to force the parts together. Also referred to as drive, or force, fit.

Prussian blue - A blue pigment; in solution, useful in determining the area of contact between two surfaces. Prussian blue is commonly used to determine the width and location of the contact area between the valve face and the valve seat.

R

Race (bearing) - The inner or outer ring that provides a contact surface for balls or rollers in bearing.

Ream - To size, enlarge or smooth a hole by using a round cutting tool with fluted edges.

Ring job - The process of reconditioning the cylinders and installing new rings.

Runout - Wobble. The amount a shaft rotates out-of-true.

S

Saddle - The upper main bearing seat.

Scored - Scratched or grooved, as a cylinder wall may be scored by abrasive particles moved up and down by the piston rings.

Scuffing - A type of wear in which there's a transfer of material between parts moving against each other; shows up as pits or grooves in the mating surfaces.

Seat - The surface upon which another part rests or seats. For example, the valve seat is the matched surface upon which the valve face rests. Also used to refer to wearing into a good fit; for example, piston rings seat after a few miles of driving.

Short block - An engine block complete with crankshaft and piston and, usually, camshaft assemblies.

Static balance - The balance of an object while it's stationary.

Step - The wear on the lower portion of a ring land caused by excessive side and back-clearance. The height of the step indicates the ring's extra side clearance and the length of the step projecting from the back wall of the groove represents the ring's back clearance.

Stroke - The distance the piston moves when traveling from top dead center to bottom dead center, or from bottom dead center to top dead center.

Stud - A metal rod with threads on both ends.

T

Tang - A lip on the end of a plain bearing used to align the bearing during assembly.

Tap - To cut threads in a hole. Also refers to the fluted tool used to cut threads.

Taper - A gradual reduction in the width of a shaft or hole; in an engine cylinder, taper usually takes the form of uneven wear, more pronounced at the top than at the bottom.

Throws - The offset portions of the crankshaft to which the connecting rods are affixed.

Thrust bearing - The main bearing that has thrust faces to prevent excessive endplay, or forward and backward movement of the crankshaft.

Thrust washer - A bronze or hardened steel washer placed between two moving parts. The washer prevents longitudinal movement and provides a bearing surface for thrust surfaces of parts.

Tolerance - The amount of variation permitted from an exact size of measurement. Actual amount from smallest acceptable dimension to largest acceptable dimension.

U

Umbrella - An oil deflector placed near the valve tip to throw oil from the valve stem area.

Undercut - A machined groove below the normal surface.

Undersize bearings - Smaller diameter bearings used with re-ground crankshaft journals.

V

Valve grinding - Refacing a valve in a valve-refacing machine.

Valve train - The valve-operating mechanism of an engine; includes all components from the camshaft to the valve.

Vibration damper - A cylindrical weight attached to the front of the crankshaft to minimize torsional vibration (the twist-untwist actions of the crankshaft caused by the cylinder firing impulses). Also called a harmonic balancer.

W

Water jacket - The spaces around the cylinders, between the inner and outer shells of the cylinder block or head, through which coolant circulates.

Web - A supporting structure across a cavity.

Woodruff key - A key with a radiused backside (viewed from the side).

Chapter 3
Cooling, heating and air conditioning systems

Contents

Specifications

Expansion tank cap pressure rating	Refer to top of cap
Thermostat rating (opening temperature range)....................................	195 degrees F
Cooling system capacity..	See Chapter 1
HVAC refrigerant type...	R-134a
Refrigerant capacity..	Refer to HVAC specification tag

Torque specifications

Ft-lbs (unless otherwise indicated)

Note: *One foot-pound (ft-lb) of torque is equivalent to 12 inch-pounds (in-lbs) of torque. Torque values below approximately 15 ft-lbs are expressed in inch-pounds, since most foot-pound torque wrenches are not accurate at these smaller values.*

Compressor mounting bolts/nuts..	37
Compressor refrigerant line fitting nut	15
Condenser-to-radiator bolts/nuts...	53 in-lbs
Condenser refrigerant line fitting bolt..	15
Radiator upper mounting bracket bolts...	89 in-lbs
Radiator lower mounting bracket bolts	44
Refrigerant line fitting to thermal expansion valve nut (at firewall)	15
Thermostat housing cover bolts	
V6 engines	18
Four-cylinder engines..	89 in-lbs
Thermostat housing mounting bolts	89 in-lbs
Water pump mounting bolts	
V6 engines	89 in-lbs
Four-cylinder engines..	18
Water pump sprocket bolts..	89 in-lbs
Water pump access cover bolts	89 in-lbs
Water pump drain plug	15

1   General information

Engine cooling system

1   The cooling system consists of a radiator, an expansion tank, a pressure cap (located on the expansion tank), a thermostat, electric cooling fans, various hoses and a belt or chain driven water pump.

2   The expansion tank is designed to store coolant and to separate any trapped air in the coolant. It has a pressure cap on top. The radiator on these models does not have a pressure cap.

Warning: *Unlike a conventional coolant recovery tank, the pressure cap on the expansion tank should never be opened after the engine has warmed up because of the danger of severe burns caused by steam or scalding coolant.*

3   When the engine is cold, the thermostat restricts the circulation of coolant from the engine. When the minimum operating temperature is reached, the thermostat begins to open, allowing coolant to return to the radiator.

Transaxle cooling systems

4   All models are equipped with a transaxle cooler, located inside the radiator, which cools the transaxle fluid. The transaxle is connected to the cooler by a pair of hoses: one delivers hot transaxle fluid to the radiator and the other brings the cooled fluid back to the transaxle.

5   For more information on transaxle oil coolers, refer to Chapter 7B.

Heating system

6   The heating system consists of the heater controls, the heater core, the heater blower assembly (which houses the blower motor and the blower motor resistor), and the hoses connecting the heater core to the engine cooling system. Hot engine coolant is circulated through the heater core. When the heater mode is activated, a flap door opens to expose the heater box to the passenger compartment. A fan switch on the heater controls activates the blower motor, which forces air through the core, heating the air.

Air conditioning system

7   The air conditioning system consists of the condenser with an integral receiver-drier, which is mounted in front of the radiator, the evaporator/heater case assembly under the dash, a compressor mounted on the engine, and the plumbing connecting all of the above components.

8   A blower fan forces the warmer air of the passenger compartment through the evaporator core (sort of a radiator-in-reverse), transferring the heat from the air to the refrigerant. The liquid refrigerant boils off into low pressure vapor, taking the heat with it when it leaves the evaporator.

2   Antifreeze - general information

Warning: *Do not allow antifreeze to come in contact with your skin or painted surfaces of the vehicle. Rinse off spills immediately with plenty of water. Antifreeze is highly toxic if ingested. Never leave antifreeze lying around in an open container or in puddles on the floor; children and pets are attracted by its sweet smell and may drink it. Check with local authorities about disposing of used antifreeze. Many communities have collection centers that will see that antifreeze is disposed of safely. Never dump used antifreeze on the ground or pour it into drains.*

Note: *Non-toxic antifreeze is now manufactured and available at local auto parts stores, but even this type must be disposed of properly.*

1   The cooling system should be filled with a water/ethylene glycol based antifreeze solution, which will prevent freezing down to at least 20-degrees F, or lower if local climate requires it. It also provides protection against corrosion and increases the coolant boiling point. The engines in these vehicles have aluminum heads. Depending on the engine and model year, the specified coolant may vary (see Chapter 1, Specifications). The manufacturer recommends that the correct type of coolant be used and strongly urges that coolant types not be mixed.

2   Drain, flush and refill the cooling system at least every other year (see Chapter 1). The use of antifreeze solutions for periods of longer than two years is likely to cause damage and encourage the formation of rust and scale in the system.

3   Before adding antifreeze to the system, inspect all hose connections. Antifreeze can leak through very minute openings.

4   The exact mixture of antifreeze to water, which you should use, depends on the relative weather conditions. The mixture should contain at least 50-percent antifreeze, but should never contain more than 70-percent antifreeze. Consult the mixture ratio chart on the container before adding coolant.

5   Hydrometers are available at most auto parts stores to test the coolant (see illustration). Use antifreeze that meets the manufacturers specifications for engines with aluminum heads. **Warning:** Do not remove the expansion tank cap, drain the coolant or replace the thermostat until the engine has cooled completely.

3   Thermostat - check and replacement

Check

1   Before assuming the thermostat is to blame for a cooling system problem, check the coolant level, drivebelt tension (see Chapter 1) and temperature gauge operation.

2   If the engine seems to be taking a long time to warm up, based on heater output or temperature gauge operation, the thermostat is probably stuck open. Replace the thermostat with a new one.

3   If the engine runs hot, use your hand to check the temperature of the lower radiator hose. If the hose isn't hot, but the engine is, the thermostat is probably stuck closed, preventing the coolant inside the engine from circulating through the radiator. Replace the thermostat.

Caution: *Don't drive the vehicle without a thermostat. The computer may stay in open loop and emissions and fuel economy will suffer.*

4   If the lower radiator hose is hot, it means that the coolant is flowing and the thermostat is open. Consult the *Troubleshooting* section at the front of this manual for cooling system diagnosis.

Replacement

Warning: *The engine must be completely cool before beginning this procedure.*

2005 and earlier four-cylinder models

Note: *The thermostat housing is located on the firewall-side of the engine block.*

5   Raise the front of the vehicle and support it securely on jackstands.

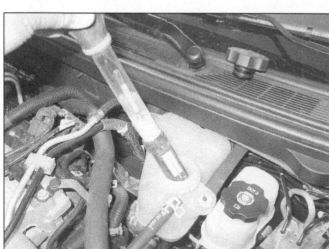

2.5 Use a hydrometer (available at auto parts stores) to test the condition of your coolant

3.7 Location of the water pump drain bolt (four-cylinder models)

3.8 Remove the bolts securing the water pump pipe to the thermostat housing

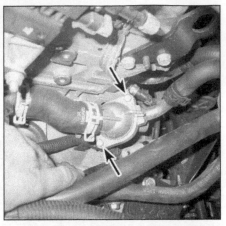

3.19 Thermostat housing cover mounting bolts (four-cylinder shown - V6 similar)

6   Drain the cooling system (see Chapter 1). If the coolant is relatively new and still in good condition, save it and reuse it.

7   Remove the water pump drain bolt and drain the remaining coolant into a container (see illustration).

8   Remove the bolts that hold the water pump pipe/thermostat housing cover to the thermostat housing (see illustration), then separate the cover from the housing and the water pipe from the water pump, using a twisting motion.

9   Remove the inner sleeve from the thermostat housing. Note the location of the notch on the lower section of the sleeve. Note how the thermostat is installed (which end is facing out), then remove the thermostat.

10   Install the thermostat cartridge in the housing, aligning the dimple on the cartridge with the slot in the housing.

11   Install a new O-ring seal onto the water pipe. Insert the water pipe into the water pump and swing the thermostat housing cover into place.

12   Install the bolts and tighten them to the torque listed in this Chapter's Specifications.

13   The remaining installation is the reverse of the removal.

14   Refill the cooling system (see Chapter 1).

15   Start the engine and allow it to reach normal operating temperature, then check for leaks and proper thermostat operation (as described in Steps 3 and 4).

2006 and later four-cylinder models

Note: *The thermostat housing on 2006 and later four-cylinder engines is located at the left rear corner of the cylinder head.*

16   Disconnect the cable from the negative terminal of the battery (see Chapter 5, Section 1).

17   Drain the cooling system (see Chapter 1). If the coolant is relatively new and still in good condition, save it and reuse it. Remove the water pump drain bolt and drain any excess coolant into a container (see illustration 3.7).

18   Loosen the hose clamps and detach

3.20 Note how the thermostat is installed, then remove it from the housing

the hoses from the fittings on the thermostat housing cover. If a hose sticks, grasp it near the end with a pair of adjustable pliers and twist it to break the seal, then pull it off. If the hose is old or if it has deteriorated, cut it off and install a new one. If the outer surface of the thermostat cover, which mates with the hose, is already corroded, pitted, or otherwise deteriorated, it might be damaged even more by hose removal. If it is, replace the thermostat cover.

19   Remove the fasteners and detach the thermostat cover (see illustration). If the cover is stuck, tap it with a soft-face hammer to jar it loose. Be prepared for some coolant to spill as the gasket seal is broken.

20   Note how it's installed, which end is facing up, or out and then remove the thermostat (see illustration).

21   Remove all traces of old gasket material and sealant from the housing and cover with a gasket scraper.

22   Install a new rubber gasket on the thermostat (see illustration) and install the thermostat in the housing, spring-end first.

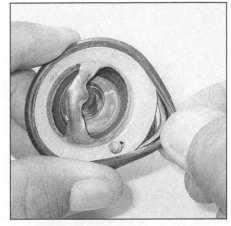

3.22 Place a new rubber seal onto the thermostat

23   Install the thermostat cover and bolts, then tighten the bolts to the torque listed in this Chapter's Specifications.

24   Reattach the radiator hose to the outlet pipe on the thermostat cover. Make sure that the hose clamp is tight. If it isn't, replace it.

25   The remaining installation is the reverse of the removal.

26   Refill the cooling system (see Chapter 1).

27   Reconnect the battery (see Chapter 5, Section 1).

28   Start the engine and allow it to reach normal operating temperature, then check for leaks and proper thermostat operation (as described in Steps 3 and 4).

3.5L and 3.9L V6 models

Note: *The thermostat housing on 2004 through 2006 3.5L V6 engines is located at the left end of the engine. On 2007 3.5L V6 engines and all 3.9L V6 engines, the thermostat housing is located at the drivebelt end of the engine, above the air conditioning compressor.*

29   Loosen the hose clamp, then detach the hose from the thermostat housing cover fit-

4.2 The location of the right cooling fan electrical connector (left fan is similar) - seen from below

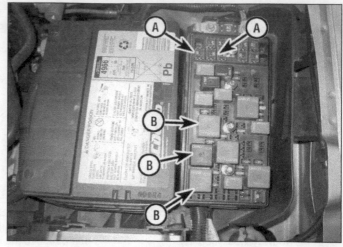

4.3 The cooling fan fuses (A) and relays (B) can be found in the engine compartment's fuse/relay box

ting. If it's stuck, grasp it near the end with a pair of adjustable pliers and twist it to break the seal, then pull it off. If the hose is old or if it has deteriorated, cut it off and install a new one.

30   If the outer surface of the fitting (where it mates with the hose) is corroded, pitted, or otherwise deteriorated, hose removal may damage it to the point where it must be replaced. Replace the thermostat housing cover if necessary.

31   Remove the thermostat housing cover fasteners and remove the cover.

Note: *If the cover is stuck, tap it with a soft-face hammer to jar it loose. Be prepared for some coolant to spill as the gasket seal is broken.*

32   Note how the thermostat is installed and which end is facing up or out, and then remove the thermostat and seal (see illustration 3.20).

33   On models so equipped, install a new rubber seal on the replacement thermostat (see illustration 3.22). Place the thermostat in the housing, spring end first.

34   If the thermostat housing was equipped with a conventional paper gasket, remove all traces of gasket material from the mating surfaces. Install the thermostat into the housing and apply a thin coat of RTV sealant to both sides of the new gasket and install it to the thermostat housing.

35   Install the housing cover and mounting bolts and tighten them to the torque listed in this Chapter's Specifications.

36   The remainder of the installation is the reverse of the removal.

37   Refill the cooling system (see Chapter 1).

38   Start the engine and allow it to reach normal operating temperature. Check for leaks and proper thermostat operation (as described in Steps 3 and 4).

3.6L V6 engine

Note: *The thermostat is located on the left side of the engine below the air cleaner ducting.*

39   Drain the cooling system. If the coolant is in good condition, save and reuse it.

40   Remove the radiator hose and the heater hose from the thermostat housing, then remove the surge tank hose.

41   Remove the thermostat housing bolts and housing, then remove the thermostat.

42   Clean all sealing surfaces. If a gasket is used, thinly coat it with RTV sealant prior to installation. If a rubber seal is used, install a new seal with the thermostat.

43   The remainder of installation is the reverse of removal. Refill the cooling system (see Chapter 1)

44   Start the engine and allow it to reach normal operating temperature. Check for leaks and proper thermostat operation as described in Steps 3 and 4.

---

4   Engine cooling fans - check and replacement

Warning: *To avoid possible injury or damage, DO NOT operate the engine with a damaged fan. Do not attempt to repair fan blades - replace a damaged fan with a new one.*

Check

1   These models are equipped with two cooling fans mounted side-by-side behind the radiator. The PCM (Powertrain Control Module - engine computer) and three relays are used to operate the fans at Low or High speeds depending on engine needs and other conditions. The fans are protected by fuses inside the engine compartment's fuse/relay box.

2   If the engine is overheating and neither of the cooling fans operate, check each fan

by unplugging the fan motor electrical connector and applying battery power directly to the motor terminals with fused jumper wires (see illustration). When done correctly, the fan should come on. If a fan motor doesn't work, replace the motor.

3   If the fan motors are okay but are still not coming on when the engine gets hot, the fan relays might be defective. Locate the relays in the engine compartment's fuse/relay box (see illustration). You can pull each relay out and test them individually (see Chapter 12).

Note: *Relays are used to control a circuit by turning it on and off in response to a signal from the PCM. The control circuits themselves are fairly complex, and checking them should be left to a dealer service department. It's also likely that a failure in one of these circuits will result in a warning light in the instrument display.*

4   Locate the fuses in the engine compartment fuse/relay box (see illustration 4.3). Check the fuses (see Chapter 12). If the fuses are okay, check all wiring and connections to the fan motors. If no obvious problems are found, have the cooling fan system and circuit diagonosed by a dealer service department or repair shop with the proper diagnostic equipment.

Replacement

Warning: *The engine must be completely cool before beginning this procedure.*

5   If you're working on a 2005 or 2006 Pontiac G6 model, have the air conditioning system discharged at a qualified service center.

6   Drain the cooling system (see Chapter 1). If the coolant is relatively new and still in good condition, it can be saved and reused.

Warning: *Be careful to clean up any spilled coolant as stated in the Warning at the beginning of Section 2.*

Note: *On four-cylinder models, there is an additional drain plug on the bottom of the water pump (see illustration 3.7).*

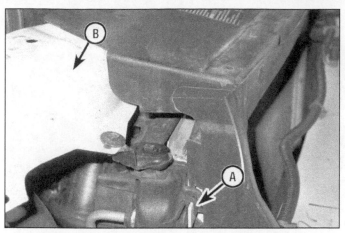

4.10 Because the radiator comes out the bottom, you'll need to suspend the condenser by tying rope around the tabs on each side (A) and the radiator support (B)

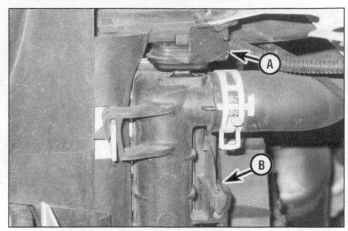

4.11 The upper radiator mounting brackets (A) and fan shroud mounting tabs (B)

2006 and earlier models

7   On Malibu models, remove both headlight housings (see Chapter 12).

8   On G6 models, remove the air filter assembly duct (see Chapter 4).

9   On four-cylinder engine models, unclip the transaxle cooler lines from the fan shroud. On all other engine models, disconnect the transaxle cooler lines from the radiator (see Chapter 7B).

10   Support the air conditioning condenser by tying it to the radiator support by its upper tabs (see illustration).

11   Remove the upper radiator mounting brackets (see illustration).

12   Pry the fan shroud upper mounting tabs upwards out of the retainers on the radiator (see illustration 4.10).

13   Remove the fasteners for the engine splash shield and the air deflectors on each side and then remove them (see illustration).

14   On V6 engine models, remove the right fender splash shield (see Chapter 11).

15   Remove the lower radiator hose from the radiator.

16   Disconnect the fan motor harness electrical connectors (see illustration 4.2).

17   Remove the lower radiator mounting brackets (see illustration).

Note: *Check the condition of the lower radiator mounting bushings and replace them if necessary.*

18   Remove the transaxle oil cooler lines from the radiator and transaxle (see Chapter 7B).

19   On G6 models, remove the air conditioning hoses to the condenser (see Section 15).

20   Make sure that no other wiring harnesses, hoses or anything else is attached to the fan shroud and that it is clear for removal.

21   Carefully tilt the bottom end of the radiator and fan shroud forward and remove the fan shroud from the bottom while supporting the radiator.

2007 and later models

22   Secure the condenser to the body structure with cord.

23   Remove the upper radiator brackets. Raise the vehicle and support it securely on jackstands.

24   Disconnect both radiator hoses from the radiator. Disconnect the cooling fan electrical connectors.

25   Remove both engine splash shields.

26   Disconnect the transmission cooler lines and seal the open ends to prevent contamination.

27   Remove the lower radiator brackets and mounts.

28   Remove the two condenser mounting bolts from the radiator. Discard them; they should be replaced with new ones on assembly.

29   Pull the radiator up while holding down on the condenser to disconnect them. Remove the condenser mounting nuts from the radiator; they will also have to be replaced.

30   Remove the radiator side seals, then remove the radiator.

31   Unclip the cooling fan shroud assembly from the radiator.

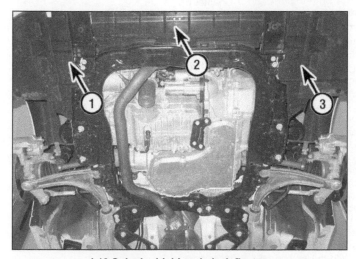

4.13 Splash shield and air deflectors

1   Right air deflector
2   Engine splash shield
3   Left air deflector

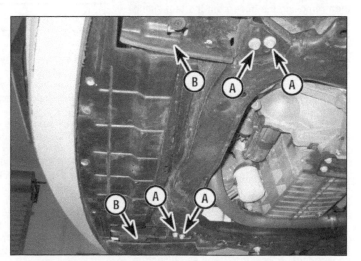

4.17 Remove the mounting fasteners (A) for the lower radiator mounting brackets (B)

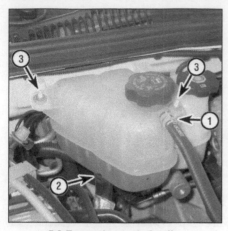

5.2 Expansion tank details:

1   *Return hose*
2   *Outlet hose*
3   *Mounting fasteners*

7.3 The weep hole on a water pump (3.5L engine model shown)

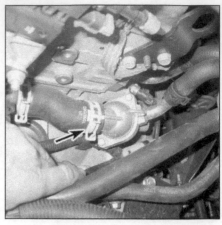

8.5 Remove the radiator hose from the thermostat housing cover

All models

32   Installation is the reverse of removal. Tighten all fasteners to the torque listed in this Chapter's Specifications. Refill the cooling system with the proper concentration of antifreeze and check the transaxle fluid level (see Chapter 1).

5   Coolant expansion tank - removal and installation

Warning: *Wait until the engine is completely cool before beginning this procedure.*
1   Drain the cooling system (see Chapter 1). If the coolant is relatively new and still in good condition, it can be saved and reused.
Note: *On four-cylinder models, there is an additional drain plug on the bottom of the water pump (see illustration 3.7).*
2   Disconnect the expansion tank hoses by sliding the hose clamp away with pliers and then removing the hose from the tank fitting (see illustration).
3   Remove the expansion tank mounting fasteners (see illustration 5.2).
4   Clean out the tank with soapy water and a brush to remove any deposits inside. Inspect the reservoir carefully for cracks. If you find a crack, replace the reservoir.
5   Installation is the reverse of removal. Refill the cooling system with the proper concentration of antifreeze (see Chapter 1).

6   Radiator - removal and installation

1   The procedure for removing and installing the radiator is nearly the same as the procedure for removing and installing the engine cooling fans. Follow the Steps as outlined in Section 4 along with the following additional instructions:

a) *It is not necessary to remove the fan shroud from the radiator for radiator removal. They can be removed together as an assembly and then separated after they are out of the vehicle.*
b) *Remove the upper radiator hose from the radiator after removing the lower hose.*
c) *The transaxle oil cooling lines may stay attached to the radiator and removed from it after the radiator is out of the vehicle.*
d) *To separate the air conditioning condenser from the radiator, remove the two mounting bolts for the condenser (one on each side) and then slide the radiator up slightly while moving the condenser down to release it from the additional retaining tabs and clips.*

Note: *Be careful not to damage the flexible upper air deflectors while removing the radiator.*

7   Water pump - check

1   A failure in the water pump can cause serious engine damage due to overheating.
2   If a failure occurs in the shaft seal, coolant will leak from the engine front cover or a hole that is exposed externally near the shaft seal.
3   Water pumps are equipped with weep or vent holes. It is possible to check the water pump weep hole using a flashlight. If a failure occurs in the shaft seal, coolant will leak from the hole. Use the flashlight to find the hole(s) on the water pump and check for leaks.
Note: *The hole(s) can be found on the top or bottom of the pump body usually near the shaft seal (see illustration).*
4   If the water pump shaft bearings fail, there may be a howling sound near the water pump while it's running. With the engine off, shaft wear can be felt if the water pump pulley is rocked up-and-down. Don't mistake drivebelt slippage, which causes a squealing

sound, for water pump bearing failure.
5   A quick water pump performance check is to turn the heater on. If the pump is failing, it might not be able to efficiently circulate hot water all the way to the heater core as it should.

8   Water pump - replacement

Warning: *The engine must be completely cool before beginning this procedure.*
1   Drain the cooling system (see Chapter 1). If the coolant is relatively new and still in good condition, it can be saved and reused.
Note: *On four-cylinder engine models, there is an additional drain plug on the bottom of the water pump (see illustration 3.7).*

Four-cylinder engines

2   Disconnect the electrical connector for the ECT sensor and then remove it (see Chapter 6).
3   Remove any harnesses attached to the thermostat housing.
4   Disconnect the outlet hose on the expansion tank (see Section 5).
5   Disconnect the radiator hose from the thermostat housing cover (see illustration).
6   Remove the exhaust manifold heat shield (see Chapter 2A).
7   Disconnect the heater hoses from the thermostat housing and then remove the housing mounting bolts (see illustration).
8   Remove the thermostat housing and the water pipe by twisting it away from the water pump to separate them. Then, separate the water pipe from the thermostat housing and discard the O-ring seals for the pipe and housing (see illustration).
9   Loosen the right front wheel lug nuts. Raise the vehicle and support it securely on jackstands.
10   Remove the right front wheel, followed by the fender splash shield (see Chapter 11).
11   Remove the cover over the water pump

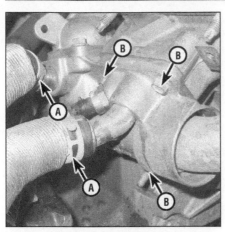

8.7 Remove the heater hoses (A) and the thermostat housing mounting bolts (B)

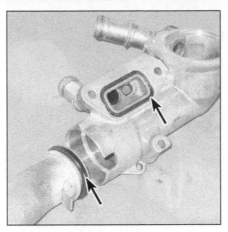

8.8 Remove the old seals and replace them with new ones

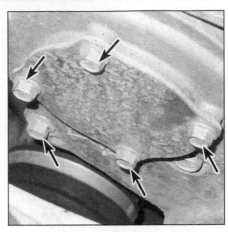

8.11 Remove the water pump access cover

8.12a Before installing the special tool onto the water pump sprocket, note the location of the bolt holes for the sprocket tool (A) and the water pump to sprocket bolts (B)

8.12b Remove one of the sprocket-to-water pump bolts, install the fabricated special tool and lock the sprocket into position. Then, remove the other two sprocket bolts

8.14a Remove the two water pump bolts on the rear of the engine block . . .

sprocket that's located left of the drivebelt tensioner (see illustration).

12   Install a special holding tool onto the water pump sprocket (tool J-43651), available from specialty tool manufacturers and some dealer service departments. A tool can be fabricated if necessary (see illustrations). Proper installation of either tool should result in absolutely no sprocket movement.

Note: *The special tool is designed to lock the sprocket into position, allowing the balance shaft chain to remain in its timed state while the water pump is being replaced. If you're using a homemade tool like the one shown in the illustration, remove one of the water pump-to-sprocket bolts before installing the tool.*

13   Remove all water pump-to-sprocket bolts.

14   Remove the two bolts attaching the water pump to the front and two bolts attaching it to the rear of the engine block and remove the

pump from the engine (see illustrations). If the water pump is stuck, gently tap it with a soft-faced hammer to break the seal.

15   Clean the bolt threads and the threaded holes and remove any corrosion or sealant. Remove all traces of old gasket material from the sealing surfaces. Remove the sealing ring from the water pump (if the same pump is to be installed).

16   Install a new sealing ring into the groove in the pump. To install the water pump, place a threaded stud loosely into the pump flange to serve as a guide pin; this will help to align the water pump sprocket with the water pump flange as the pump is installed.

17   Install the water pump mounting bolts and tighten them loosely. Install two bolts into the water pump sprocket and tighten them loosely. Remove the threaded stud (guide pin) and install the third water pump sprocket bolt.

18   Tighten the water pump mounting bolts

8.14b . . . and on the front of the engine block

8.24 Use a prybar or long screwdriver to hold the pulley in place while loosening the water pump pulley bolts

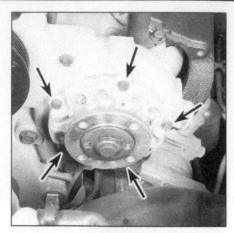

8.27 Water pump mounting bolts (3.5L and 3.9L V6 engines)

10.2 Remove the insulating panel by pulling it down against the tension of the retaining clips

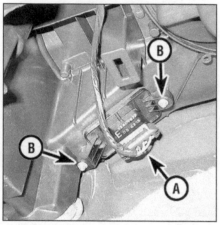

10.3 The blower motor resistor electrical connector (A) and mounting screws (B)

(two in the front and two in the rear of the engine block) to the torque listed in this Chapter's Specifications.

19   Tighten the water pump sprocket bolts to the torque listed in this Chapter's Specifications.

20   Install the water pump access cover and tighten the bolts to the torque listed in this Chapter's Specifications.

21   Install the thermostat housing and water pipe and tighten the bolts to the torque listed in this Chapter's Specifications. Lubricate the seal lightly with silicon gel before installing the water pipe into the water pump and the thermostat housing.

22   The remaining installation is the reverse of removal. Refill the cooling system with the proper concentration of antifreeze (see Chapter 1). Start the engine and allow it to reach normal operating temperature while inspecting the system for leaks.

V6 engines

23   Remove the air duct between the air cleaner housing and the throttle body if nec-

essary (see Chapter 4)

24   Loosen, but do not remove, the water pump pulley bolts (see illustration).

25   Remove the drivebelt (see Chapter 1).

26   Remove the pulley bolts and the pulley from the water pump.

27   Remove the water pump mounting bolts and then remove the pump (see illustration).

28   Clean the mounting bolt threads and any threaded holes in the engine to remove corrosion and sealant.

29   Compare the replacement pump with the old one to make sure that they're identical.

30   Remove all traces of old gasket material from the engine with the appropriate tools.

31   Thoroughly clean the mating surfaces of the engine and water pump with lacquer thinner or acetone.

32   Apply a thin film of RTV sealant to both sides of the new gasket.

33   Place the gasket and water pump into position. Install the mounting bolts until they are all finger tight.

34   With everything correctly in place, tighten the water pump mounting bolts to the torque listed in this Chapter's Specifications.

35   The remaining installation is the reverse of the removal. Tighten the water pump pulley bolts to the torque listed in this Chapter's Specifications. Refill the cooling system with the proper concentration of antifreeze (see Chapter 1). Start the engine and allow it to reach normal operating temperature while inspecting the system for leaks.

9   Coolant temperature sending unit - check

Warning: *Wait until the engine is completely cool before beginning this procedure.*

1   The coolant temperature indicator system consists of a temperature gauge and warning light on the dash and a sensor mounted on the engine. The Engine Coolant Temperature (ECT) sensor (see Chapter 6), which is an

information sensor for the Powertrain Control Module (PCM), provides a signal to the PCM which operates the temperature gauge.

2   If an overheating condition has occurred, first check the coolant level in the system (see Chapter 1) and that the coolant mixture is correct (see Section 2). Also, refer to the *Troubleshooting* section at the beginning of this book before assuming that the temperature indicator is faulty.

3   If the temperature gauge does not move from the C position, check the wiring harness connections going to the instrument cluster.

4   If there is a problem with the ECT sensor, it is very likely that the Check Engine light will come on and the sensor or circuit will need repair (see Chapter 6).

10   Blower motor resistor and blower motor - replacement

Warning: *The models covered by this manual are equipped with a Supplemental Restraint System (SRS), more commonly known as airbags. Always disarm the airbag system before working in the vicinity of any airbag system component to avoid the possibility of accidental deployment of the airbag, which could cause personal injury (see Chapter 12). Do not use a memory saving device to preserve the PCM's memory when working on or near airbag system components.*

1   Disconnect the cable from the negative battery terminal (see Chapter 5, Section 1).

2   Remove the insulating panel from below the dash and under the glove box in the passenger compartment, if equipped (see illustration).

Blower motor resistor

3   Disconnect the electrical connector at the blower motor resistor (see illustration).

4   Remove the mounting screws securing the resistor and pull it out of the evaporator/heater core housing.

5   Installation is the reverse of removal.

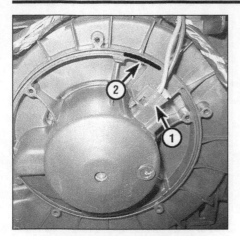

10.8 Blower motor mounting details:

1   Blower motor electrical connector.
2   Using a sharp utility knife, carefully cut the plastic in a complete circle (between the rings on the blower motor housing) to remove the blower motor.

11.3 The heater/air conditioner control assembly mounting screws - 2007 and earlier models

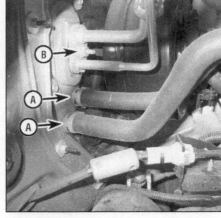

12.4 The heater hoses (A) are connected to the heater core pipes next to the brake booster and the refrigerant line fittings (B) are directly above

6   Reconnect the battery (see Chapter 5, Section 1).

Blower motor

7   To replace the blower motor, part of the plastic housing must be cut with a sharp utility knife and the motor must be reinstalled with a factory mounting kit.
8   Disconnect the blower motor electrical connector (see illustration).
9   Cut the plastic blower motor housing between the two concentric rings (see illustration 10.8).
10   Lower the blower motor.
11   Installation of replacement blower motor requires a factory installation kit that includes screws and a metal ring to secure the blower motor to the housing.
Note: *Before installing the screws, make sure that the motor is placed in the original position.*
12   Reconnect the battery (see Chapter 5, Section 1).

---

11   Heater/air conditioner control assembly - removal and installation

---

Warning: *The models covered by this manual are equipped with a Supplemental Restraint System (SRS), more commonly known as airbags. Always disarm the airbag system before working in the vicinity of any airbag system component to avoid the possibility of accidental deployment of the airbag, which could cause personal injury (see Chapter 12). Do not use a memory saving device to preserve the PCM's memory when working on or near airbag system components.*

1   Disconnect the cable from the negative battery terminal (see Chapter 5, Section 1).
2   Remove the center trim bezel (see Chapter 11).

3   Remove the heater/air conditioner control assembly retaining screws (see illustration) and pull the unit out.
4   Disconnect the electrical connectors on the back of the control assembly.
5   Installation is the reverse of removal.
6   Reconnect the battery (see Chapter 5, Section 1).
7   Start the vehicle and allow it to idle for one minute. Caution: The control assembly will go through a calibration when the power is restored to it. DO NOT make any adjustments to the control assembly during this time or it will not work properly.

---

12   Heater core - replacement

Warning: *The models covered by this manual are equipped with a Supplemental Restraint System (SRS), more commonly known as airbags. Always disarm the airbag system before working in the vicinity of any airbag system component to avoid the possibility of accidental deployment of the airbag, which could cause personal injury (see Chapter 12). Do not use a memory saving device to preserve the PCM's memory when working on or near airbag system components.*
Warning: *The air conditioning system is under high pressure. DO NOT loosen any fittings or remove any components until after the system has been discharged. Air conditioning refrigerant must be properly discharged into an EPA-approved container at a dealer service department or an automotive air conditioning repair facility. Always wear eye protection when disconnecting air conditioning system fittings.*
Note: *Heater core removal on these models is quite involved and requires the removal of the entire dash (instrument panel) and the evaporator/heater core housing.*

1   Have the air conditioning system discharged by a dealer service department or an

automotive air conditioning shop before proceeding (see Warning above).
2   Disconnect the cable from the negative battery terminal (see Chapter 5, Section 1).
3   Drain the cooling system (see Chapter 1). If the coolant is relatively new and still in good condition, it can be saved and reused.
Note: *On four-cylinder engine models, there is an additional drain plug on the bottom of the water pump(see illustration 3.7). Be careful to clean up any spilled coolant as stated in the Warning at the beginning of Section 2.*
4   Disconnect the heater hoses from the heater core fittings at the firewall (see illustration).
5   Disconnect the refrigerant lines to the evaporator core at the firewall (see illustration 12.4). Plug all open lines and fittings to prevent contamination of the air conditioning system.
6   Remove the instrument panel (see Chapter 11). Set the panel upside-down on a protective surface (with the top of the dash facing down.
7   Remove the all of the harness electrical connectors to the evaporator/heater core housing.
8   Remove the plastic air ducts that are in front of the housing that lead to the floor. They are held on by two screws at the top and a small push-type fastener at the bottom.
9   Remove the mounting bolts that hold the evaporator/heater core housing to the instrument panel.
10   Make sure that nothing else is attached to the housing, then separate it from the panel by pulling is straight up. Keep the housing upside down to get to the heater core cover.
11   The heater core cover on the evaporator/heater core housing is the section of plastic that is closest to the heater core pipes. There are ten mounting points that hold it onto the housing. These mounting points are heat-staked (plastic) and must be drilled out

13.1 Look for the evaporator drain hose on the firewall towards the middle-right side (passenger's side)

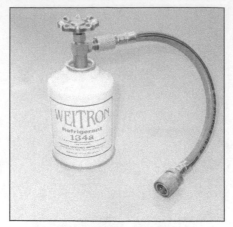

13.9 A basic R-134a charging kit is available at most auto parts stores - it must say R-134a (not R-12) on the kit and the can of refrigerant

13.12 Cans of R-134A refrigerant (available at auto parts stores) can be added to the low side of the air conditioningsystem with a simple recharging kit

to remove the cover.

Note: *The cover is designed for mounting screws to be used when the cover is placed back onto the housing. You will need to get these screws from an authorized dealership parts department.*

12   With the cover removed, pull the heater core out of the housing. Place rags under the heater core pipes to catch any excess coolant that may spill.

13   Installation is the reverse of removal. Don't forget to reconnect the heater hoses and refrigerant lines at the firewall using new sealing washers for the refrigerant line fittings.

14   Refill the cooling system when you're done (see Chapter 1).

15   Reconnect the battery (see Chapter 5, Section 1).

16   Have the air conditioning system evacuated, recharged and leak tested by the shop that discharged it.

13   Air conditioning and heating system - check and maintenance

Warning: *The air conditioning system is under high pressure. Do not loosen any hose fittings or remove any components until after the system has been discharged by an air conditioning technician. Always wear eye protection when disconnecting air conditioning system fittings.*

1   The following maintenance checks should be performed on a regular basis to ensure the air conditioner continues to operate at peak efficiency.

a) *Check the compressor drivebelt. If it's worn or deteriorated, replace it (see Chapter 1).*
b) *Check the drivebelt tension and, if necessary, adjust it (see Chapter 1).*
c) *Check the system hoses. Look for cracks, bubbles, hard spots and deterioration. Inspect the hoses and all fittings*

for oil bubbles and seepage. If there's any evidence of wear, damage or leaks, replace the hose(s).
d) *Inspect the condenser fins for leaves, bugs and other debris. Use a "fin comb" or compressed air to clean the condenser.*
e) *Make sure the system has the correct refrigerant charge.*
f) *Check the evaporator housing drain tube (see illustration) for blockage.*

2   It's a good idea to operate the system for about 10 minutes at least once a month, particularly during the winter. Long term non-use can cause hardening, and subsequent failure, of the seals.

3   Because of the complexity of the air conditioning system and the special equipment necessary to service it, in-depth troubleshooting and repairs are not included in this manual (refer to the *Haynes Automotive Heating and Air Conditioning Repair Manual*). However, simple checks and component replacement procedures are provided in this Chapter.

4   The most common cause of poor cooling is simply a low system refrigerant charge. If a noticeable drop in cool air output occurs, the following quick check will help you determine if the refrigerant level is low.

Checking the refrigerant charge

5   Warm the engine up to normal operating temperature.

6   Place the air conditioning temperature selector at the coldest setting and the blower at the highest setting. Open the vehicle doors (to make sure the air conditioning system doesn't cycle off as soon as it cools the passenger compartment).

7   If the compressor discharge line feels warm and the compressor inlet pipe feels cool, the system is probably adequately charged.

8   Place a thermometer in the dashboard vent nearest the evaporator and operate

the system until the indicated temperature is around 40 to 45-degrees F. If the ambient (outside) air temperature is very high, say 110-degrees F, the duct air temperature may be as high as 60-degrees F, but generally the air conditioning is 30 to 40-degrees F cooler than the ambient air.

Note: *Humidity of the ambient air also affects the cooling capacity of the system. Higher ambient humidity lowers the effectiveness of the air conditioning system.*

Adding refrigerant

9   Buy an automotive charging kit at an auto parts store. A charging kit includes a can of refrigerant, a tap valve and a short section of hose that can be attached between the tap valve and the system low side service valve (see illustration).

Caution: *There are two types of refrigerant used in automotive systems; R-12 - which has been widely used on earlier models - and the more environmentally-friendly R-134a used in all models covered by this manual. These two refrigerants (and their appropriate refrigerant oils) are not compatible and must never be mixed or components will be damaged. Use only R-134a refrigerant in the models covered by this manual.*

10   Hook up the charging kit by following the manufacturer's instructions.

Warning: *DO NOT attempt to hook the charging kit hose to the system high side! The fittings on the charging kit are designed to fit only on the low side of the system.*

11   Back off the valve handle on the charging kit and screw the kit onto the refrigerant can, making sure first that the O-ring or rubber seal inside the threaded portion of the kit is in place.

Warning: *Wear protective eyewear when dealing with pressurized refrigerant cans.*

12   Remove the dust cap from the low-side charging connection and attach the quick-connect fitting on the kit hose (see illustration).

13.15 Insert a thermometer in the center vent, turn on the air conditioning system and wait for it to cool down; depending on the humidity, the output air should be 30 to 40 degrees cooler than the ambient air temperature

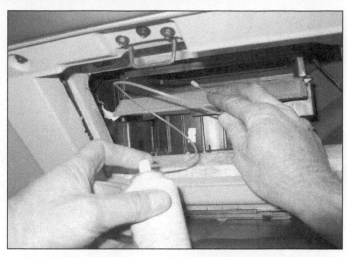

13.23 Pull the glove box door down, open the access door and insert the nozzle of the disinfectant can into the evaporator housing

13   Warm up the engine and turn on the air conditioner. Keep the charging kit hose away from the fan and other moving parts.

14   Turn the valve handle on the kit until the stem pierces the can, then back the handle out to release the refrigerant. You should be able to hear the rush of gas. Add refrigerant to the low side of the system until the compressor discharge line feels warm and the compressor inlet pipe feels cool. Allow stabilization time between each addition.

15   If you have an accurate thermometer, place it in the center air conditioning vent (see illustration) and then note the temperature of the air coming out of the vent. A fully-charged system which is working correctly should cool down to about 40-degrees F. Generally, an air conditioning system will put out air that is 30 to 40-degrees F cooler than the ambient air. For example, if the ambient (outside) air temperature is very high (over 100 degrees F), the temperature of air coming out of the registers should be 60 to 70 degrees F.

16   When the can is empty, turn the valve handle to the closed position and release the connection from the low-side port. Replace the dust cap.

Caution: *Never add more than one can of refrigerant to the system. If more refrigerant than that is required, the system should be evacuated and leak tested.*

17   Remove the charging kit from the can and store the kit for future use with the piercing valve in the UP position, to prevent inadvertently piercing the can on the next use.

Heating systems

18   If the carpet under the heater core is damp, or if antifreeze vapor or steam is coming through the vents, the heater core is leaking. Remove it (see Section 12) and install a new unit (most radiator shops will not repair a leaking heater core).

19   If the air coming out of the heater vents

isn't hot, the problem could stem from any of the following causes:

a) *The thermostat is stuck open, preventing the engine coolant from warming up enough to carry heat to the heater core. Replace the thermostat (see Section 3).*

b) *There is a blockage in the system, preventing the flow of coolant through the heater core. Feel both heater hoses at the firewall. They should be hot. If one of them is cold, there is an obstruction in one of the hoses or in the heater core, or the heater control valve is shut. Detach the hoses and back flush the heater core with a water hose. If the heater core is clear but circulation is impeded, remove the two hoses and flush them out with a water hose.*

c) *If flushing fails to remove the blockage from the heater core, the core must be replaced (see Section 12).*

Eliminating air conditioning odors

20   Unpleasant odors that often develop in air conditioning systems are caused by the growth of a fungus, usually on the surface of the evaporator core. The warm, humid environment there is a perfect breeding ground for mildew to develop.

21   The evaporator core on most vehicles is difficult to access, and factory dealerships have a lengthy, expensive process for eliminating the fungus by opening up the evaporator case and using a powerful disinfectant and rinse on the core until the fungus is gone. You can service your own system at home, but it takes something much stronger than basic household germ-killers or deodorizers.

22   Aerosol disinfectants for automotive air conditioning systems are available in most auto parts stores, but remember when shopping for them that the most effective treatments are also the most expensive. The basic

procedure for using these sprays is to start by running the system in the RECIRC mode for ten minutes with the blower on its highest speed. Use the highest heat mode to dry out the system and keep the compressor from engaging by disconnecting the wiring connector at the compressor (see Section 14).

23   Make sure that the disinfectant can comes with a long spray hose. Point the nozzle through the interior ventilation filter chamber allowing the nozzle to protrude inside the evaporator housing (see illustration), and then spray according to the manufacturer's recommendations. Follow the manufacturer's recommendations for the length of spray and waiting time between applications.

Note: *Although the manufacturer states that these vehicles are not equipped with an interior ventilation filter, we've have found that the evaporator/heater core housing is equipped with an access door like vehicles that have a filter. Use this door for applying the disinfectant, if equipped. Alternatively, if your model is not equipped with an access door, remove the blower motor resistor (see Section 10) and insert the disinfectant nozzle there instead.*

24   Once the evaporator has been cleaned, the best way to prevent the mildew from coming back again is to make sure your evaporator housing drain tube is clear (see illustration 13.1).

14   Air conditioning compressor - removal and installation

Warning: *The air conditioning system is under high pressure. DO NOT loosen any fittings or remove any components until after the system has been discharged. Air conditioning refrigerant must be properly discharged into an EPA-approved container at a dealer service department or an automotive air conditioning repair facility. Always wear eye protection when dis-*

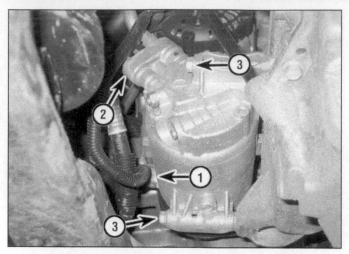

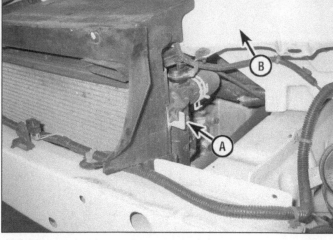

**14.5 Air conditioning compressor mounting details
(four-cylinder model shown - V6 similar):**

1   *Clutch field coil electrical connector*
2   *Refrigerant line fitting*
3   *Mounting fasteners (one hidden)*

**15.3 Tie a rope through the radiator tabs (A) and secure it to the
radiator support (B)**

connecting air conditioning system fittings.
Caution: *If you are replacing the compressor
due to major internal damage, you must also
replace the refrigerant filter (see Section 17).
Additionally, there is a small screen filter designed to be installed in the suction line fitting
for the compressor under the same condition.
It is available in three sizes and requires a
special tool for installation. We recommend
that this part and the special tool for installation be obtained by an authorized dealership
parts department.*

Removal

1   Have the air conditioning system discharged by a dealer service department or
by an automotive air conditioning shop before
proceeding (see Warning above).
2   Remove the drivebelt (see Chapter 1).
3   Raise the vehicle and secure it on
jackstands. Remove the right front wheel.
4   Remove the right front fender splash

15.7 Location of the refrigerant line fittings

shield (see Chapter 11). If you're working on
a 2005 or 2006 G6 model, with a 3.5L engine,
remove the right radiator air deflector.
5   Disconnect the electrical connector from
the compressor clutch field coil (see illustration).
6   Disconnect the compressor refrigerant
line fitting from the compressor. Discard the
sealing washers (see illustration 14.5).
7   Remove the compressor mounting fasteners and remove the compressor (see illustration 14.5).

Installation

8   If a new compressor is being installed,
follow the directions with the compressor
regarding the draining of excess oil prior to
installation.
9   The clutch may have to be transferred
from the original to the new compressor.
10   Before connecting the refrigerant line
fitting to the compressor, replace all sealing
washers.
11   The remainder of installation is the
reverse of removal. Tighten the line fitting fastener and compressor mounting fasteners to
the torque listed in this Chapter's Specifications.
12   Have the system evacuated, recharged
and leak tested by the shop that discharged it.

---

15 Air conditioning condenser -
removal and installation

Warning: *The air conditioning system is under
high pressure. DO NOT loosen any fittings or
remove any components until after the system
has been discharged. Air conditioning refrigerant must be properly discharged into an EPA-
approved container at a dealer service department or an automotive air conditioning repair*

facility. Always wear eye protection when disconnecting air conditioning system fittings.
Note: *The air conditioning condenser has
an integral desiccant bag mounted inside of
a housing on the left side of the condenser.
These components cannot be separated and
are serviced as an assembly.*
1   Have the air conditioning system discharged by a dealer service department or
by an automotive air conditioning shop before
proceeding (see Warning above).
2   Remove both headlight housings (see
Chapter 12).
3   Remove the upper radiator mounting
brackets (see illustration 4.10). Secure the
radiator to the radiator support by tying a rope
through the tabs on the radiator (see illustration).
4   Raise the front of the vehicle and support it securely on jackstands.
5   Remove the fasteners for the engine
splash shield and the air deflectors on each side
and then remove them (see illustration 4.12).
6   Remove the right fender splash shield
(see Chapter 11).
7   Disconnect the refrigerant lines from the
condenser (see illustration). Plug all open
lines and fittings to prevent contamination of
the air conditioning system.
8   Remove the lower radiator mounting
brackets (see illustration 4.16).
9   Remove the two mounting bolts (one on
each side) that fasten the condenser to the
radiator. On 2008 and later models, the nuts
and bolts will have to be replaced with new
ones. Also on 2008 and later models, remove
both lower radiator brackets.
10   Slide the radiator up slightly while moving the condenser down to release it from the
additional retaining tabs and clips.
Note: *Be careful not to damage the flexible
upper air deflectors while removing the condenser.*

16.1 The location of the air conditioning pressure sensor - (G6 model shown, other models similar)

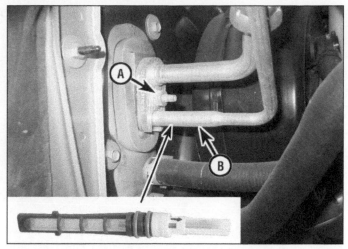

17.2 Remove the refrigerant line fitting nut (A). The filter is inside the lower refrigerant line (B)

11   Lower the condenser from the engine compartment. If you're going to reinstall the same condenser, plug the openings and store it upright.

12   If you're going to install a new condenser, pour 1 ounce of refrigerant oil of the correct type into it prior to installation.

Note: *If the old condenser had more than this amount of refrigerant oil in it, put that amount of new oil in the new condenser.*

13   Before reconnecting the refrigerant lines to the condenser, be sure to use new sealing washers. Tighten the condenser line fitting fastener to the torque listed in this Chapter's Specifications.

14   Installation is otherwise the reverse of removal.

15   Have the system evacuated, recharged and leak tested by the shop that discharged it.

16   Air conditioning pressure sensor - replacement

Note: *The air conditioning pressure sensor is threaded onto a Schrader valve. Therefore, it is not necessary to discharge the air conditioning system to replace it.*

Note: *The air conditioning pressure sensor is used to monitor excessively high and low system pressures and will prevent the system from operating if either condition occurs.*

1   Unplug the electrical connector from the sensor (see illustration).

2   Unscrew the sensor from the Schrader valve. Use a back-up wrench on the fitting for the sensor to prevent damaging the refrigerant line.

3   Lubricate the switch O-ring with clean refrigerant oil of the correct type.

4   Screw the new sensor onto the threads until hand tight, and then tighten it securely.

5   Reconnect the electrical connector.

17   Air conditioning refrigerant filter - replacement

Warning: *The air conditioning system is under high pressure. DO NOT loosen any fittings or remove any components until after the system has been discharged. Air conditioning refrigerant must be properly discharged into an EPA-approved container at a dealer service department or an automotive air conditioning repair facility. Always wear eye protection when disconnecting air conditioning system fittings.*

Note: *The refrigerant filter, which is small and tube shaped, is located in the refrigerant line connected to the thermal expansion valve on the firewall. When major air conditioning components are replaced, this filter should be inspected or replaced also.*

1   Have the air conditioning system discharged by a dealer service department or by an automotive air conditioning shop before proceeding (see Warning above).

2   Remove the nut securing the refrigerant lines to the thermal expansion valve on the firewall (see illustration). Plug all open lines and fittings to prevent contamination of the air conditioning system.

3   Discard the sealing washers.

4   The refrigerant filter is located inside the refrigerant line from the condenser (see illustration 17.2). Care must taken when removing it, as it can easily be broken off inside the line. You can use a pair of needle-nose pliers to remove it. If it's really stuck, special extractor tools are available at most auto parts stores.

5   Install a new filter into the line.

6   Install the lines to the expansion valve with new sealing washers.

7   Have the system evacuated, recharged and leak tested by the shop that discharged it.

Notes

Chapter 4
Fuel and exhaust systems

Contents

Specifications

Fuel pressure (ignition key turned to ON, engine not running)
Four cylinder engines.. 50 to 60 psi
V6 engines
2007 and earlier models, except 2007 3.5L V6 .... 50 to 60 psi
2007 3.5L V6 and 2008 and later models.................................... 56 to 62 psi
Fuel pressure leakdown (5 minutes after turning key to OFF) .... No more than 5 psi drop
Fuel injector coil resistance (all engines, all years) .... 11 to 14 ohms

Torque specifications

Ft-lbs (unless otherwise indicated)

Note: *One foot-pound (ft-lb) of torque is equivalent to 12 inch-pounds (in-lbs) of torque. Torque values below approximately 15 ft-lbs are expressed in inch-pounds, because most foot-pound torque wrenches are not accurate at these smaller values.*

Throttle body mounting bolts/nuts (all engines, all years) .... 89 in-lbs

1   General information

1   This Chapter covers the removal and installation procedures for the important parts of the air intake, fuel and exhaust systems. Because emission-control systems are integral parts of the engine management system, there are many cross-references to Chapter 6. Information on the engine management system, information sensors and output actuators is in Chapter 6.

2   The air intake system consists of the air filter housing, the air intake duct, the throttle body and the intake manifold. Incoming air passes through the air filter element, the Mass Air Flow (MAF) sensor (V6 models only), the air intake duct, the throttle body, the intake manifold plenum and the intake manifold runners before being mixed with fuel sprayed into the intake ports by the fuel injectors.

3   The Sequential Fuel Injection (SFI) system consists of the fuel tank, an electric fuel pump/fuel level sending unit module mounted inside the tank, the fuel rail, the fuel injectors, the fuel pressure regulator and the metal and flexible fuel lines that connect the various components of the SFI system.

4   The exhaust system consists of the exhaust manifold(s), the catalytic converter(s), the resonator, the muffler and the exhaust pipes connecting these components. The system is suspended from the vehicle pan by rubber hangers. You'll find the removal and installation procedures for the exhaust manifold(s) in Chapter 2, and for the rest of the exhaust system in this Chapter. There is more information about - and the replacement procedures for - the catalytic converter(s) in Chapter 6.

2   Fuel pressure relief procedure

Warning: *Gasoline is extremely flammable, so take extra precautions when you work on any part of the fuel system. Don't smoke or allow open flames or bare light bulbs near the work area, and don't work in a garage where a gas-type appliance (such as a water heater or a clothes dryer) is present. Since gasoline is carcinogenic, wear fuel-resistant gloves when there's a possibility of being exposed to fuel, and, if you spill any fuel on your skin, rinse it off immediately with soap and water. Mop up any spills immediately and do not store fuel-soaked rags where they could ignite. The fuel system is under constant pressure, so, if any fuel lines are to be disconnected, the fuel pressure in the system must be relieved first. When you perform any kind of work on the fuel system, wear safety glasses and have a Class B type fire extinguisher on hand.*

Caution: *After the fuel pressure has been relieved, it's a good idea to lay a shop towel over any fuel connection to be disassembled, to absorb the residual fuel that may leak out when servicing the fuel system.*

1   The fuel system referred to in this Chapter is defined as the fuel tank and tank-mounted fuel pump/fuel gauge sender unit, the fuel filter, the fuel injectors and the metal pipes and flexible hoses of the fuel lines between these components. All these components contain fuel, which is pressurized as soon as the ignition key is turned to ON, and remains pressurized while the engine is running (and even after the ignition is switched off). And because the pressure remains for some time after the ignition has been switched off, it must be relieved before any fuel lines are disconnected.

2   Remove the fuel filler cap to relieve any pressure built-up in the fuel tank.

3   Open the trunk and locate the fuse and relay panel inside the left side of the trunk and pull the fuel pump fuse (see illustration). Start the engine and allow it to run until it stalls (it may not even start). Once it has stalled (or has failed to start), crank the starter for three more seconds, then turn off the ignition key.

4   The fuel pressure is now relieved. You may now open up the fuel system to service any component. Warning: This procedure merely relieves the pressure that the engine needs to run. But remember that fuel is still present in the system components, and take precautions accordingly before disconnecting any of them.

5   After relieving the fuel pressure, be sure to disconnect the cable from the negative terminal of the battery before working on any fuel system component (see Chapter 5).

3   Fuel pump/fuel pressure - check

Warning: *Gasoline is extremely flammable, so take extra precautions when you work on any part of the fuel system. See the Warning in Section 2.*

Fuel pump operation check

1   The fuel pump is located inside the fuel tank, which muffles its sound when the engine is running. But you can actually hear the fuel pump. Sit inside the vehicle with the

2.3 The fuel pump fuse is located on the fuse and relay panel in the left side of the trunk

windows closed, turn the ignition key to ON (not START) and listen carefully for the soft whirring sound made by the fuel pump as it's briefly turned on by the PCM to pressurize the fuel system prior to starting the engine. You will only hear a soft whirring sound for a second or two, but that sound tells you that the pump is working. If you can't hear the pump, remove the fuel filler cap, then have an assistant turn the ignition switch to ON while you listen for the sound of the pump operating for a couple of seconds.

2   If the pump does not come on when the ignition key is turned to ON, check the fuel pump fuse and relay (both of which are located in the rear compartment fuse and relay box - see illustration 2.3). If the fuse and relay are okay, check the wiring back to the fuel pump. If the fuse, relay and wiring are okay, the fuel pump is probably defective. If the pump runs continuously with the ignition key in its ON position, the Powertrain Control Module (PCM) is probably defective. Have the PCM checked by a dealer service department or other qualified repair shop.

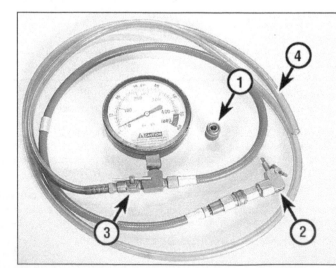

3.3 A typical fuel pressure gauge set up

1   *Screw-on adapter for the Schrader valve on the fuel rail*
2   *Hose with fitting to connect to the adapter*
3   *Bleeder valve (optional)*
4   *Bleeder hose (optional)*

3.5a The fuel pressure test port on four-cylinder models is located at the left end of the fuel rail

3.5b The fuel pressure test port on V6 models is located on the right end of the fuel rail

Fuel pressure check

3   To measure the fuel pressure you'll need a fuel pressure gauge compatible with high-pressure fuel injection systems, and a hose and fitting suitable for connecting the gauge to the Schrader valve-type test port on the fuel feed line (see illustration).

4   Relieve the fuel pressure (see Section 2).

5   Locate the fuel pressure test port (see illustrations), unscrew the cap and connect a fuel pressure gauge.

6   Start the engine and allow it to idle. Note the gauge reading as soon as the pressure stabilizes, and compare it with the pressure listed in this Chapter's Specifications.

a)   *If the pressure is lower than specified, suspect a restricted fuel filter (the fuel filter is an integral part of the fuel pump/ fuel level sensor module; to replace it you must replace the fuel pump). Also check the supply side fuel lines and hoses for kinks, blockages and leaks.*

b)   *If there are no restrictions and you changed the fuel filter but the pressure is still lower than specified, the regulator (which is an integral component of the fuel pump module) might be stuck in the open position, which will prevent the fuel system from reaching its normal operating pressure range. The regulator is not serviceable separately; if it's defective you must replace the fuel pump module.*

c)   *Remove the fuel pump (see Section 7) and inspect the fuel inlet strainer for restrictions. If the fuel strainer is okay, the regulator or the fuel pump could be clogged or defective. Replace the fuel pump module (which includes the regulator).*

d)   *It's possible that one or more of the fuel injectors might be leaking or stuck open (but if this is happening the engine will be running very poorly and will most likely have set a trouble code). Remove the fuel rail and inspect the injectors (see Section 13).*

e)   *If the fuel pressure is higher than specified, the fuel pressure regulator might be stuck in the closed position. The regulator is an integral component of the fuel pump module, and is not serviceable separately; if it's defective you must replace the fuel pump module.*

7   Turn off the engine. Verify that the fuel pressure loses no more than 8 psi for five minutes after the engine is turned off.

8   Relieve the fuel pressure (see Section 2), then disconnect the fuel pressure gauge and screw on the test port cap. Clean up any spilled gasoline.

9   Start the engine and verify that there are no fuel leaks.

4   Fuel lines and fittings - general information

Warning: *Gasoline is extremely flammable, so take extra precautions when you work on any part of the fuel system. See the Warning in Section 2.*

Warning: *Before disconnecting any fuel line fittings, relieve the fuel system pressure (see Section 2) and equalize tank pressure by removing the fuel filler cap. This procedure will merely relieve the increased pressure necessary for the engine to run - remember that fuel will still be present in the system components, so you should be ready to mop up fuel spills when disconnecting fuel line fittings.*

1   Always relieve the fuel pressure (see Section 2) before servicing fuel lines or fittings, then disconnect the cable from the negative battery terminal (see Chapter 5, Section 1) before proceeding.

2   Whenever you're working under the vehicle, be sure to inspect all fuel and evaporative emission lines for leaks, kinks, dents and other damage. Always replace a damaged fuel or EVAP line immediately. Leaking fuel and EVAP lines will result in loss of fuel and excessive air pollution (the leaking raw fuel emits unburned hydrocarbon vapors into

the atmosphere).

3   If you find signs of dirt in the lines during disassembly, disconnect all lines and blow them out with compressed air. Inspect the fuel strainer on the fuel pump pick-up unit (see Section 7) for damage and deterioration. (There is also a fuel filter inside the fuel pump/ fuel level sending unit but you cannot replace it without replacing the pump module, so consider this option only if, after cleaning the strainer and all the fuel lines, the fuel system is still clogged.)

4   The fuel supply line connects the fuel pump in the fuel tank to the fuel rail on the engine. The Evaporative Emission (EVAP) system vapor lines connect the fuel tank to the EVAP canister and the canister to the intake manifold. The fuel and EVAP lines are secured to the underbody with plastic clips (see illustration). These clips are attached to the vehicle floorpan by pushing them onto a threaded stud that's welded to the pan. To remove one of these clips, pry it open, disengage the fuel line and EVAP lines, then unscrew the clip or simply pull it straight down.

4.4 Plastic clips like this secure the fuel and EVAP lines to the underside of the vehicle

4.12 Pull the end of the retainer off the fuel line, then disengage the other end from the female side of the fitting

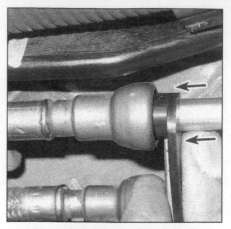

4.13a Insert the fuel line separator tool into the female side of the fitting, push it into the fitting until it releases the locking tabs inside the fitting . . .

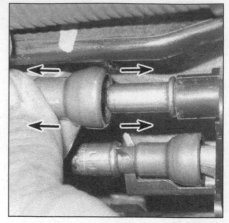

4.13b . . . then pull the two halves of the fitting apart

Steel tubing

5   Because fuel lines used on fuel-injected vehicles are under fairly high pressure, it is critical that they be replaced with lines of equivalent specification. Never use copper or aluminum tubing to replace steel tubing. These materials cannot withstand normal vehicle vibration.

6   Some steel fuel lines have threaded fittings. When loosening these fittings to service or replace components:

a) *Hold the stationary fitting with one wrench while loosening or tightening the tubing nut with another.*

b) *If you're going to replace one of these fittings, use original equipment parts or parts that meet original equipment standards.*

Plastic tubing

7   Most of the fuel (and EVAP) lines on the vehicles covered in this manual are plastic. If you ever have to replace a plastic line, use only plastic tubing meeting original equipment standards.

Caution: *When removing or installing plastic fuel line tubing, be careful not to bend or twist it too much, which can damage it. And damaged fuel lines MUST be replaced! Also, be aware that the plastic fuel tubing is NOT heat resistant, so keep it away from excessive heat. Nor is it acid-proof, so don't wipe it off with a shop rag that has been used to wipe off battery electrolyte. If you accidentally spill or wipe electrolyte on plastic fuel tubing, replace the tubing.*

Flexible hoses

Warning: *Use only original equipment replacement hoses or their equivalent. Unapproved hoses might fail when subjected to the high operating pressures of the fuel system.*

8   Don't route fuel hoses within four inches of exhaust system components or within ten inches of a catalytic converter. Make sure that

no rubber hoses are installed directly against the vehicle, particularly in places where there is any vibration. If allowed to touch some vibrating part of the vehicle, a hose can easily become chafed and it might start leaking. A good rule of thumb is to maintain a minimum of 1/4-inch clearance around a hose (or metal line) to prevent contact with the vehicle underbody.

Fuel line and EVAP line fittings

9   The vehicles covered in this manual use two kinds of fuel line quick-connect fittings (metal or plastic) for most connections at the fuel pump, the fuel tank, under the vehicle and in the engine compartment. (A third type of plastic quick-connect fitting is used only at the EVAP canister and on the vent hose connection at the fuel tank for the EVAP canister vent solenoid.)

10   The procedure for releasing each type of fuel line fitting is different. But a few rules of thumb apply to all fittings:

a) *Inspect the fitting for dirt. If the fitting is dirty, clean it off before disassembling it. The seals in the fitting will stick to the fuel line as they age. Twist the fitting on the line, then push and pull the fitting until it moves freely.*

b) *Always disconnect all fuel line fittings from a fuel system component before removing the component.*

c) *When disconnecting a quick-connect fitting, inspect the condition of the retainer before reconnecting the fitting. The best strategy with respect to retainers is to simply replace the retainer every time that you disconnect the fitting.*

d) *When you disconnect a fitting with an O-ring inside, inspect the O-ring before reconnecting the fitting. Fuel line fittings are under the same pressure as the rest of the fuel system, so to avoid leaks (and fires!) make VERY SURE that the O-ring is in good condition. Even better, simply replace it.*

e) *In most cases, the fitting itself is a non-removable part of the fuel line, so you might have to replace an entire fuel line if a fitting is damaged or defective.*

Metal collar quick-connect fittings

Disconnection

Note: *You'll find these fittings at the connections between the fuel supply and return lines in the engine compartment.*

Note: *You'll need a special tool set (available at most auto parts stores) to disconnect these fittings.*

Note: *The photos accompanying the disconnection procedure depicted here shows the metal collar quick-connect fittings at a four-cylinder fuel rail, but the metal quick-connect fittings on a V6 fuel rail are identical.*

11   Relieve the fuel system pressure (see Section 2).

12   Pull off the clip end of the retainer, then remove it from the fitting (see illustration).

13   Using a fuel line separator tool of the proper size (available at most auto parts stores), insert the tool into the female side of the fitting, then push it into the fitting to release the locking tabs and pull the fitting apart (see illustrations).

Reconnection

14   Inspect the O-ring (see illustration). If it's dried out, cracked, torn or otherwise deteriorated, replace it.

15   Apply a few drops of clean engine oil to the male pipe end.

16   Push both sides of the fitting together until the retaining tabs snap into place. Pull on both sides of the fitting to verify that it's securely connected.

17   Install the retainer, making sure it clips into place (see illustration).

18   Start the engine and check for fuel leaks.

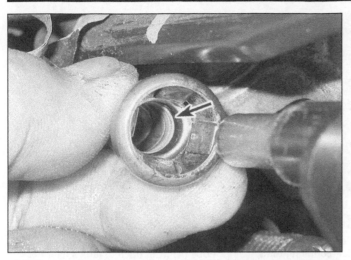

4.14 Inspect the old O-ring inside the female side of the fitting; if it's cracked, torn or deteriorated, replace it

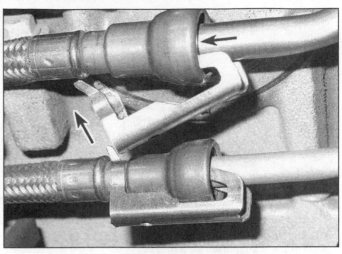

4.17 To install a retainer, insert the hooked end into the female side of the fitting, then push the clip end onto the fuel line until it snaps into place

Plastic collar quick-connect fittings

Bartholomew-type

Disconnection

19 To release a Bartholomew-type quick-connect fitting, depress the tabs of the retainer (see illustration). Once the retainer is released, continue pressing on the tabs while pulling the two fuel lines apart (see illustration).

20 Remove and discard the old retainer from the male side of the fitting.

21 Remove and discard the indicator ring from the male side of the fitting.

Reconnection

22 Inspect the old O-ring inside the female side of the fitting (see illustration). If it's dried out, cracked, torn or deteriorated, replace it.

4.19a To release a Bartholomew-type plastic quick-connect fitting, depress the tabs on the connector housing with a small screwdriver, then continue pressing on them . . .

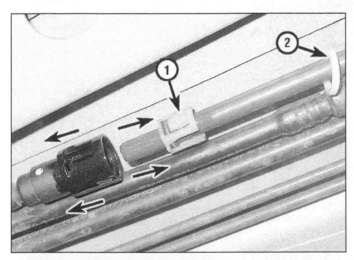

4.19b . . . until the two fuel lines are disconnected, then remove and discard the old retainer (1) and the indicator ring (2) (the indicator ring is used only during factory assembly; there is no need to reinstall it)

4.22 Inspect the old O-ring inside the female side of the fitting; if it's cracked, torn or deteriorated, replace it

4.23 Install a new retainer in the female side of the fitting; make sure that the release tabs are aligned with the windows in the connector

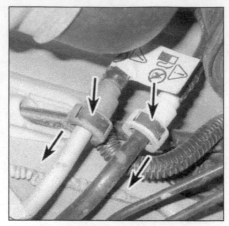

4.29 To disconnect a Push Down TI type plastic collar quick-connect fitting, depress the button on the side of the collar and pull the pipe out of the collar. To reconnect, simply push the pipe into the collar until it clicks into place

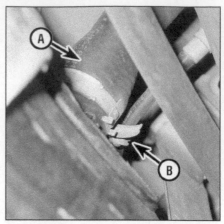

5.5 To disconnect the fuel filler neck hose (A) from the fuel tank's filler neck pipe, loosen this clamp screw (B) and pull off the hose. If there's still fuel in the tank, siphon the excess fuel out of the tank through the filler neck pipe into an approved container

23   Insert a new retainer in the female side of the fitting. Make sure that the release tabs are aligned with the "windows" of the connector (see illustration).
24   Apply a few drops of engine oil to the tip of the male fuel line.
25   Push both sides of the fitting together until the retainer release tabs snap into place.
26   Pull on both sides of the fitting to verify that it's securely connected.
27   Start the engine and check for fuel leaks.

Push Down TI type

28   Push Down TI type fittings are typically used as chassis fuel and EVAP line connections. They're frequently used to connect the fuel and EVAP lines that are coming from components in, on or near the fuel tank to the fuel/EVAP lines going forward to the engine compartment.
29   To disconnect a Push Down TI type fitting, simply depress the button on the female, or collar, side of the fitting (see illustration) and

pull the fuel or EVAP pipe out of the collar.
30   Be sure to wipe off the end of the pipe with a clean cloth before reconnecting it to a Push Down TI fitting, then simply push it into the collar until it clicks into place.

5   Fuel tank - removal and installation

Warning: *Gasoline is extremely flammable, so take extra precautions when you work on any part of the fuel system. See the Warning in Section 2.*
Warning: *Before disconnecting or opening any part of the fuel system, relieve the fuel system pressure (see Section 2), and equalize the pressure inside the fuel tank by removing the fuel filler cap.*
1   Relieve the fuel system pressure (see Section 2).
2   Disconnect the cable from the negative battery terminal (see Chapter 5, Section 1).
3   Raise the vehicle and place it securely

on jackstands.
4   Remove the part of the exhaust system that's underneath the fuel tank (see Section 14).
5   Loosen the hose clamp that secures the fuel filler neck hose to the fuel tank (see illustration) and disconnect the hose from the tank.
6   It's easier to remove the fuel tank when it's nearly empty. But there is no fuel tank drain plug, so if there's still a lot of fuel in the tank, siphon or hand-pump the remaining fuel from the tank through the tank filler neck pipe.
Warning: *Don't start the siphoning action by mouth! Use a siphoning kit (available at most auto parts stores).*
7   Loosen the hose clamp screw and disconnect the EVAP recirculation hose from the fuel tank.
8   Detach the electrical connector from the side of the EVAP canister (see illustration) and disconnect it. Set the harness aside.
9   Disconnect the quick-connect fittings for the fuel and EVAP lines (see illustration 5.8). If you're unfamiliar with quick-connect fittings, refer to Section 4.
10   Support the fuel tank.
11   Remove the fuel tank strap bolts (see illustration) and remove the straps.
12   Lower the tank enough to access the top of the tank and disconnect the electrical connectors from the fuel pump/fuel level sensor module and from the fuel tank pressure sensor.
13   Lower the tank the rest of the way.
14   Installation is the reverse of removal.

6   Fuel tank cleaning and repair - general information

Warning: *Gasoline is extremely flammable, so take extra precautions when you work on any part of the fuel system. See the Warning in Section 2.*

5.8 Before lowering the fuel tank, detach this electrical connector (1) from the EVAP canister, disconnect it and set the harness aside, disconnect the quick-connect fittings (2) for the fuel and EVAP lines and remove the EVAP canister (3)

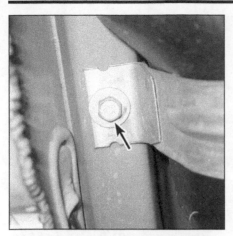

5.11 To detach the fuel tank straps from the underside of the vehicle, remove the strap bolts

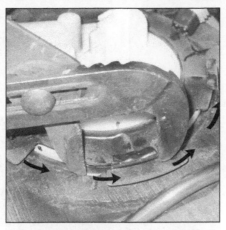

7.6 Use a large pair of water pump pliers to loosen and unscrew the fuel pump locknut; if the locknut is too tight to loosen this way, carefully tap it loose with a hammer and a brass punch

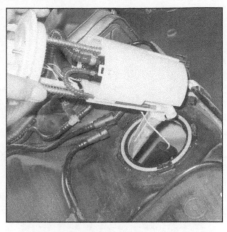

7.7 Carefully remove the fuel pump/ fuel level sensor module from the fuel tank; once the module has cleared the mounting hole in the tank, angle it as shown to work the fuel pump inlet strainer and the fuel level sensor float and float arm through the hole without damaging anything

1   The fuel tank is plastic and cannot be repaired. No reliable repair procedures are available to correct leaks or damage. Fuel tank replacement is the only approved service.
2   To remove sediment from the bottom of the tank, have the fuel tank steam-cleaned. Remove the fuel pump/level sensor module (see Section 7) and all EVAP system components (see Chapter 6) prior to cleaning. Allow plenty of time for the tank to air-dry before returning it to service.

7   Fuel pump/fuel level sensor module - removal and installation

Warning: *Gasoline is extremely flammable, so take extra precautions when you work on any part of the fuel system. See the Warning in Section 2.*
1   Relieve the system fuel pressure (see Section 2), and equalize tank pressure by removing the fuel filler cap.
2   Disconnect the cable from the negative battery terminal (see Chapter 5, Section 1).
3   Remove the fuel tank (see Section 5).
4   Disconnect the fuel supply line from the fuel pump/fuel level sending unit module. Use a shop rag to soak up any spilled fuel.
5   Mark the orientation of the fuel pump in relation to the fuel tank to ensure that the fuel pump is correctly realigned when you install it again. (If you're going to install a new pump, note the location of your alignment mark on the old pump and make a mark at the same spot on the new unit.)
6   Using a pair of large water pump pliers, unscrew the fuel pump/fuel level sending unit module locknut by turning it counterclockwise (see illustration). If the locknut is tight, use a hammer and a brass punch to loosen it (don't use a steel punch, which could produce

sparks when struck by the hammer).
7   Raise the fuel pump/fuel level sensor module high enough to disconnect the quick-connect fitting for the ventilation harness, then remove the pump assembly (see illustration), taking care not to damage the fuel inlet strainer or the fuel level sensor float arm and float.
8   Before installing the pump, inspect the O-ring (see illustration) for cracks, tears and deterioration. If it's worn or damaged, replace it. Also inspect the fuel pump inlet strainer. Make sure that it's clean and free of debris and dirt. If it's dirty, try washing it with carburetor cleaner spray. If this filter is seriously damaged, you'll have to replace the fuel pump/ fuel level sending unit module. The filter is not available separately.
9   Insert the fuel pump/fuel level sensor module into the fuel tank and connect the ventilation hose to the fuel pump assembly before locking the pump into place. Align the fuel pump/fuel level sending unit module with

its hole in the tank and carefully insert it into the tank. Make sure that you don't damage the fuel inlet strainer, the float arm or the float during installation. If the float arm is bent, the fuel level that is indicated on the fuel level gauge on the instrument cluster will be incorrect.
10   Installation is otherwise the reverse of removal.

8   Fuel pump/fuel level sensor - component replacement

Note: *You can purchase the complete fuel pump/fuel level sensor module, or you can purchase either the fuel pump or the fuel level sensor separately. If only one component fails, use this procedure to separate the two components, then reassemble the good component and the new replacement component.*
1   Remove the fuel pump/fuel level sensor module (see Section 7).

7.8 Remove and inspect the O-ring seal for the fuel pump mounting flange; if it's cracked, torn or deteriorated, replace it

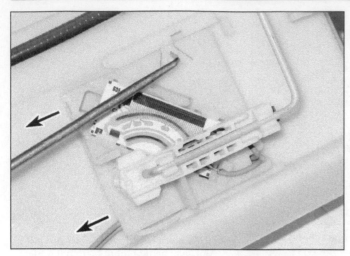

8.5 To detach the fuel level sensor from the fuel pump module, disengage the locking tab with a pick (shown) or with a small screwdriver, then slide the sensor unit off the retaining rails (this is a typical setup; not all units are identical to this one, but they're similar)

9.2 To disconnect the PCV hose from the air intake duct on a four-cylinder model, loosen this hose clamp and pull off the hose

2   Place the fuel pump/fuel level sensor module on a clean workbench surface.

3   Disengage the fuel level sensor wiring harness from the molded-in harness guide.

4   Disconnect the fuel level sensor electrical connector from the pump module cover.

5   To remove the fuel level sensor unit, depress the locking tab (see illustration) and slide off the sensor.

6   No further disassembly of the fuel pump/ fuel level sensor assembly is possible.

7   To install the fuel level sensor on the fuel pump module, slide the sensor unit into place until you hear a click, then gently pull on the sensor unit to verify that it's locked into place.

8   Installation is otherwise the reverse of removal.

9   Air filter housing - removal and installation

Four-cylinder models
Air intake duct

1   Make sure that the ignition key is turned to OFF.

2   Disconnect the PCV hose from the air intake duct (see illustration).

3   Loosen the hose clamp screw at the air filter housing (see illustration) and pull back the clamp.

4   Loosen the hose clamp screw at the throttle body (see illustration), pull back the clamp and remove the duct.

5   Installation is the reverse of removal.

Air filter housing

6   Remove the air intake duct (see Steps 1 through 4).

7   Disconnect the electrical connector from the Mass Air Flow (MAF) sensor (see Chapter 6).

8   Remove the two air filter housing mounting bolts (see illustration) and remove the filter housing.

9   Inspect all rubber mounting grommets while the air filter housing is out. If a grommet is cracked, torn or otherwise damaged, replace it.

10   Installation is the reverse of removal.

9.3 To disconnect the air intake duct from the air filter housing on a four-cylinder model, loosen this hose clamp and pull off the duct

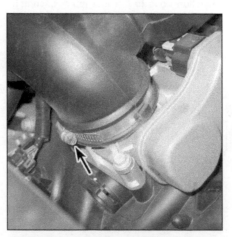

9.4 To disconnect the air intake duct from the throttle body on a four-cylinder model, loosen this hose clamp screw, pull back the clamp and pull off the duct

9.8 To detach the air filter housing from a four-cylinder model, remove these two bolts

9.11 To disconnect the air intake duct from the Mass Air Flow (MAF) sensor on a V6 model, loosen this hose clamp screw, pull back the clamp and pull off the duct

9.12 To disconnect the air intake duct from the throttle body on a V6 model, disconnect the PCV fresh air inlet hose (1), then loosen the hose clamp screw (2), pull back the clamp and pull off the duct

V6 models

Air intake duct

11   Loosen the hose clamp screw and disconnect the air intake duct from the Mass Air Flow (MAF) sensor (see illustration).

12   Disconnect the PCV fresh air inlet hose from the air intake duct (see illustration).

13   Loosen the hose clamp screw, disconnect the air intake duct from the throttle body and remove the duct.

14   Installation is the reverse of removal.

Air filter housing

15   Remove the air intake duct (see Steps 11 through 13).

16   Remove the Mass Air Flow (MAF) sensor from the air filter housing (see Chapter 6).

17   The remainder of the air filter housing removal procedure is similar to the procedure for removing the air filter housing on four-cylinder models (see Steps 6 through 9).

18   Installation is the reverse of removal.

---

10   Sequential Fuel Injection (SFI) system - general information

1   These models are equipped with a Sequential Fuel Injection (SFI) system. The SFI system consists of three basic sub-systems: the air induction system, the fuel system and the electronic control system.
Note: *Refer to Chapter 6 for more information on the components of the electronic control system.*

Air induction system

2   The air induction system consists of the air filter housing, the Mass Air Flow (MAF) sensor, the Intake Air Temperature (IAT) sensor (an integral part of the MAF sensor), the

air intake duct, the throttle body, the Throttle Position (TP) sensor, the air intake plenum and the intake manifold. The MAF/IAT sensor is an information sensor for the Powertrain Control Module (PCM). The MAF sensor uses a heated wire system to send the PCM an analog (constantly variable) voltage signal corresponding to the volume of air passing into the engine. The IAT sensor measures the temperature of the intake air. The PCM uses these signals to calculate the mass (density) of air entering the engine.
Note: *For more information about the MAF/IAT sensor, refer to Chapter 6. The TP sensor is an integral component of the throttle body and cannot be serviced separately.*

3   All models are equipped with an electronic throttle body. The electronic throttle control system consists of the Accelerator Pedal Position (APP) sensor, the Powertrain Control Module (PCM) and the solenoid motor inside the throttle body that controls the angle of the throttle plate. The APP sensor, which is an integral component of the accelerator pedal assembly, is a potentiometer (similar to a fuel level sensor or a throttle position sensor) that monitors the position (or angle) of the accelerator pedal. As you depress the accelerator pedal, the APP sensor outputs a variable voltage signal to the PCM, which sends a command to the throttle body's solenoid motor, which opens the throttle plate in proportion to the position (angle) of the accelerator pedal. As the throttle plate opens or closes, the amount of air that can pass through the system increases or decreases accordingly. As the throttle plate opens or closes, the Throttle Position (TP) sensor, which is located on the end of the throttle plate shaft, opens or closes with it. And as more or less air enters the engine, the MAF/IAT sensor signal to the PCM also changes. In response to these two signals from the TP and MAF/IAT sensors,

the PCM opens each injector for a longer or shorter duration to increase the amount of fuel delivered to the inlet ports. The interval during which an injector is open is known as pulse width.

Fuel system

4   An electric fuel pump located inside the fuel tank supplies fuel under pressure to the fuel rail, which distributes fuel evenly to all injectors. A filter between the fuel pump and the fuel rail protects the components of the system. From the fuel rail, fuel is injected into the intake ports, just above the intake valves, by a fuel injector.

5   The amount of fuel supplied by the injectors is precisely controlled by injector "drivers" inside the PCM. The injector drivers, which are turned on and off by the PCM, control the ground side of each injector circuit: When the ground path is closed, the injectors are on; when the ground path is open, the injectors are off. The PCM uses signals from the Crankshaft Position (CKP) sensor and (on V6 models and 2007 four-cylinder models) the Camshaft Position (CMP) sensor to determine when to trigger each injector in cylinder firing order (hence the term "sequential injection"). This precise control of injector timing produces more power, better fuel economy and lower exhaust emissions.

6   To prevent fuel starvation, the fuel pump delivers more fuel to the fuel rail than the injectors can use under most circumstances. When the pressure exceeds a certain threshold, the excess fuel is returned to the fuel tank by the fuel pressure regulator, which is an integral component of the fuel pump/fuel level sensor module. The fuel pump and the fuel level sensor are separately available, but the fuel pressure regulator is not. If the fuel pressure regulator fails, you must replace the fuel pump.

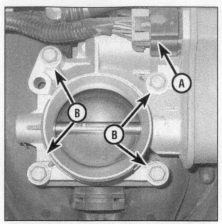

12.5a To remove the throttle body from the intake manifold on a four-cylinder model, disconnect the electrical connector (A) and remove the four mounting bolts (B)

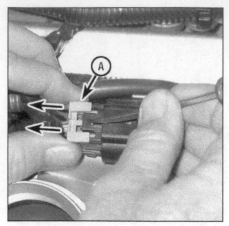

12.5b To disconnect the throttle body electrical connector on a four-cylinder model, carefully pry the lock loose with a small screwdriver, then depress the release tab (A) and pull off the connector

Electronic control system

7   The PCM controls the SFI system and the engine management system. It receives signals from an array of information sensors that monitor such variables as intake air mass and temperature, coolant temperature, engine speed, crankshaft position, acceleration/deceleration, and exhaust gas oxygen content. These signals help the PCM determine the injection duration necessary for the optimal air/fuel ratio. These sensors and various PCM-controlled output actuators (relays, solenoids, etc.) are located throughout the engine compartment. For further information regarding the PCM, the engine management system, the information sensors and the output actuators, see Chapter 6.

11   Sequential Fuel Injection (SFI) system - general check

Warning: *Gasoline is extremely flammable, so take extra precautions when you work on any part of the fuel system. See the Warning in Section 2.*

1   Inspect the SFI system electrical connectors. Verify that all ground wire connections are tight. Loose connectors and poor grounds can cause many problems that resemble more serious malfunctions.
2   Verify that the battery is fully charged (see Chapters 1 and 5 for help with the battery). The PCM, information sensors and output actuators depend on a steady and adequate voltage to function correctly.
3   Inspect the air filter element (see Chapter 1). A dirty or partially blocked filter will severely impede performance and economy.
4   Inspect any fuses for the circuit you're checking. If you find a blown fuse, replace it and note whether it blows again. If it does, look for a short in the circuit.
5   Inspect the air intake duct, the throttle

body and the intake manifold for leaks, which will cause an excessively lean mixture. Also inspect all vacuum hoses connected to the intake manifold and to the throttle body.
6   Remove the air intake duct (see Section 9) and inspect the throttle body for dirt, carbon, varnish or other residue inside the bore of the throttle body, particularly around the throttle plate. If it's dirty, clean it with carburetor cleaner spray and a shop towel.
7   With the engine running, place an automotive stethoscope against each injector, one at a time, and listen for a clicking sound that indicates operation. If you don't have a stethoscope, you can place the tip of a long screwdriver against the injector and listen through the handle. If you can hear all of the injectors operating, but there is a misfire condition present, the electrical circuits are functioning, but the injectors might be dirty or fouled from carbon deposits. Commercial cleaning products might help. If not, then you might have to replace the injectors (see Section 13).
8   If you can't hear an injector operating, disconnect its electrical connector and note whether it makes any difference in the way the engine runs (engine rpm should drop if the injector is working). Unplugging a non-operational injector shouldn't make any difference in engine rpm. If the engine rpm does drop, or if the engine stalls, when you disconnect an injector, then the injector is probably still operating when it's connected (though it might be so dirty that it's not operating well).
9   If the rpm doesn't drop when you disconnect an injector, the injector is probably defective (provided that the cylinder has compression and spark is present). Disconnect the electrical connector and measure the resistance of the injector coil with an ohmmeter, then compare your measurement with the injector resistance listed in this Chapter's Specifications.
Note: *On V6 models you will have to remove the upper intake manifold (see Chapter 2B or*

Chapter 2C) to access some of the injectors. If the indicated injector resistance is out of the specified range of resistance, replace the injector.
10   If the injector coil resistance is okay, but the injector is not operating, the circuit between the PCM and the injector might be open somewhere. It might also indicate that the driver inside the PCM is defective. The signal to the injector can be tested with a "noid" light, available at most auto parts stores. If the light doesn't flash when connected to the injector harness connector and the engine is cranked, have the circuit between the PCM and the injectors tested by a dealer service department before replacing the PCM or the injector. If the service department determines that an injector driver is bad, you'll have to replace the PCM (see Chapter 6).

12   Throttle body - inspection, removal and installation

Inspection

1   Verify that the throttle linkage operates smoothly.
2   Remove the air intake duct from the throttle body, open the throttle plate and inspect the throttle body bore for carbon and residue build-up. If it's dirty, clean it with solvent or carburetor cleaner. Make sure that the solvent or carb cleaner is safe for oxygen sensor systems and catalytic converters.
Caution: *Do not clean the Throttle Position (TP) sensor or the solenoid motor with solvent. Also, do NOT use a metal brush to clean the bore of the throttle body, which is protected by a special coating. Scrubbing the bore with a stiff brush could ruin the coating. Instead, wipe out the bore with a clean shop rag and a little carburetor cleaner.*

Removal and installation

Four-cylinder models
Warning: *Wait until the engine is completely cool before beginning this procedure.*
3   Disconnect the cable from the negative battery terminal (see Chapter 5, Section 1).
4   Disconnect the air intake duct from the throttle body.
5   Disconnect the throttle body electrical connector (see illustrations).
6   Remove the four throttle body mounting bolts (see illustration 12.5a) and remove the throttle body.
7   Remove the O-ring type throttle body gasket (see illustration) and inspect it. If the gasket isn't cracked, torn or otherwise deteriorated, it's okay to reuse it. But if it's damaged or worn, replace it. (If the vehicle is fairly old, it's a good idea to replace this gasket regardless of its apparent condition.)
8   If necessary, clean the throttle body as outlined in Step 2.
9   Installation is the reverse of removal. Be sure to install the throttle body gasket, then install the throttle body and tighten the throttle

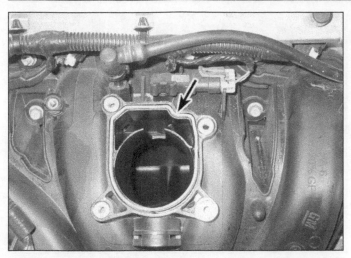

12.7 Remove and inspect the throttle body's O-ring type gasket; if it's in good condition it's okay to reuse it, but if it's cracked, torn or otherwise deteriorated, replace it

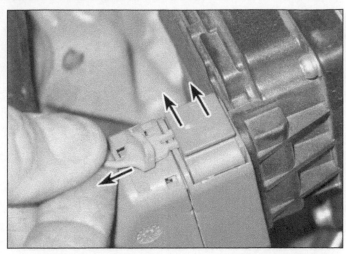

12.13 To disconnect the throttle body electrical connector on V6 models, pull out the lock, then carefully pry up the release tab with a small screwdriver and pull off the connector

body mounting bolts to the torque listed in this Chapter's Specifications.

10   Start the engine, then verify that the throttle body operates correctly and that there are no air leaks.

V6 models

11   Disconnect the cable from the negative battery terminal (see Chapter 5, Section 1).
12   Remove the air intake duct (see Section 9).
13   Disconnect the throttle body electrical connector (see illustration).
14   Remove the nut or bolt that secures the heater hose clamp to the throttle body, then set the heater hoses aside.
15   Remove the two upper mounting bolts and the two lower mounting nuts (see illustra-

tion) and remove the throttle body.
16   Remove the O-ring type throttle body gasket and inspect it. If the gasket isn't cracked, torn or otherwise deteriorated, it's okay to reuse it. But if it's damaged or worn, replace it. (If the vehicle is fairly old, it's a good idea to replace this gasket regardless of its apparent condition.)
17   If necessary, clean the throttle body as outlined in Step 2.
18   Installation is the reverse of removal. Be sure to install the throttle body gasket, then install the throttle body and tighten the throttle body mounting bolts and nuts to the torque listed in this Chapter's Specifications.
19   Start the engine, then verify that the throttle body operates correctly and that there are no air leaks.

13   Fuel rail and injectors - removal and installation

1   Relieve the fuel system pressure (see Section 2).
2   Equalize tank pressure by removing the fuel filler cap.
3   Disconnect the cable from the negative battery terminal (see Chapter 5, Section 1).

Four-cylinder models

4   Remove the air intake duct (see Section 9).
5   Remove the PCV hose (see illustration).
6   Detach the coolant pipe from the clips (see illustration 13.5).
7   Disconnect the electrical connectors

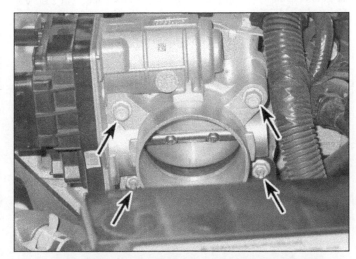

12.15 To detach the throttle body from the intake manifold, remove the two upper bolts and the two lower nuts

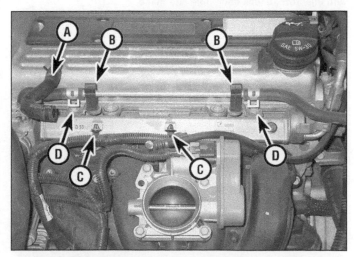

13.5 To remove the fuel rail, disconnect the PCV hose (A), detach the coolant pipe from its clips (B), detach the harness clips (C), and remove the two mounting bolts (D)

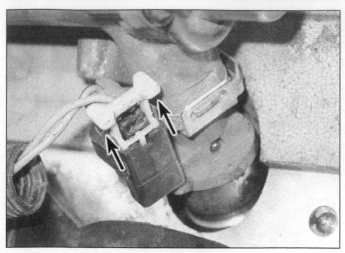

13.7 To disconnect the electrical connector from an injector on a four-cylinder model, pull up on this lock, then unplug the connector (this design is typical on these models; GM uses several variations of this design)

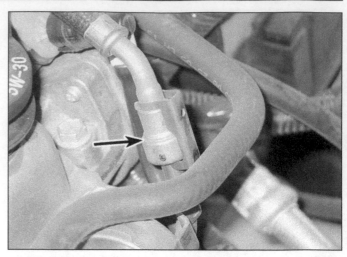

13.8 On four-cylinder models, the fuel supply line is connected to the fuel rail with this metal collar type quick-connect fitting at the left end of the fuel rail. If you're unfamiliar with this type of fitting, refer to Section 4

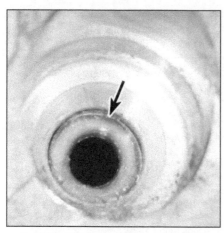

13.10 When an injector is removed, the lower O-ring on the injector sometimes comes off. So after removing the fuel rail and injectors, check each hole and make sure that there is no O-ring in any hole

from the fuel injector (see illustration), then detach the injector wiring harness clips from the fuel rail (see illustration 13.5) and set the harness aside.

8   Disconnect the fuel supply line from the fuel rail (see illustration). Remove and discard the old O-rings.

9   Remove the fuel rail mounting bolts (see illustration 13.5).

10   Remove the fuel rail and injectors as a single assembly. After removing the fuel rail and injectors, look inside each injector hole and make sure that no O-rings remain in the holes (see illustration).

11   Remove the retainer that secures each fuel injector to the fuel rail and pull out the injector (see illustration).

12   Remove the old O-rings from each injector (see illustration) and discard them. Always install new O-rings on the injectors before reassembling the injectors and the fuel rail.

13   To ensure that the new injector O-rings

are not damaged when the injectors are installed into the fuel rail and into the intake manifold, lubricate them with clean engine oil.

14   Installation is otherwise the reverse of removal. Be sure to tighten the fuel rail mounting bolts securely.

15   Start the engine and verify that there are no fuel leaks.

V6 models

16   Disconnect the fuel line from the fuel rail (see illustration).

17   Remove the upper intake manifold (see "Intake manifold - removal and installation" in Chapter 2B or 2C).

18   On 2004 through 2006 3.5L V6 models, disconnect the main fuel injector harness electrical connector (connects the fuel injector harness to the main engine harness), then disconnect the electrical connectors from the injectors and set the harness aside. On 3.9L V6 models and 2007 3.5L V6 models, disconnect

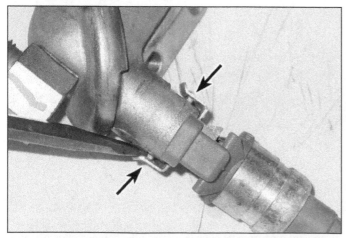

13.11 To release an injector retainer, free it from the small lugs on each side of the injector and pull it off

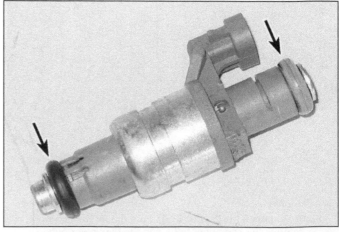

13.12 Remove and discard the old O-rings from each injector and install new ones

13.16 On V6 models, the fuel supply line is connected to the fuel rail with this metal collar type quick-connect fitting at the right end of the fuel rail. If you're unfamiliar with this type of fitting, refer to Section 4

13.19 To detach the fuel rail from the lower intake manifold, remove these two bolts

the electrical connectors from the Engine Coolant Temperature (ECT) and from the Camshaft Position (CMP) sensors (both are part of the injector harness), remove the bolt that secures the main injector harness electrical connector to the intake manifold, disconnect the electrical connectors from the fuel injectors and set the injector harness aside.

19   Remove the fuel rail mounting bolts (see illustration).

20   Carefully disengage the injectors from the lower intake manifold and lift the fuel rail and all six injectors from the engine as a single assembly.

21   Remove the injector retainers (see illustration) and remove the injectors from the fuel rail.

22   Remove and discard the old injector O-rings (see illustration). Always install new O-rings on the injectors before reassembling the injectors and the fuel rail.

23   To ensure that the new injector O-rings

are not damaged when the injectors are installed into the fuel rail and into the intake manifold, lubricate them with clean engine oil.

24   Installation is otherwise the reverse of removal. Be sure to tighten the fuel rail mounting bolts securely.

25   Start the engine and verify that there are no fuel leaks.

14   Exhaust system servicing - general information

Inspection

Warning: *Inspect and repair exhaust system components only after allowing the exhaust components to cool completely. This applies particularly to the catalytic converter, which operates at very high temperatures. Also, when working under the vehicle, make sure it is securely supported on jackstands.*

1   The exhaust system consists of the exhaust manifold(s), the catalytic converter(s), the exhaust pipes, the muffler, and all brackets, hangers and clamps that support the exhaust system. Inspect the exhaust system regularly to ensure that it remains safe and quiet. Look for any damaged or bent parts, open seams, holes, loose connections, excessive corrosion or other defects which could allow exhaust fumes to enter the vehicle. Also check the catalytic converter(s) when you inspect the exhaust system. Inspect the catalytic converter heat shield(s) for cracks, dents and loose or missing fasteners. If a heat shield is damaged, the converter might also be damaged. Damaged or deteriorated exhaust system components should not be repaired; they should be replaced with new parts.

2   Before trying to disassemble any exhaust components, spray the fasteners with a penetrating oil to help ease removal. If the exhaust system components are extremely corroded

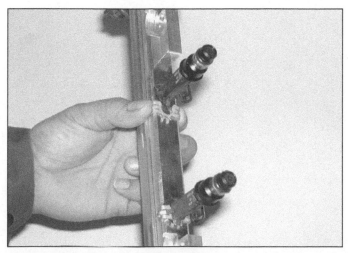

13.21 Remove each injector retaining clip and pull the injector out of the fuel rail

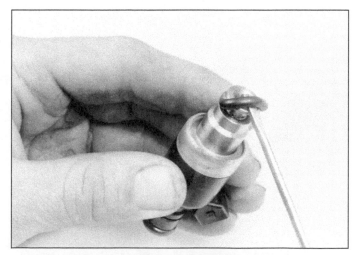

13.22 Carefully remove the O-rings from each fuel injector, then replace them with new ones

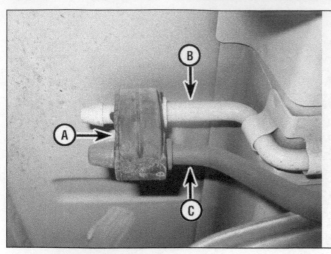

14.4 A typical rubber exhaust hanger (this one supports the muffler):

A)   Rubber exhaust hanger
B)   Support bracket (welded or bolted to vehicle)
C)   Muffler support bracket (welded to exhaust pipe)

or rusted together, welding equipment will probably be required to remove them. The convenient way to accomplish this is to have a muffler repair shop remove the corroded sections with a cutting torch. If, however, you want to save money by doing it yourself (and you don't have a welding outfit with a cutting torch), simply cut off the old components with a hacksaw. If you have compressed air, special pneumatic cutting chisels can also be used. If you decide to tackle the job at home,

be sure to wear safety goggles to protect your eyes from metal chips and work gloves to protect your hands.

3   Here are some simple guidelines to follow when repairing the exhaust system:

a) Work from the back to the front when removing exhaust system components.
b) Apply penetrating oil to the exhaust system component fasteners to make them easier to remove.
c) Use new gaskets, hangers and clamps when installing exhaust systems components.
d) Apply anti-seize compound to the threads of all exhaust system fasteners at reassembly.
e) Be sure to allow sufficient clearance between newly installed parts and all points on the underbody to avoid overheating the floor pan and possibly damaging the interior carpet and insulation. Pay particularly close attention to the catalytic converter and heat shield.

Component replacement
Rubber exhaust hangers
4   The exhaust system is attached to the body with mounting brackets and rubber hangers (see illustration). Anytime you must raise the vehicle to perform any under-vehicle service, make sure that you inspect the exhaust hangers. Look for cracks, tears and deterioration. If a hanger is worn or damaged, replace it.

Catalytic converters
5   The procedures for replacing the catalytic converter(s) are in Chapter 6.

Chapter 5
Engine electrical systems

Contents

Specifications

General

Firing order
Four-cylinder engine 1-3-4-2
V6 engine 1-2-3-4-5-6
Cylinder numbering (from drivebelt end to transaxle end)
Four-cylinder engine 1-2-3-4
V6 engine
Front bank 2-4-6
Rear bank 1-3-5
Ignition timing .... Not adjustable

Torque specifications   **Ft-lbs** (unless otherwise indicated)

Note: *One foot-pound (ft-lb) of torque is equivalent to 12 inch-pounds (in-lbs) of torque. Torque values below approximately 15 ft-lbs are expressed in inch-pounds, because most foot-pound torque wrenches are not accurate at these smaller values.*

Starter motor mounting bolts
2006 and earlier models 30
2007 models
Four-cylinder engines 39
V6 engines 32
2008 and later models
Four-cylinder engines 30
V6 engines 37

---

1   General information, precautions and battery disconnection

1   The engine electrical systems include all ignition, charging and starting components. Because of their engine-related functions, these components are discussed separately from chassis electrical devices such as the lights, the instruments, etc. (see Chapter 12).

Precautions

2   Always observe the following precautions when working on the electrical system: Be extremely careful when servicing engine electrical components. They are easily damaged if checked, connected or handled improperly.

a) *Never leave the ignition switched on for long periods of time when the engine is not running.*

b) *Never disconnect the battery cables while the engine is running.*

c) *Maintain correct polarity when connecting battery cables from another vehicle during jump starting - see the Booster battery (jump) starting Section at the front of this manual.*

d) *Always disconnect the negative cable from the battery before working on the electrical system, but read the following battery disconnection procedure first.*

3   It's also a good idea to review the safety-related information regarding the engine electrical systems located in the *Safety first!* Section at the front of this manual, before beginning any operation included in this Chapter.

Battery disconnection

Warning: *On 2006 and later models with On-Star, make absolutely sure the ignition key is in the Off position and Retained Accessory Power (RAP) has been depleted before disconnecting the cable from the negative battery terminal. Also, never remove the OnStar fuse with the ignition key in any position other than Off. If these precautions are not taken, the OnStar system's back-up battery will be activated, and remain activated, until it goes dead. If this happens, the OnStar system will not function as it should in the event that the main vehicle battery power is cut off (as might happen during a collision).*

4   The battery is located in the engine compartment on all vehicles covered by this manual. To disconnect the battery for service procedures that require battery disconnection, simply disconnect the cable from the negative battery terminal. Make sure that you isolate the cable to prevent it from coming into contact with the battery negative terminal.

5   Some vehicle systems (radio, alarm system, power door locks, etc.) require battery power all the time, either to enable their operation or to maintain control unit memory (Powertrain Control Module, automatic transaxle control module, etc.), which would be lost if the battery were to be disconnected. So before you disconnect the battery, note the following points:

a)   *Before connecting or disconnecting the cable from the negative battery terminal, make sure that you turn the ignition key and the lighting switch to their OFF positions. Failure to do so could damage semiconductor components.*

b)   *On a vehicle with power door locks, it is a wise precaution to remove the key from the ignition and to keep it with you, so that it does not get locked inside if the power door locks should engage accidentally when the battery is reconnected! To disconnect the battery for service procedures requiring power to be cut from the vehicle, first open the driver's door to disable Retained Accessory Power (RAP), then loosen the cable end bolt and disconnect the cable from the negative battery terminal. Isolate the cable end to prevent it from coming into accidental contact with the battery terminal.*

c)   *After the battery has been disconnected, then reconnected (or a new battery has been installed) on vehicles with an automatic transaxle, the Transaxle Control Module (TCM) will need some time to relearn its adaptive strategy. As a result, shifting might feel firmer than usual. This is a normal condition and will not*

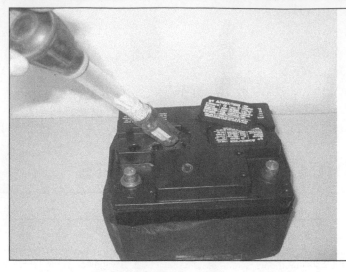

3.1a Use a battery hydrometer to draw electrolyte from the battery cell; this hydrometer is equipped with a thermometer to make temperature corrections

adversely affect the operation or service life of the transaxle. Eventually, the TCM will complete its adaptive learning process and the shift feel of the transaxle will return to normal.

d)   *The engine management system's PCM has some learning capabilities that allow it to adapt or make corrections in response to minor variations in the fuel system in order to optimize driveability and idle characteristics. However, the PCM might lose some or all of this information when the battery is disconnected. The PCM must go through a relearning process before it can regain its former driveability and performance characteristics. Until it relearns this lost data, you might notice a difference in driveability, idle and/or (if you have an automatic) shift "feel."*

Memory savers

6   Devices known as "memory savers" (typically, small 9-volt batteries) can be used to avoid some of the above problems. A memory saver is usually plugged into the cigarette lighter, and then you can disconnect the vehicle battery from the electrical system. The memory saver will deliver sufficient current to maintain security alarm codes and - maybe, but don't count on it! - PCM memory. It will also run "unswitched" (always on) circuits such as the clock and radio memory, while isolating the car battery in the event that a short circuit occurs while the vehicle is being serviced.

Warning: *If you're going to work around any airbag system components, disconnect the battery and do not use a memory saver. If you do, the airbag could accidentally deploy and cause personal injury.*

Caution: *Because memory savers deliver current to operate unswitched circuits when the battery is disconnected, make sure that the circuit that you're going to service is actually open before working on it!*

2   Battery - emergency jump starting

1   Refer to the *Booster battery (jump) starting* procedure at the front of this manual.

3   Battery - check, removal and installation

Warning: *Hydrogen gas is produced by the battery, so keep open flames and lighted cigarettes away from it at all times. Always wear eye protection when working around a battery. Rinse off spilled electrolyte immediately with large amounts of water.*

Check

1   A battery cannot be accurately tested until it is at or near a fully charged state. Disconnect the negative battery cable from the battery and perform the following tests:

a)   ***Battery state of charge test*** *- Visually inspect the indicator eye (if equipped) on the top of the battery. If the indicator eye is dark in color, charge the battery as described in Chapter 1. If the battery is equipped with removable caps, check the battery electrolyte. The electrolyte level should be above the upper edge of the plates. If the level is low, add distilled water. DO NOT OVERFILL. The excess electrolyte may spill over during periods of heavy charging. Test the specific gravity of the electrolyte using a hydrometer (see illustration). Remove the caps and extract a sample of the electrolyte and observe the float inside the barrel of the hydrometer. Follow the instructions from the tool manufacturer and determine the specific gravity of the electrolyte for each cell. A fully charged battery will indicate approximately 1.270 (green zone) at 68-degrees F (20-degrees C). If the specific gravity of the electrolyte is low (red zone), charge the battery as described in Chapter 1.*

3.1b To test the open-circuit voltage of the battery, connect the black probe of a voltmeter to the negative terminal and the red probe to the positive terminal of the battery; if the battery is fully charged, the voltmeter should indicate about 12.5 volts (depending on the outside air temperature)

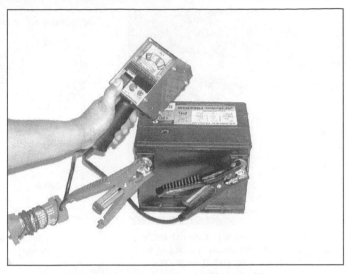

3.1c Some battery load testers are equipped with an ammeter, which enables you to impose a precise load on the battery (less expensive testers, like this one, have only a load switch and a voltmeter)

b) *Open circuit voltage test - Using a digital voltmeter, perform an open circuit voltage test (see illustration). Connect the negative probe of the voltmeter to the negative battery post and the positive probe to the positive battery post. The battery voltage should be greater than 12.5 volts. If the battery is less than the specified voltage, charge the battery before proceeding to the next test. Do not proceed with the battery load test until the battery is fully charged.*

c) *Battery load test - An accurate check of the battery condition can only be performed with a load tester (available at most auto parts stores). This test evaluates the ability of the battery to operate the starter and other accessories during periods of heavy amperage draw (load). Connect a battery load-testing tool to the battery terminals (see illustration). Load test the battery according to the tool manufacturer's instructions. This tool increases the load demand (amperage draw) on the battery. Maintain the load on the battery for 15 seconds and observe that the battery voltage does not drop below 9.6 volts. If the battery condition is weak or defective, the tool will indicate this condition immediately.*
Note: Cold temperatures will cause the minimum voltage reading to drop slightly. Follow the chart given in the tool manufacturer's instructions to compensate for cold climates. Minimum load voltage for freezing temperatures (32-degrees F/0-degrees C) should be approximately 9.1 volts.

d) *Battery drain test - This test will indicate whether there's a constant drain on the vehicle's electrical system that can cause the battery to discharge. Make sure all accessories are turned off. If the vehicle has an underhood light, verify that it's working properly, then disconnect it. Connect one lead of a digital ammeter to the disconnected negative battery cable clamp and the other lead to the negative battery post. A drain of approximately 100 milliamps or less is considered normal (due to the engine control computers, clocks, digital radios and other components that normally cause a key-off battery drain). An excessive drain (approximately 500 milliamps or more) will cause the battery to discharge. The problem circuit or component can be located by removing the fuses, one at a time, until the excessive drain stops and normal drain is indicated on the meter.*

Replacement
Battery
2   Remove the battery's protective cover (see illustration).
3   Disconnect both cables from the battery terminals (see illustration).
Warning: *Always disconnect the negative cable first and connect it last, or you might accidentally short the battery with the tool you're using to loosen the cable clamps.*

3.2 To access the battery terminals and the battery hold-down bolt, remove the battery's protective cover

3.3 Whenever you must disconnect the cables from the battery terminals, ALWAYS disconnect the cable from the negative terminal (1) FIRST, then disconnect the cable from the positive terminal (2)

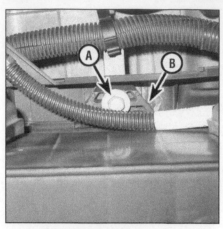

3.4 To detach the battery from the battery tray, remove the hold-down bolt (A) and hold-down clamp (B), then lift the battery out

4   Remove the battery hold-down clamp bolt (see illustration) and remove the hold-down clamp.
5   Lift out the battery. Use a battery lifting strap that attaches to the battery posts to lift the battery safely and easily.
6   Installation is the reverse of removal.
Warning: *When connecting the battery cables, always connect the positive cable first and the negative cable last to avoid a short circuit caused by the tool used to tighten the cable clamps.*

---

4   Battery cables - replacement

---

1   Periodically inspect the entire length of each battery cable for damage, cracked or burned insulation and corrosion. Poor battery cable connections can cause starting problems and decreased engine performance.
2   Check the cable-to-terminal connections at the ends of the cables for cracks, loose

Terminal end corrosion or damage.

Insulation cracks.

Chafed insulation or exposed wires.

Burned or melted insulation.

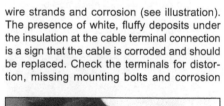

4.2 Typical battery cable problems

wire strands and corrosion (see illustration). The presence of white, fluffy deposits under the insulation at the cable terminal connection is a sign that the cable is corroded and should be replaced. Check the terminals for distortion, missing mounting bolts and corrosion

(see Chapter 1 for further information regarding battery cable maintenance).
3   When removing the cables always disconnect the negative cable from the negative battery post first and hook it up last or the tool used to loosen the cable clamps may short the battery. Even if only the positive cable is being replaced, be sure to disconnect the negative cable from the negative battery post first.
4   Before disconnecting any cables, note the routing of both cables to ensure correct installation. Disconnect the old cables from the battery terminals (see Section 3), then disconnect them at the other end(s). Trace each cable from the battery down to its lower end and disconnect it. The smaller ground cables are attached to the upper radiator crossmember (see illustration) and the larger ground cable is attached to the transaxle (see illustration). The positive cable is connected to the starter motor solenoid (see illustration). Some of these cables are taped and/or cable-tied together, so before you can remove these cables you'll have to cut off the electrical tape and/or cable ties. Be extremely careful when cutting off electrical tape; you might create driveability problems if you accidentally cut or damage one of the small wires in some of the harnesses. Instead, remove the battery to give yourself some room to work (see Section 3). Then locate the point at which the cables are bundled into another harness and the point at which they emerge from the harness down below, then cut off whatever holds the cables together. Some cables are also secured by clips to the battery tray or to other engine compartment components. You can usually detach these clips without damaging them. If you do damage a clip when disengaging it, replace it with a new clip when you install the new cable(s).
5   Positive cables are almost always red and larger in cross-section; ground cables are usually black and smaller in cross-section. But if you're replacing either or both of the

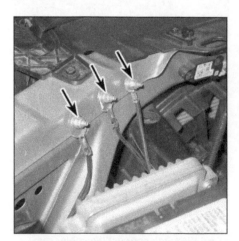

4.4a To detach the smaller ground cables from the upper radiator crossmember, remove these nuts

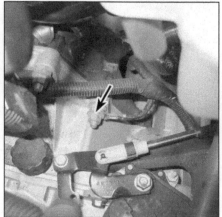

4.4b To detach the larger ground cable from the transaxle, remove this nut

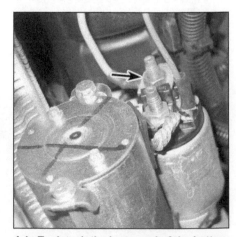

4.4c To detach the lower end of the battery cable from the starter solenoid, remove this nut (four-cylinder model shown, V6 models similar)

cables, take them with you when buying new cables. It is vitally important that you replace the cables with the identical parts.

6   Clean the threads of the starter solenoid and/or ground connection with a wire brush to remove rust and corrosion. Apply a light coat of battery terminal corrosion inhibitor or petroleum jelly to the threads to prevent future corrosion.

7   Attach the lower ends of the cables first, then connect the positive cable to the positive battery post (don't reconnect the ground cable to the negative battery post until you're completely finished). Before connecting a new cable to the battery, make sure that it reaches the battery post without having to be stretched.

8   Reattach the new cables to whatever harness they were previously attached to with new cables ties and/or electrical tape. If either cable is supposed to be secured by any brackets or clips, make sure that you reattach them.

9   After both cables are completely installed, reconnect the ground cable to the negative battery post.

---

5   Ignition system - general information

1   The ignition system consists of the ignition control module(s), the ignition coil pack(s), the spark plugs, the Camshaft Position (CMP) sensor (V6 models only), the Crankshaft Position (CKP) sensor, the knock sensor(s) and the Powertrain Control Module (PCM).

2006 and earlier four-cylinder models

2   2006 and earlier four-cylinder models are equipped with a single "coil-over-plug" coil pack/ignition control module assembly that fits directly onto the four spark plugs. The ignition control module can be replaced separately. The coil pack assembly consists of two ignition coils. Each coil is connected to the spark plugs for two cylinders by a short boot, which contains a coil spring that carries voltage to the plugs. And each coil fires the two plugs to which it's connected simultaneously. One coil fires the plugs for cylinders 1 and 4; the other coil fires the plugs for cylinders 2 and 3. When the piston in cylinder No. 1 is on the compression stroke, the piston in cylinder No. 4 is on the exhaust stroke. When the coil fires cylinder No. 1, most of the spark voltage goes to that cylinder because the pressure - and therefore the resistance - is high in that cylinder (the higher the resistance, the higher the voltage needed to jump the gap from the spark plug's center electrode to ground). Conversely, the piston in cylinder No. 4, which is on the exhaust stroke, produces no pressure, and therefore no resistance, so little voltage is needed to jump the gap from

the spark plug's center electrode to ground. The ignition coil for cylinders 2 and 3 works the same way. This design is known as a "waste spark" ignition. The ignition control module on four-cylinder models houses the "driver modules" that turn the ignition coils on and off by closing and opening their ground paths. The timing of these drivers is controlled by the PCM. The ignition control module on 2004 through 2006 four-cylinder coil packs is serviceable separately from the coil pack.

2007 and later four-cylinder models

3   On these models there are four individual coil-over-plug type ignition coils. Each coil can be removed separately. There is no removable ignition control module. The module's functions are controlled inside the Powertrain Control Module (PCM).

3.5L and 3.9L V6 models

4   V6 models use three ignition coils; each coil fires two spark plugs via spark plug wires. The three coils and the ignition control module are combined into one integral coil pack unit mounted on the left end of the rear valve cover. V6 models do not have a separate ignition control module; it's an integral part of the coil pack. If a coil pack or its ignition module fails, you must replace the entire unit. In this design, each coil fires two spark plugs. One coil fires the No. 1 and No. 4 cylinders; the second coil fires the No. 2 and No. 5 cylinders; the third coil fires the No. 3 and No. 6 cylinders. When the piston in cylinder No. 1 is on the compression stroke, the piston in cylinder No. 4 is on the exhaust stroke. When the coil fires cylinder No. 1, most of the spark voltage goes to that cylinder because the pressure - and therefore the resistance - is high in that cylinder (the higher the resistance, the higher the voltage needed to jump the gap from the spark plug's center electrode to ground). Conversely, the piston in cylinder No. 4, which is on the exhaust stroke, produces no pressure, and therefore no resistance, so little voltage is needed to jump the gap from the spark plug's center electrode to ground. The ignition coils for cylinders 2 and 5 and 3 and 6 work the same way. This design is known as a "waste spark" ignition.

3.6L V6 models

5   The 3.6L engine uses an individual coil for each cylinder. The coils are mounted along the center of the camshaft cover and include the coils, driver module and spark plug boots. The Powertrain Control Module commands the coil driver module on or off based on signals from engine speed, MAF sensor signal, and position information from the crankshaft position and camshaft position sensors. The coil modules can be moved from cylinder to cylinder to aid in misfire diagnosis.

All models

6   The CKP, CMP (V6 models) and knock sensor(s) are information sensors used by the PCM to control ignition timing and other engine operating parameters. The PCM also uses a number of other information sensors to make decisions regarding the correct ignition timing. These other sensors include the Throttle Position (TP) sensor, the Engine Coolant Temperature (ECT) sensor, the Mass Air Flow (MAF) sensor, the Intake Air Temperature (IAT) sensor, the Vehicle Speed Sensor (VSS) and the transmission gear position sensor or Transmission Range (TR) switch. For more information on the CKP, CMP, knock and these other sensors, refer to Chapter 6.

---

6   Ignition system - check

Warning: *Because of the very high voltage generated by the ignition system (as much as 40,000 volts), use extreme care when you're servicing ignition components such as the ignition coil pack and spark plugs.*

Note: *The ignition system components are difficult to diagnose. In the event of ignition system failure, if the checks do not clearly indicate the source of the ignition system problem, have the vehicle tested by a dealer service department or other qualified auto repair facility.*

Note: *For the following test, you'll need a calibrated spark tester (available at auto parts stores). If you're going to check the ignition system on a four-cylinder engine, you'll also need three short spark plug wires to connect the (removed) coil pack to the other three spark plugs.*

1   If a malfunction occurs and the vehicle won't start, do not immediately assume that the ignition system is causing the problem. First, check the following items:

a) Make sure the battery cable clamps, where they connect to the battery, are clean and tight.

b) Test the condition of the battery (see Section 3). If it does not pass all the tests, replace it with a new battery.

c) Check the wiring and connections for the ignition control module and for the ignition coil pack. On V6 models, inspect the spark plug wires; make sure that they're in good shape and tightly connected to the coil pack and to the spark plugs.

d) Check the related fuses inside the fuse box (see Chapter 9). If they're burned, determine the cause and repair the circuit.

2   If the engine turns over but won't start, verify that there is sufficient secondary ignition voltage to fire the spark plug as follows.

3   If you're testing for spark on a 2004 through 2006 four-cylinder engine, remove the ignition coil pack (see Section 8), then remove the spark plug boots from the coil pack. Connect the calibrated spark tester to the spark

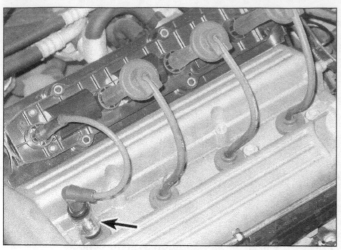

6.3 Here's the setup used for checking to see if the ignition coil is sending power to the spark plug on a 2004 through 2006 four-cylinder engine. If the coil is delivering power to the plug, the tester will flash

7.2 To remove the ignition control module, disconnect the electrical connector and remove the mounting screws

plug wire for the No. 1 cylinder and connect the other three spark plug wires between the coil high-tension terminals and the other three spark plugs (see illustration).

4   On 2007 and later four-cylinder models, remove the ignition coil for the No. 1 cylinder. On V6 models, disconnect the spark plug wire from the spark plug for the No 1 cylinder. Connect the spark tester between the coil (four-cylinder models) or spark plug wire (V6 models) and the spark plug.

5   Crank the engine while watching the tester. If the tester flashes, sufficient voltage is reaching the spark plug to fire it.

Caution: *Do NOT crank the engine or allow it to run for more than five seconds; running the engine for more than five seconds might set a Diagnostic Trouble Code (DTC) for a cylinder misfire.*

6   Repeat this test for each cylinder.

7   Proceed on this basis until you have verified that there's a good spark from each coil terminal. If there is, then you have verified that the coil(s) are functioning correctly.

8   If all the coils are firing correctly but the engine has a misfire, then one or more of the plugs might be fouled. Remove and check the spark plugs or install new ones (see Chapter 1). Also inspect the boots carefully for corrosion (high resistance) or deterioration of the insulation (low resistance). If any of the boots look damaged or deteriorated, replace them as a set.

9   No further testing of the ignition system is possible without special tools. If the problem persists, have the ignition system tested by a dealer service department.

7   Ignition control module (four-cylinder models) - replacement

Note: *This procedure does NOT apply to 2007 and later four-cylinder models, which do not have a removable ignition control module.*

1   Make sure that the ignition key is turned to OFF.

7.4 To remove the ignition control module, pull it straight up; to remove the "interconnect," simply disconnect it from the module and plug it into the new module

2   Disconnect the electrical connector from the ignition control module (see illustration).

3   Remove the ignition control module mounting screws.

4   Remove the ignition control module and the "interconnect" (see illustration).

5   If you're replacing the ignition control module, disconnect the "interconnect" from the module and plug it into the new module. (The interconnect is an adapter plug that connects the terminals on the ignition control module to the terminals on the coil pack assembly.) You'll have to swap it to the new module if you're replacing the old module. Either end of the interconnect can be plugged into the ignition control module or the ignition coil pack. But pay attention to how the plug is oriented in relation to the terminals because it only goes in one way. One side of the interconnect - and one side of the terminals on the ignition control module and on the ignition coil pack - has rounded corners and the other side has square corners. The interconnect is equipped with a weather-resistant grommet. Make sure that this grommet is in good shape. If it's cracked, torn or deteriorated, replace it.

6   Installation is the reverse of removal. Tighten the ignition control module mounting screws to the torque listed in this Chapter's Specifications.

8   Ignition coil pack - removal and installation

Four-cylinder engines
2006 and earlier models

1   Lift the battery box cover up and off of the ball studs, if equipped then disconnect the cable from the negative battery terminal (see Section 1).

2   Disconnect the electrical connector from the ignition control module (see illustration 7.2).

8.3 To detach the ignition coil pack assembly from the valve cover on a 2004 through 2006 four-cylinder model, remove these four bolts (the ignition control module is already removed in this photo, but it's not necessary to do so unless you're planning to replace the coil pack)

8.4 To remove the ignition coil pack assembly from the valve cover on a 2004 through 2006 four-cylinder model, grasp it firmly and pull straight up; the boots should come off with the coil pack (if any of them stay with the spark plugs, simply pull them off the plugs)

It's not necessary to remove the ignition control module from the ignition coil pack in order to remove the coil pack assembly, which is something you must do in order to remove the valve cover or to service the cylinder head components. However, if you're going to replace the ignition coil pack, you'll have to remove the ignition control module in order to remove the cover from the ignition coil pack.

3   Remove the four ignition coil pack mounting bolts (see illustration).

4   Pull the ignition coil pack straight up, detaching the spark plug boots (see illustration).

5   If you're replacing the ignition coil pack,

remove the cover (see illustration) and install it on the new coil pack.

6   If you're replacing the ignition coil pack, remove the four boots from the coil pack and inspect them for cracks, tears and deterioration. If any of the boots are damaged, replace them.

7   Before installing the boots on the ignition coil pack, coat the interior of each boot with silicone dielectric compound.

8   Installation is otherwise the reverse of removal. Tighten the ignition coil pack mounting bolts to the torque listed in this Chapter's Specifications.

2007 and later models

Note: *There are four separate, coil-over-plug ignition coils on these models. The following procedure applies to all four coils.*

9   Disconnect the cable from the negative battery terminal (see Section 1).

10   Disconnect the electrical connector from the ignition coil (see illustration).

11   Remove the ignition coil mounting bolt.

12   To disengage the ignition coil from the spark plug, grasp it firmly and pull it straight up.

13   Installation is the reverse of removal.

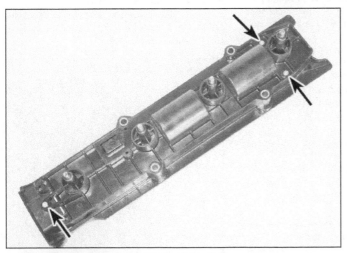

8.5 To separate the cover from ignition coil pack, remove these three screws (2004 through 2006 four-cylinder models)

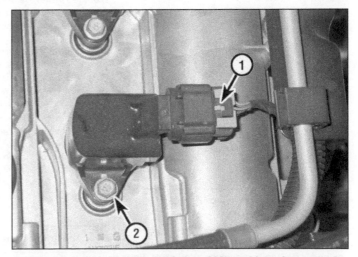

8.10 To remove an ignition coil from a 2007 and later four-cylinder model, disconnect the electrical connector (1), remove the coil mounting bolt (2), then grasp the coil firmly and pull it straight up

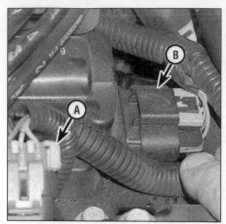

8.14 On 3.5L models, disconnect (if applicable) the electrical connector (A) next to the ignition coil, then disconnect the electrical connector from the coil pack (B). On 2007 and later models, also disconnect the MAP sensor connector (not shown)

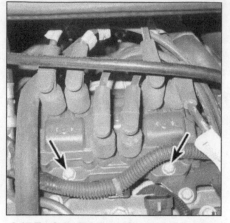

8.17 To detach the ignition coil on a 2004 through 2006 V6 model, remove the four coil mounting bolts (other two bolts, on rear side of coil, not visible in this photograph). To detach the coil on 2007 and later V6 models (not shown), remove the two coil mounting bolts and the two mounting nuts

10.2 To measure standing voltage, put your multimeter in the volts mode, connect the positive probe of the meter to the positive battery terminal and the negative probe to the negative terminal and note the reading, which should be at least 12.6 volts

3.5L and 3.9L V6 models

14   On 2004 through 2006 3.5L models, disconnect the electrical connector that overlaps the rear edge of the coil pack (see illustration). On 2007 and later models, disconnect the Manifold Absolute Pressure (MAP) sensor electrical connector.

15   Disconnect the electrical connector from the ignition coil pack.

16   Disconnect the spark plug wires from the ignition coil (see Chapter 1).

17   Remove the ignition coil pack mounting bolts (see illustration) and remove the ignition coil pack assembly.

18   Before installing the boots on the ignition coil pack, coat the interior of each boot with silicone dielectric compound.

19   Installation is otherwise the reverse of removal.

3.6L V6 engine

20   Disconnect the cable from the negative battery terminal (see Section 1).

21   Remove the engine trim cover.

22   Disconnect the coil driver module electrical connector. On the odd cylinders, it may be necessary to disconnect the evaporative emissions purge line and solenoid.

23   Unbolt the coil driver module(s), then twist and remove the module(s).

24   Installation is the reverse of removal.

---

9   Charging system - general information and precautions

1   The charging system supplies electrical power for the ignition system, the lights, the radio, the electronic control systems and all other electrical components on the car. The charging system consists of the battery, the alternator (with an integral voltage regulator),

the Powertrain Control Module (PCM), the charge indicator lamp on the instrument panel cluster, a fusible link (located inline between the starter solenoid terminal and the alternator) and the wiring between all the components.

2   The alternator generates alternating current (AC), which is rectified to direct current (DC) to charge the battery and supply power to other electrical systems. The alternator is driven by a drivebelt at the front of the engine (right side of the vehicle). The alternator is located on the front side of four-cylinder engines and on the rear side of V6 engines. The voltage regulator limits the alternator charging voltage by regulating the current supplied to the alternator field circuit. The regulator is a solid-state electronic assembly mounted inside the alternator. The regulator is not separately replaceable on these vehicles. If it's defective, you must replace the alternator.

3   Inspect the alternator drivebelt, battery and all charging system wires and connections at the intervals listed in Chapter 1.

4   Be very careful when making any circuit connections and note the following:

a) *Never start the engine with a battery charger connected.*

b) *Never disconnect a battery cable with the engine running.*

c) *Always disconnect both battery cables before using a battery charger: negative cable first, positive cable last.*

---

10   Charging system - check

1   If the charging system malfunctions, don't immediately assume that the alternator is causing the problem. First check the following items:

a) *Ensure that the battery cable connections at the battery are clean and tight.*

b) *If the battery is not a maintenance-free type, check the electrolyte level and specific gravity. If the electrolyte level is low, add clean, mineral-free tap water. If the specific gravity is low, charge the battery.*

c) *Check the alternator wiring and connections.*

d) *Check the drivebelt condition and tension (see Chapter 1).*

e) *Check the alternator mounting bolts for looseness.*

f) *Run the engine and check the alternator for abnormal noise.*

2   Use a voltmeter to check the battery voltage with the engine off. It should be at least 12.6 volts (see illustration).

3   Start the engine and check the battery voltage again. It should now be approximately 13.5 to 14.5 volts (see illustration)

10.3 To measure charging voltage, turn on the engine and note the reading again, which should then be about 13.5 volts to 14.5 volts

11.5 To disconnect the electrical connector from the alternator on a four-cylinder engine, pry up the release tab and pull the connector straight out

11.6 To detach the alternator from a four-cylinder engine, remove these three mounting bolts

4   If the charging voltage reading is no greater than the standing voltage, inspect the condition of the fusible link that's located in the wire between the alternator and the starter solenoid terminals. If a fusible link is badly blown, it will be obvious that a meltdown has occurred. If a visual inspection is inconclusive, use a continuity tester or ohmmeter to determine whether there is continuity through the fusible link. If the fusible link is blown, replace the fusible link and the wire in which it's located as a single assembly (available at dealer parts departments). The correct size of the fusible link should be printed on the outside of the link. Make sure that you obtain the correct fusible link and wire for the application. After replacing the fusible link, check the charging voltage again.

5   If the voltage reading is more or less than the specified charging voltage, the voltage regulator is defective. Replace the alternator (the voltage regulator cannot be replaced separately).

6   The charging system (battery) light on the instrument cluster lights up when the ignition key is turned to ON, but it should go out when the engine starts.

7   If the charging system light stays on after the engine has been started, there is a problem with the charging system. Before replacing the alternator, check the battery condition, alternator belt tension and electrical cable connections.

8   If replacing the alternator doesn't restore voltage to the specified range, have the charging system tested by a dealer service department or other qualified repair shop.

11   Alternator - removal and installation

Four-cylinder models

1   Lift the battery box cover up and off of the ball studs, if equipped then disconnect the cable from the negative battery terminal (see Section 1).

2   Remove the air intake duct (see Air filter housing - removal and installation in Chapter 4).

3   Unbolt the dipstick tube and set it aside.

4   Remove the accessory drivebelt (see Chapter 1).

5   Disconnect the electrical connectors from the alternator (see illustration).

6   Remove the alternator mounting bolts (see illustration).

7   Installation is the reverse of removal. Be sure to tighten the alternator mounting bolts securely.

V6 models

8   Disconnect the cable from the negative battery terminal (see Section 1).

9   Remove the accessory drivebelt (see Chapter 1).

10   Disconnect the electrical connectors from the alternator (see illustration).

11   Remove the front alternator mounting bolt. On 3.6L engines, remove the idler pulley.

12   Remove the alternator rear mounting bolts (see illustration). On 3.9L models, the right bolt (the one nearest the pulley) is a stud bolt. A clip for the power steering line is secured to this stud bolt by a nut. On these models you'll have to remove the nut and detach the power steering line clip from the stud before you can remove the stud.

13   Remove the alternator.

14   Installation is the reverse of removal. Be sure to tighten the alternator mounting bolts securely.

11.10 To remove the alternator from a V6 engine, disconnect the electrical connector (A) and the battery cable (B) from the alternator, remove the front mounting bolt (C) . . .

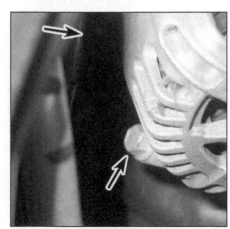

11.12 . . . then remove the two rear mounting bolts (one bolt not visible)

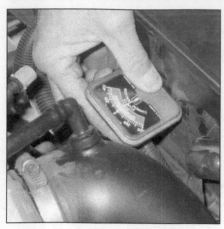

13.3 Use an inductive-type ammeter to measure the current draw

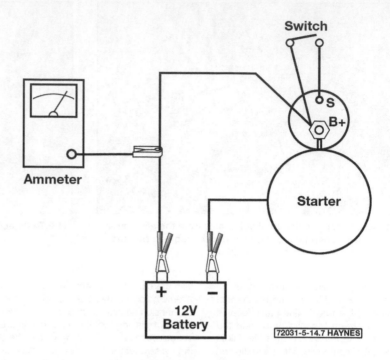

13.4 Starter motor bench testing details

12 Starting system - general information and precautions

1   The starting system consists of the battery, the ignition switch, the Transmission Range (TR) switch, the starter motor solenoid, the starter motor and the wires that connect these components. The solenoid is located on top of and is an integral part of the starter motor. The starter is located on the front left side of the engine block on four-cylinder models and on the left rear side of the block on V6 models.

2   The starter motor can be operated only when the shift lever is in PARK or NEUTRAL. When the ignition key is turned to the START position, it closes the starter control circuit, which sends battery voltage to the TR switch. If the shift lever is in PARK or NEUTRAL, battery voltage is sent to the starter solenoid terminal, which energizes the solenoid, which moves a lever that engages the starter pinion gear with the flywheel ring gear to crank the engine.

3   Always observe the following precautions when working on the starting system:

a)   *Excessive cranking of the starter motor can overheat it and cause serious damage. Never operate the starter motor for more than 15 seconds at a time without pausing for at least two minutes to allow it to cool.*

b)   *The starter is connected directly to the battery and could arc or cause a fire if mishandled, overloaded or short-circuited.*

c)   *Always detach the cable from the negative battery terminal before working on the starting system.*

13 Starter motor and circuit - check

1   If a malfunction occurs in the starting circuit, do not immediately assume that the starter is causing the problem. First, check the following items:

a)   *Make sure that the battery cable clamps are clean and tight where they connect to the battery.*

b)   *Check the condition of the battery cables (see Section 4). Replace any defective battery cables with new parts.*

c)   *Test the condition of the battery (see Section 3). If it does not pass all the tests, replace it with a new battery.*

d)   *Check the starter solenoid wiring and connections. Refer to the wiring diagrams at the end of Chapter 13.*

e)   *Check the starter mounting bolts for tightness.*

f)   *Make sure that the shift lever is in PARK or NEUTRAL (automatic transaxle) or the clutch pedal is pressed (manual transaxle).*

g)   *Check the adjustment of the Transmission Range (TR) switch (see Chapter 6).*

2   If the starter motor does not operate when the ignition switch is turned to the START position, check for battery voltage to the solenoid. Connect a test light or voltmeter to the starter solenoid switched terminal (the small wire) while an assistant turns the ignition switch to the START position. If voltage is not available, check the starting system circuit (see the wiring diagrams at the end of Chapter 12). If voltage is available but the starter motor does not operate, remove the starter from the engine compartment (see Section 14) and bench test the starter (see Step 4).

3   If the starter turns over slowly, check the starter cranking voltage and the current draw from the battery. This test must be performed with the starter assembly on the engine. Crank the engine over (for 10 seconds or less) and observe the battery voltage. It should not drop below 8.0 volts on manual transaxle models or 8.5 volts on automatic transaxle models. Also, observe the current draw using an inductive type ammeter (see illustration). It should not exceed 400 amps or drop below 250 amps.

Caution: *The battery cables might overheat because of the large amount of current being drawn from the battery. Discontinue the testing until the starting system has cooled down. If the starter motor cranking amp values are not within the correct range, replace it with a new unit. There are several conditions that may affect the starter cranking potential. The battery must be in good condition and the battery cold-cranking rating must not be underrated for the particular application. Be sure to check the battery specifications carefully. The battery terminals and cables must be clean and not corroded. Also, in cases of extreme cold temperatures, make sure the battery and/ or engine block is warmed before performing the tests.*

4   If the starter is receiving voltage but does not activate, remove and check the starter/ solenoid assembly on the bench (see illustration). Most likely the solenoid is defective. In some rare cases, the engine may be seized so be sure to try and rotate the crankshaft pulley (see Chapter 2A) before proceeding. With the starter/solenoid assembly mounted in a vise on the bench, install one jumper cable from the negative battery terminal to the body of the

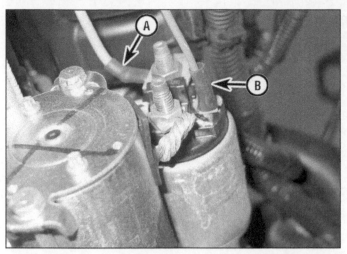

14.4 Remove the nut and disconnect the battery cable (A) from the B+ terminal, then disconnect the starter control cable (B) from the starter solenoid terminal (four-cylinder models)

14.5 To detach the starter motor from a four-cylinder engine, use a socket and extension to remove the upper mounting bolt (not shown), then remove the lower mounting bolt

starter. Install the other jumper cable from the positive battery terminal to the B+ terminal on the starter. Install a starter switch and apply battery voltage to the solenoid S terminal (for 10 seconds or less) and see if the solenoid plunger, shift lever and overrunning clutch extends and rotates the pinion drive. If the pinion drive extends but does not rotate, the solenoid is operating but the starter motor is defective. If there is no movement but the solenoid clicks, the solenoid and/or the starter motor is defective. If the solenoid plunger extends and rotates the pinion drive, the starter/solenoid assembly is working properly.

14 Starter motor - removal and installation

Four-cylinder models

1   Lift the battery box cover up and off of the ball studs, if equipped then disconnect the cable from the negative battery terminal (see Section 1).
2   Raise the front of the vehicle and place it securely on jackstands.
3   On four-cylinder models equipped with an electric coolant pump, disconnect the electrical connector and harness to the pump then remove the pump-to-oil pan mounting bolts and move the pump out of the way without disconnecting the coolant hoses.
4   Disconnect the battery cable (the larger

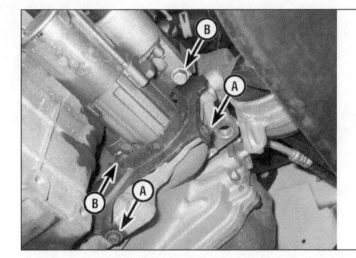

14.9 Starter motor mounting details (V6 models)

A Driveplate access cover bolts
B Starter mounting bolts

cable) and the starter control cable (the smaller cable) from the starter motor solenoid terminals (see illustration).
5   Remove the starter motor mounting bolts (see illustration) and remove the starter motor.
6   Installation is the reverse of removal. Be sure to tighten the starter motor mounting bolts to the torque listed in this Chapter's Specifications.

V6 models

7   Disconnect the cable from the negative battery terminal (see Section 1).

8   Disconnect the battery cable (the larger cable) and the starter control cable (the smaller cable) from the starter motor solenoid terminals (see illustration 14.4).
9   Remove the driveplate/torque converter access cover (see illustration).
10   Remove the starter motor mounting bolts and remove the starter motor.
11   Installation is the reverse of removal. Be sure to tighten the starter motor mounting bolts to the torque listed in this Chapter's Specifications.

Notes

Chapter 6
Emissions and engine control systems

Contents

Specifications

Torque specifications   Ft-lbs (unless otherwise indicated)

Note: *One foot-pound (ft-lb) of torque is equivalent to 12 inch-pounds (in-lbs) of torque. Torque values below approximately 15 ft-lbs are expressed in inch-pounds, since most foot-pound torque wrenches are not accurate at these smaller values.*

Accelerator pedal position nuts	89 in-lbs
Camshaft actuator solenoid valve or magnet screws	89 in-lbs
Engine coolant temperature sensor	
Four-cylinder and 3.5L and 3.9L V6 models	15
3.6L V6 models	18
Intake manifold tuning valve solenoid screws	89 in-lbs
Knock sensor retaining bolt	18
Oxygen sensors	
Four-cylinder models	
Upstream sensor	22
Downstream sensor	30
V6 models	30
Transmission Range (TR) switch	
TR switch mounting bolts	180 in-lbs
TR switch lever retaining nut	26

1   General information

1   The emission control systems and components are an integral part of the engine management system, which is called the Sequential Fuel Injection (SFI) system (see Chapter 4 for more information on the SFI system). The SFI system also includes all the government-mandated diagnostic features of the second generation of on-board diagnostics, which is known as On-Board Diagnostics II (OBD-II).

2   At the center of the SFI and OBD-II systems is the on-board computer, which is known as the Powertrain Control Module (PCM). Using a variety of information sensors, the PCM monitors all of the important engine operating parameters (temperature, speed, load, etc.). It also uses an array of output actuators (such as the ignition coils, fuel injectors and various solenoids and relays) to respond to and alter these parameters as necessary to maintain optimal performance,

economy and emissions. The principal emission control systems used on the vehicles covered in this manual include the:

- *Catalytic converters*
- *Evaporative Emission Control (EVAP) system*
- *Exhaust Gas Recirculation (EGR) system*
- *Intake Manifold Tuning (IMT) Valve system (3.9L V6 models)*
- *Positive Crankcase Ventilation (PCV) system*

3   The Sections in this Chapter include

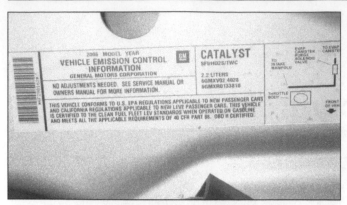

1.7 The Vehicle Emissions Control Information (VECI) label, which is located on the radiator crossmember, specifies critical tune-up and/or adjustment procedures, if applicable (there aren't any on this vehicle), and provides a vacuum hose diagram that shows how vacuum-controlled emission components are connected

2.1 Simple code readers are an economical way to extract trouble codes when the CHECK ENGINE light comes on

general descriptions and component replacement procedures for most of the information sensors and output actuators, as well as the important components that are part of the systems listed above. Refer to Chapter 4 for more information on the air intake, fuel and exhaust systems, and to Chapter 5 for information on the ignition system. Refer to Chapter 1 for any scheduled maintenance for emission-related systems and components.

4   The procedures in this Chapter are intended to be practical, affordable and within the capabilities of the home mechanic. The diagnosis of most engine and emission control functions and driveability problems requires specialized tools, equipment and training. When servicing emission devices or systems becomes too difficult or requires special test equipment, consult a dealer service department.

5   Although engine and emission control systems are very sophisticated on late-model vehicles, you can do most of the regular maintenance and some servicing at home with common tune-up and hand tools and relatively inexpensive meters. Because of the Federally mandated warranty that covers

the emission control system, check with a dealer about warranty coverage before working on any emission-related systems. After the warranty has expired, you may wish to perform some of the component replacement procedures in this Chapter to save money. Remember that the most frequent cause of emission and driveability problems is a loose electrical connector or a broken wire or vacuum hose, so always check the electrical connections, the electrical wiring and the vacuum hoses first.

6   Pay close attention to any special precautions given in this Chapter. Remember that illustrations of various systems might not exactly match the system installed on the vehicle on which you're working because of changes made by the manufacturer during production or from year to year.

7   A Vehicle Emission Control Information (VECI) label (see illustration) is located in the engine compartment. This label contains emission-control and engine tune-up specifications and adjustment information. It also includes a vacuum hose routing diagram for emission-control components. When servicing the engine or emission systems, always check the VECI label in your vehicle. If any

information in this manual contradicts what you read on the VECI label on your vehicle, always defer to the information on the VECI label.

2   On-Board Diagnostic (OBD) system and Diagnostic Trouble Codes (DTCs)

Scan tool information

1   Hand-held scanners are handy for analyzing the engine management systems used on late-model vehicles. Because extracting the Diagnostic Trouble Codes (DTCs) from an engine management system is now the first step in troubleshooting many computer-controlled systems and components, even the most basic generic code readers are capable of accessing a computer's DTCs (see illustration). More powerful scan tools can also perform many of the diagnostics once associated with expensive factory scan tools. If you're planning to obtain a generic scan tool for your vehicle, make sure that it's compatible with OBD-II systems. If you don't plan to purchase a code reader or scan tool and don't have access to one, you can have the codes extracted by a dealer service department or by an independent repair shop.

2   With the advent of the Federally mandated emission control system known as On-Board Diagnostics-II (OBD-II), specially designed scanners were developed. Several tool manufacturers have released OBD-II scan tools for the home mechanic (see illustration).

Note: *An aftermarket generic scanner should work with any model covered by this manual. Before purchasing a generic scan tool, verify that it will work properly with the OBD-II system you want to scan. If necessary, of course, you can always have the codes extracted by a dealer service department or an independent repair shop with a professional scan tool. Some auto parts stores even provide this service for free.*

2.2 Hand-held scan tools can extract trouble codes and also perform diagnostics

Information Sensors

Accelerator Pedal Position (APP) sensor - as you press the accelerator pedal, the APP sensor alters its voltage signal to the PCM in proportion to the angle of the pedal, and the PCM commands a motor inside the throttle body to open or close the throttle plate accordingly

Camshaft Position (CMP) sensor - produces a signal that the PCM uses to identify the number 1 cylinder and to time the firing sequence of the fuel injectors

Crankshaft Position (CKP) sensor - produces a signal that the PCM uses to calculate engine speed and crankshaft position, which enables it to synchronize ignition timing with fuel injector timing, and to detect misfires

Engine Coolant Temperature (ECT) sensor - a thermistor (temperature-sensitive variable resistor) that sends a voltage signal to the PCM, which uses this data to determine the temperature of the engine coolant

Fuel tank pressure sensor - measures the fuel tank pressure and controls fuel tank pressure by signaling the EVAP system to purge the fuel tank vapors when the pressure becomes excessive

Intake Air Temperature (IAT) sensor - monitors the temperature of the air entering the engine and sends a signal to the PCM to determine injector pulse-width (the duration of each injector's on-time) and to adjust spark timing (to prevent spark knock)

Knock sensor - a piezoelectric crystal that oscillates in proportion to engine vibration which produces a voltage output that is monitored by the PCM. This retards the ignition timing when the oscillation exceeds a certain threshold

Manifold Absolute Pressure (MAP) sensor - monitors the pressure or vacuum inside the intake manifold. The PCM uses this data to determine engine load so that it can alter the ignition advance and fuel enrichment

Mass Air Flow (MAF) sensor - measures the amount of intake air drawn into the engine. It uses a hot-wire sensing element to measure the amount of air entering the engine

Oxygen sensors - generates a small variable voltage signal in proportion to the difference between the oxygen content in the exhaust stream and the oxygen content in the ambient air. The PCM uses this information to maintain the proper air/fuel ratio. A second oxygen sensor monitors the efficiency of the catalytic converter

Throttle Position (TP) sensor - a potentiometer that generates a voltage signal that varies in relation to the opening angle of the throttle plate inside the throttle body. Works with the PCM and other sensors to calculate injector pulse width (the duration of each injector's on-time)

Photos courtesy of Wells Manufacturing, except APP and MAF sensors.

OBD-II system general description

3   All vehicles covered by this manual are equipped with the OBD-II system. This system consists of the on-board computer, known as the Powertrain Control Module (PCM), and information sensors that monitor various functions of the engine and send a constant stream of data to the PCM during engine operation. Unlike earlier on-board diagnostics systems, the OBD-II system doesn't just monitor everything, store Diagnostic Trouble Codes (DTCs) and illuminate a Malfunction Indicator Light (MIL) when there's a problem. It even predicts the probable failure of systems and components when their data starts to become suspicious!

4   The PCM is the "brain" of the electronically controlled OBD-II system. It receives data from a number of information sensors and switches. Based on the data that it receives from the sensors, the PCM constantly alters engine operating conditions to optimize driveability, performance, emissions and fuel economy. It does so by turning on and off and by controlling various output actuators such as relays, solenoids, valves and other devices. The PCM is mounted in the left front corner of the engine compartment.

5   If your vehicle is still under warranty, virtually every fuel, ignition and emission control component in the OBD-II system is covered by a Federally mandated extended emissions warranty that is longer than the warranty covering the rest of the vehicle. Vehicles sold in California and in some other states have even longer emissions warranties than other states. Read your owner's manual for the terms of the warranty protecting the emission-control systems on your vehicle. It isn't a good idea to "do-it-yourself" at home while the vehicle emission systems are still under warranty because owner-induced damage to the PCM, the sensors and/or the control devices might VOID this warranty. So as long as the emission systems are still warranted, take the vehicle to a dealer service department if there's a problem.

Information sensors

6   **Accelerator Pedal Position (APP) sensor** - The APP sensor, which is located at the top of the accelerator pedal, is part of the electronic throttle control system. None of the vehicles covered in this manual use a conventional accelerator cable. The APP sensor constantly monitors the angle of the accelerator pedal and sends this data to the PCM, which controls an output actuator known as the throttle motor (located in the throttle body) to open or close the throttle plate inside the throttle body to the correct position. The APP sensor consists of a couple of identical "potentiometers," variable resistors that receive a reference voltage from the PCM and return a signal voltage to the PCM that's proportional to the angle of the accelerator pedal. One of the potentiometers is redundant, and serves as a back-up in the event that the primary

potentiometer fails. The PCM compares the signal outputs from both potentiometers to assess the accuracy of the primary potentiometer's signal. The APP sensor is an integral component of the accelerator pedal assembly. If it's defective, replace the accelerator pedal assembly.

7   **Camshaft Position (CMP) sensor** - The CMP sensor, which is used on V6 models and 2007 and later four-cylinder models, is a Hall effect switching device that produces a signal that the PCM uses to monitor the position of the intake camshaft (four-cylinder models) or the exhaust camshaft on the front cylinder head (V6 models). This data enables the CMP sensor to identify the number 1 cylinder so that it can time the firing order of the spark plugs and the firing sequence of the fuel injectors. On four-cylinder engines the CMP sensor is located at the left front of the cylinder head. On V6 engines the CMP sensor is located at the front and on top of the block, between the timing belt and the intake manifold.

8   **Crankshaft Position (CKP) sensor** - The CKP sensor is a permanent magnet generator (also known as a variable reluctance sensor) that produces a variable AC voltage signal that the PCM uses to determine the position of the crankshaft. The PCM uses data from the CKP sensor (and from the CMP sensor on V6 models) to synchronize ignition timing with fuel injector timing, to control spark knock and to detect misfires. On four-cylinder models, the CKP sensor is located on the front of the block, right above the starter motor. On V6 models, the CKP sensor is located on the backside (the firewall side) of the block.

9   **Engine Coolant Temperature (ECT) sensor** - The ECT sensor is a thermistor (temperature-sensitive variable resistor) that sends a voltage signal to the PCM, which uses this data to determine the temperature of the engine coolant. The ECT sensor tells the PCM when the engine is sufficiently warmed up to go into closed loop, helps the PCM control the air/fuel mixture ratio and ignition timing, and also helps the PCM determine when to turn the Exhaust Gas Recirculation (EGR) system on and off. On four-cylinder models, the ECT sensor is located on the thermostat housing, which is located on the left rear corner of the cylinder head. On V6 models, the ECT sensor is located at the left end of the rear cylinder head.

10   **Fuel tank pressure sensor** - The fuel tank pressure sensor, which is located on top of the fuel tank (on the mounting flange of the fuel pump/fuel level sending unit module), measures the fuel tank pressure when the PCM tests the EVAP system. It's also used to control fuel tank pressure by signaling the EVAP system to purge the tank when the pressure becomes excessive.

11   **Intake Air Temperature (IAT) sensor** - The IAT sensor monitors the temperature of the air entering the engine and sends a signal to the PCM. The IAT sensor is an integral component of the Mass Air Flow (MAF) sensor, which is located on the air filter housing, near the air intake duct (four-cylinder models)

or between the air filter housing and the air intake duct (V6 models).

12   **Knock sensor** - The knock sensor is a "piezoelectric" crystal that oscillates in proportion to engine vibration. (The term piezoelectric refers to the property of certain crystals that produce a voltage when subjected to a mechanical stress.) The oscillation of the piezoelectric crystal produces a voltage output that is monitored by the PCM, which retards the ignition timing when the oscillation exceeds a certain threshold. When the engine is operating normally, the knock sensor oscillates consistently and its voltage signal is steady. When detonation occurs, engine vibration increases, and the oscillation of the knock sensor exceeds a design threshold. (Detonation is an uncontrolled explosion, after the spark occurs at the spark plug, which spontaneously combusts the remaining air/fuel mixture, resulting in a "pinging" or "knocking" sound.) If allowed to continue, the engine can be damaged. On four-cylinder models, the knock sensor is located on the block, near the starter motor. On V6 models, there are two knock sensors. They're located on the upper middle part of each side of the block, right below the exhaust manifolds.

13   **Manifold Absolute Pressure (MAP) sensor** - The MAP sensor monitors the pressure or vacuum downstream from the throttle plate, inside the intake manifold. The MAP sensor measures intake manifold pressure and vacuum on the absolute scale, i.e. from zero instead of from sea-level atmospheric pressure (14.7 psi). The MAP sensor converts the absolute pressure into a variable voltage signal that changes with the pressure. The PCM uses this data to determine engine load so that it can alter the ignition advance and fuel enrichment. On four-cylinder models, the MAP sensor is located on the intake manifold, under the throttle body (which you must remove to access it). On V6 models, the MAP sensor is located on top and at the left rear corner of the intake manifold.

14   **Mass Air Flow (MAF) sensor** - The MAF sensor is the means by which the PCM measures the amount of intake air drawn into the engine. It uses a hot-wire sensing element to measure the amount of air entering the engine. The wire is constantly maintained at a specified temperature above the ambient temperature of the incoming air by electrical current. As intake air passes through the MAF sensor and over the hot wire, it cools the wire, and the control system immediately corrects the temperature back to its constant value. The current required to maintain the constant value is used by the PCM to determine the amount of air flowing through the MAF sensor. The MAF sensor also includes an integral Intake Air Temperature (IAT) sensor. The two components cannot be serviced separately; if either sensor is defective, replace the MAF sensor. On four-cylinder models, the MAF sensor is located on the air filter housing, near the air intake duct. On V6 models, the MAF sensor is located between the air filter housing and the air intake duct. The MAF sensor is

also referred to an MAF/IAT sensor because it also incorporates the Intake Air Temperature (IAT) sensor.

15   **Oxygen sensors** - An oxygen sensor is a galvanic battery that generates a small variable voltage signal in proportion to the difference between the oxygen content in the exhaust stream and the oxygen content in the ambient air. The PCM uses the voltage signal from the upstream oxygen sensor to maintain a "stoichiometric" air/fuel ratio of 14.7:1 by constantly adjusting the "on-time" of the fuel injectors. On four-cylinder models, there are two oxygen sensors: the upstream sensor is located on the exhaust manifold, right above the flange, and the downstream sensor is located behind the catalyst. On V6 models, there are four oxygen sensors: one upstream sensor in each exhaust manifold, right above the flanges, and a downstream sensor below each upstream catalyst.

16   **Throttle Position (TP) sensor** - The TP sensor is a potentiometer that receives a constant voltage input from the PCM and sends back a voltage signal that varies in relation to the opening angle of the throttle plate inside the throttle body. This voltage signal tells the PCM when the throttle is closed, half-open, wide open or anywhere in between. The PCM uses this data, along with information from other sensors, to calculate injector "pulse width" (the interval of time during which an injector solenoid is energized by the PCM). The TP sensor is located on the throttle body, on the end of the throttle plate shaft. There are actually two TP sensors. One of them is redundant, and serves as a back-up in the event that the primary TP sensor fails. The PCM also compares the voltage signals from the two TP sensors in order to gauge the accuracy of the primary sensor and to predict its demise. The TP sensors are an integral part of the electronic throttle body and cannot be serviced separately.

17   **Transmission Range (TR) switch** - The TR switch is located on top of the transaxle. The TR switch functions like a conventional Park/Neutral Position (PNP) switch: it prevents the engine from starting in any gear other than Park or Neutral, and it closes the circuit for the back-up lights when the shift lever is moved to Reverse. The PCM also sends a voltage signal to the transmission range switch, which uses a series of step-down resistors that act as a voltage divider. The PCM monitors the voltage output signal from the switch, which corresponds to the position of the manual lever. Thus the PCM is able to determine the gear selected and is able to determine the correct pressure for the electronic pressure control system of the transaxle.

18   **Transmission speed sensors** - The Input Shaft Speed (ISS) sensor is located inside the transaxle, so you cannot replace it at home. The Output Speed Shaft (OSS) sensor is a magnetic pick-up coil located on the right side of the automatic transaxle, near the inner CV joint. The OSS sensor provides the PCM with information about the rotational

speed of the output shaft in the transmission. The PCM uses this information to control the torque converter and to calculate speed scheduling and the correct operating pressure for the transaxle.

Powertrain Control Module (PCM)

19   The PCM, which is located at the left front corner of the engine compartment, is the "brain" of the engine management system. It monitors the data input of all the information sensors, processes this information by comparing it to the operational map (program) for the engine management system, then sends command decisions to the output actuators, which execute those commands. These decisions are made so quickly (in milliseconds) that the entire process is virtually undetectable. As a result, driveability is smooth and seamless.

Output actuators

20   **Camshaft actuator solenoid** - The PCM-controlled camshaft actuator solenoid controls the flow of oil into the camshaft actuator system. When the PCM sends a pulse-width modulated voltage signal to the camshaft actuator solenoid, oil flows through one of two passages into the cam phaser. When oil is directed through one passage the cam phaser advances the camshaft; when it's directed through the other passage the cam phaser retards the camshaft. The camshaft actuator solenoid and the cam phaser are components of the camshaft actuator system, which lowers emissions, increases fuel economy and improves engine idle stability. For more information about the camshaft actuator system, see Section 20.

21   **EVAP canister purge solenoid** - The EVAP canister purge solenoid (or purge valve) controls the flow of fuel vapors (unburned hydrocarbons) from the EVAP canister to the intake manifold. The EVAP canister purge solenoid is normally closed. But when ordered to do so by the PCM, it allows intake manifold vacuum to draw the fuel vapors that are stored in the EVAP canister into the intake manifold, where they're mixed with intake air, then burned along with the normal air/fuel mixture. On four-cylinder models, the purge valve is located at the left end of the valve cover. On V6 models, the purge valve is located at the left end of the intake manifold, right behind the throttle body.

22   **EVAP canister vent solenoid** - The EVAP canister vent solenoid is normally open, to allow outside air to flow through the vent, through the EVAP canister and into the fuel tank, which maintains atmospheric pressure inside the fuel tank. But when energized by the PCM, the vent solenoid closes and seals off the EVAP system for inspection and maintenance tests and for OBD-II leak and pressure tests. The EVAP canister vent solenoid is located on top of the EVAP canister (which is located under the vehicle, at the right front corner of the fuel tank).

23   **Exhaust Gas Recirculation (EGR) valve** - When the engine is put under a load (hard acceleration, passing, going up a steep hill, pulling a trailer, etc.), combustion chamber temperature increases. When combustion chamber temperature exceeds 2500-degrees F (1380-degrees C), excessive amounts of oxides of nitrogen (NOx) are produced. NOx is a precursor of photochemical smog. When combined with hydrocarbons (HC), other "reactive organic compounds" (ROCs) and sunlight, it forms ozone, nitrogen dioxide and nitrogen nitrate and other nasty stuff. The PCM-controlled EGR valve allows exhaust gases to be recirculated back to the intake manifold where they dilute the incoming air/fuel mixture, which lowers the combustion chamber temperature and decreases the amount of NOx produced during high-load conditions. The EGR valve, which is used only on 2004 through 2006 V6 models, is located at the left rear corner of the engine, behind the throttle body.

24   **Fuel injectors** - The fuel injectors, which spray a fine mist of fuel into the intake ports, where it is mixed with incoming air, are inductive coils under PCM control. The injectors are installed in the intake manifold. For more information about the injectors, see Chapter 4.

25   **Ignition coils** - The ignition coils are under the control of the Powertrain Control Module (PCM). Four-cylinder models use a coil pack assembly consisting of two ignition coils, each of which fires two cylinders. V6 models use a coil pack, which consists of three ignition coils. For more information about the ignition coils, see Chapter 5.

26   **Intake Manifold Tuning (IMT) valve solenoid** - The IMT valve solenoid is part of the IMT valve system, a variable air induction tuning system that improves performance and efficiency over a wider operating range on 3.9L V6 models. The PCM uses the IMT valve motor to alter the configuration of the intake manifold. When the IMT valve is closed, the intake manifold is divided into two smaller plenums; when the IMT valve is open the intake manifold is one large plenum. The two smaller plenums improve torque during low-speed, high-load conditions. The single larger plenum increases horsepower during high-speed, high-load conditions. For more information on the IMT valve system, refer to Section 21.

Obtaining and clearing Diagnostic Trouble Codes (DTCs)

27   All models covered by this manual are equipped with on-board diagnostics. When the PCM recognizes a malfunction in a monitored emission control system, component or circuit, it turns on the Malfunction Indicator Light (MIL) on the dash. The PCM will continue to display the MIL until the problem is fixed and the Diagnostic Trouble Code (DTC) is cleared from the PCM's memory. You'll need a scan tool to access any DTCs stored in the PCM.

28   Before outputting any DTCs stored in the

2.29 The 16-pin Data Link Connector (DLC), also referred to as the diagnostic connector, is located under the left part of the dash

PCM, thoroughly inspect ALL electrical connectors and hoses. Make sure that all electrical connections are tight, clean and free of corrosion. And make sure that all hoses are correctly connected, fit tightly and are in good condition (no cracks or tears). Also, make sure that the engine is tuned up. A poorly running engine is probably one of the biggest causes of emission-related malfunctions. Often, simply giving the engine a good tune-up will correct the problem.

Accessing the DTCs
29   On these models, all of which are equipped with On-Board Diagnostic II (OBD-II) systems, the Diagnostic Trouble Codes (DTCs) can only be accessed with a code reader or a scan tool (see illustrations 2.1 and 2.2). Simply plug the connector of the tool into the Data Link Connector (DLC) or diagnostic connector (see illustration), which is located under the lower edge of the dash, just to the right of the steering column. Then follow the instructions included with the scan tool to extract the DTCs.

30   Once you have outputted all of the stored DTCs look them up on the accompanying DTC chart.
31   After troubleshooting the source of each DTC make any necessary repairs or replace the defective component(s).

Clearing the DTCs
32   Clear the DTCs with the scan tool in accordance with the instructions provided by the scan tool's manufacturer.

Diagnostic Trouble Codes
33   The accompanying tables are a list of the Diagnostic Trouble Codes (DTCs) that can be accessed by a do-it-yourselfer working at home (there are many, many more DTCs available to dealerships with proprietary scan tools and software, but those codes cannot be accessed by a generic scan tool). If, after you have checked and repaired the connectors, wire harness and vacuum hoses (if applicable) for an emission-related system, component or circuit, the problem persists, have the vehicle checked by a dealer service department.

OBD-II trouble codes
Note: *Not all trouble codes apply to all models.*

Code	Probable cause
P0010	Camshaft Position Actuator Solenoid Control Circuit
P0011	Camshaft Position System Performance
P0013	Exhaust Camshaft Position Actuator Solenoid Valve Control Circuit
P0014	Exhaust Camshaft Position System Performance
P0016	Crankshaft Position Camshaft Position Correlation
P0017	Crankshaft Position - Exhaust Camshaft Position Correlation
P0018	Crankshaft Position - Intake Camshaft Position Correlation Bank 2
P0019	Crankshaft Position - Exhaust Camshaft Position Correlation Bank 2
P0020	Intake Camshaft Position Actuator Solenoid Control Circuit Bank 2
P0021	Intake Camshaft Position System Performance Bank 2
P0023	Exhaust Camshaft Position Actuator Solenoid Control Circuit Bank 2
P0024	Exhaust Camshaft Position System Performance Bank 2
P0030	Oxygen sensor heater control circuit (bank 1, sensor 1)
P0036	Oxygen sensor heater control circuit (bank 1, sensor 2)
P0050	Oxygen sensor heater control circuit (bank 2, sensor 1)

Code	Probable cause
P0053	Oxygen sensor heater resistance, (bank 1, sensor 1)
P0054	Oxygen sensor heater resistance, (bank 1, sensor 2)
P0056	Oxygen sensor heater control circuit (bank 2, sensor 2)
P0059	Oxygen sensor heater resistance, (bank 2, sensor 1)
P0060	Oxygen sensor heater resistance, (bank 2, sensor 2)
P0068	Throttle body air flow performance
P0101	Mass Air Flow (MAF) sensor performance
P0102	Mass Air Flow (MAF) sensor circuit, low frequency
P0106	Manifold Absolute Pressure (MAP) sensor performance
P0107	Manifold Absolute Pressure (MAP) sensor circuit, low voltage
P0108	Manifold Absolute Pressure (MAP) sensor circuit, high voltage
P0112	Intake Air Temperature (IAT) sensor circuit, low voltage
P0113	Intake Air Temperature (IAT) sensor circuit, high voltage
P0116	Engine Coolant Temperature (ECT) sensor performance
P0117	Engine Coolant Temperature (ECT) sensor circuit, low voltage
P0118	Engine Coolant Temperature (ECT) sensor circuit, high voltage
P0120	Throttle Position (TP) sensor 1 circuit
P0121	Throttle Position (TP) sensor 1 performance
P0122	Throttle Position (TP) sensor 1 circuit, low voltage
P0123	Throttle Position (TP) sensor 1 circuit, high voltage
P0125	Engine coolant temperature insufficient for closed loop fuel control
P0128	Engine coolant temperature below thermostat-regulated temperature
P0130	Oxygen sensor circuit performance (bank 1, sensor 1)
P0131	Oxygen sensor circuit, low voltage (bank 1, sensor 1)
P0132	Oxygen sensor circuit, high voltage (bank 1, sensor 1)
P0133	Oxygen sensor circuit, slow response (bank 1, sensor 1)
P0134	Oxygen sensor circuit, insufficient activity (bank 1, sensor 1)
P0135	Oxygen sensor heater performance (bank 1, sensor 1)
P0136	Oxygen sensor circuit (bank 1, sensor 2)
P0137	Oxygen sensor circuit, low voltage (bank 1, sensor 2)

OBD-II trouble codes (continued)

Note: *Not all trouble codes apply to all models.*

Code	Probable cause
P0138	Oxygen sensor circuit, high voltage (bank 1, sensor 2)
P0140	Oxygen sensor circuit, insufficient activity (bank 1, sensor 2)
P0141	Oxygen sensor heater performance (bank 1, sensor 2)
P0150	Oxygen sensor circuit, closed loop performance (bank 2, sensor 1)
P0151	Oxygen sensor circuit, low voltage (bank 2, sensor 1)
P0152	Oxygen sensor circuit, high voltage (bank 2, sensor 1)
P0153	Oxygen sensor circuit, slow response (bank 2, sensor 1)
P0154	Oxygen sensor circuit, insufficient activity (bank 2, sensor 1)
P0155	Oxygen sensor heater performance (bank 2, sensor 1)
P0157	Oxygen sensor circuit, low voltage (bank 2, sensor 2)
P0158	Oxygen sensor circuit, high voltage (bank 2, sensor 2)
P0160	Oxygen sensor circuit, insufficient activity (bank 2, sensor 2)
P0161	Oxygen sensor heater performance (bank 2, sensor 2)
P0171	Fuel trim system lean (bank 1)
P0172	Fuel trim system rich (bank 1)
P0174	Fuel trim system lean (bank 2)
P0175	Fuel trim system rich (bank 2)
P0201	Injector no. 1 circuit malfunction
P0202	Injector no. 2 circuit malfunction
P0203	Injector no. 3 circuit malfunction
P0204	Injector no. 4 circuit malfunction
P0205	Injector no. 5 circuit malfunction
P0206	Injector no. 6 circuit malfunction
P0218	Transmission fluid temperature too high
P0220	Throttle Position (TP) sensor 2 circuit
P0222	Throttle Position (TP) sensor 2 circuit, low voltage
P0223	Throttle Position (TP) sensor 2 circuit, high voltage
P0230	Fuel pump relay control circuit
P0300	Engine misfire detected

Code	Probable cause
P0301	Engine misfire detected (cylinder 1)
P0302	Engine misfire detected (cylinder 2)
P0303	Engine misfire detected (cylinder 3)
P0304	Engine misfire detected (cylinder 4)
P0315	Crankshaft Position (CKP) sensor circuit, system variation not learned
P0324	Knock Sensor Module Performance
P0325	Knock sensor (KS) module performance
P0326	Knock sensor (KS) performance
P0327	Knock sensor (KS) circuit, low input (bank 1)
P0328	Knock Sensor Circuit High Bank 1
P0330	Knock Sensor Circuit Bank 2
P0332	Knock sensor (KS) circuit, bank 2
P0333	Knock Sensor Circuit High Bank 2
P0335	Crankshaft Position (CKP) sensor circuit
P0336	Crankshaft Position (CKP) sensor performance
P0340	Camshaft Position (CMP) sensor circuit
P0341	Camshaft Position (CMP) sensor performance
P0345	Intake Camshaft Position Sensor Circuit Bank 2
P0346	Intake Camshaft Position Sensor Performance Bank 2
P0351	Ignition coil 1 control circuit
P0352	Ignition coil 2 control circuit
P0353	Ignition coil 3 control circuit
P0354	Ignition Coil 4 Control Circuit
P0355	Ignition Coil 5 Control Circuit
P0356	Ignition Coil 6 Control Circuit
P0365	Exhaust Camshaft Position Sensor Circuit
P0366	Exhaust Camshaft Position Sensor Performance
P0390	Exhaust Camshaft Position Sensor Circuit Bank 2
P0391	Exhaust Camshaft Position Sensor Performance Bank 2
P0401	Exhaust Gas Recirculation (EGR) sensor, insufficient flow
P0403	Exhaust Gas Recirculation (EGR) sensor solenoid control circuit

OBD-II trouble codes (continued)

Note: *Not all trouble codes apply to all models.*

Code	Probable cause
P0404	Exhaust Gas Recirculation (EGR) sensor, open position performance
P0405	Exhaust Gas Recirculation (EGR) sensor circuit, low voltage
P0406	Exhaust Gas Recirculation (EGR) sensor circuit, high voltage
P0411	Secondary Air Injection System Incorrect Air Flow Detected
P0412	Secondary Air Injection Solenoid Control Circuit
P0418	Secondary Air Injection Pump Control Circuit
P0420	Catalyst system, low efficiency (bank 1)
P0430	Catalyst system, low efficiency (bank 2)
P0442	Evaporative Emission (EVAP) system, small leak detected
P0443	Evaporative Emission (EVAP) system, purge solenoid control circuit
P0446	Evaporative Emission (EVAP) vent system performance
P0449	Evaporative Emission (EVAP) vent solenoid control circuit
P0451	Fuel Tank Pressure (FTP) sensor performance
P0452	Fuel Tank Pressure (FTP) sensor circuit, low voltage
P0453	Fuel Tank Pressure (FTP) sensor circuit, high voltage
P0454	Fuel Tank Pressure (FTP) sensor circuit intermittent
P0455	Evaporative Emission (EVAP) system, large leak detected
P0461	Fuel level sensor performance
P0462	Fuel level sensor circuit, low voltage
P0463	Fuel level sensor circuit, high voltage
P0464	Fuel level sensor circuit intermittent
P0480	Cooling fan relay 1 control circuit
P0481	Cooling fan relays 2 and 3 control circuit
P0496	EVAP system flow during non-purge
P0502	Vehicle Speed Sensor (VSS) circuit, low voltage
P0503	Vehicle Speed Sensor (VSS) circuit, intermittent
P0506	Idle speed low
P0507	Idle speed high
P0520	Engine oil pressure (EOP) sensor circuit

Code	Probable cause
P0521	Engine oil pressure (EOP) sensor circuit, performance
P0522	Engine oil pressure (EOP) sensor circuit, low voltage
P0523	Engine oil pressure (EOP) sensor circuit, high voltage
P0530	Air conditioning refrigerant pressure sensor circuit
P0532	Air conditioning refrigerant pressure sensor circuit, low voltage
P0533	Air conditioning refrigerant pressure sensor circuit, high voltage
P0556	Brake Booster Pressure Sensor Performance
P0557	Brake Booster Pressure Sensor Circuit Low Voltage
P0558	Brake Booster Pressure Sensor Circuit High Voltage
P0560	System voltage
P0562	System voltage low
P0563	System voltage high
P0572	Cruise control brake switch circuit, low voltage
P0573	Cruise control brake switch circuit, high voltage
P0575	Cruise control switch signal circuit
P0601	Powertrain or Transmission Control Module Read Only Memory (ROM)
P0602	Powertrain or Transmission Control Module not programmed
P0603	Powertrain or Transmission Control Module long-term memory reset
P0604	Powertrain or Transmission Control Module Random Access Memory
P0606	Powertrain Control Module (PCM) internal performance
P0607	Powertrain Control Module (PCM) performance
P0608	Vehicle Speed Sensor (VSS) output circuit
P0615	Starter relay control circuit
P0621	Alternator L-terminal circuit
P0622	Alternator F-terminal circuit
P062F	Powertrain Control Module (PCM) long-term memory performance
P062F	Transmission Control Module (TCM) EEPROM error
P0634	Transmission Control Module (TCM) Over temperature
P0641	5-volt reference circuit
P0645	Air conditioning clutch relay control circuit

OBD-II trouble codes (continued)

Note: *Not all trouble codes apply to all models.*

Code	Probable cause
P0646	Air conditioning clutch relay control circuit, low voltage
P0647	Air conditioning clutch relay control circuit, high voltage
P0650	Malfunction Indicator Light (MIL) control circuit
P0651	5-volt reference 2 circuit
P0658	Pressure Control/Shift Lock Solenoid Control Circuit Low Voltage
P0659	Pressure Control/Shift Lock Solenoid Control Circuit High Voltage
P0660	Intake Manifold Tuning (IMT) valve solenoid control circuit
P0667	Transmission Control Module Temperature Sensor Performance
P0668	Transmission Control Module Temperature Sensor Circuit Low Voltage
P0669	Transmission Control Module Temperature Sensor Circuit High Voltage
P0685	Engine controls ignition relay control circuit
P0689	Engine controls ignition relay feedback circuit, low voltage
P0690	Engine controls ignition relay feedback circuit, high voltage
P0700	Transmission Control Module (TCM) requested MIL illumination
P0703	Brake switch circuit 2
P0705	Transmission Range (TR) switch circuit
P0711	Transmission Fluid Temperature (TFT) sensor performance
P0712	Transmission Fluid Temperature (TFT) sensor circuit, low voltage
P0713	Transmission Fluid Temperature (TFT) sensor circuit, high voltage
P0716	Input speed sensor performance
P0717	Input Speed Sensor (ISS) circuit, low voltage
P0722	Output Speed Sensor (OSS) circuit, low voltage
P0723	Output Speed Sensor (OSS) circuit, intermittent
P0730	Incorrect gear ratio
P0731	Incorrect 1st gear ratio
P0732	Incorrect 2nd gear ratio
P0733	Incorrect 3rd gear ratio
P0734	Incorrect 4th gear ratio
P0741	Torque Converter Clutch (TCC) system, stuck off
P0742	Torque Converter Clutch (TCC) system, stuck on
P0748	Pressure Control (PC) solenoid control circuit
P0751	1-2 Shift Solenoid (SS) valve performance, no 1st or 4th gear

Code	Probable cause
P0752	1-2 Shift Solenoid (SS) valve performance, no 2nd or 3rd gear
P0753	1-2 Shift Solenoid (SS) control circuit
P0756	2-3 Shift Solenoid (SS) valve performance, no 1st or 2nd gear
P0757	2-3 Shift Solenoid (SS) valve performance, no 3rd, 4th or 5th gear
P0758	2-3 Shift Solenoid (SS) control circuit
P0815	Upshift switch circuit
P0816	Downshift switch circuit
P0826	Up and down shift switch circuit
P0842	Transmission Fluid Temperature (TFT) sensor circuit, low voltage
P0843	Transmission Fluid Temperature (TFT) sensor circuit, high voltage
P0850	Park/Neutral Position Switch Circuit
P0851	Park/Neutral Position Switch Circuit Low Voltage
P0852	Park/Neutral Position Switch Circuit High Voltage
P0961	Line Pressure Control (PC) solenoid system performance
P0973	1-2 Shift Solenoid (SS) control circuit, low voltage
P0974	1-2 Shift Solenoid (SS) control circuit, high voltage
P0976	2-3 Shift Solenoid (SS) control circuit, low voltage
P0977	2-3 Shift Solenoid (SS) control circuit, high voltage

3   Accelerator Pedal Position (APP) sensor - replacement

Note: *The APP sensor is located at the top of the accelerator pedal arm. The APP sensor and the accelerator pedal are a one-piece assembly and are replaced as a unit.*

1   Remove the knee bolster (see Chapter 11).

2   Locate the APP sensor at the top of the accelerator pedal assembly (see illustration).

3   Disconnect the electrical connector from the APP sensor.

4   Remove the three APP sensor mounting nuts and detach the APP sensor from the sensor mounting bracket.

5   Installation is the reverse of removal. Tighten the APP sensor mounting nuts to the torque listed in this Chapter's Specifications.

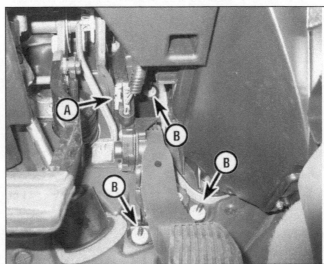

3.2 The Accelerator Pedal Position (APP) sensor is located at the top of the accelerator pedal arm. The accelerator pedal and the APP sensor are a one-piece assembly. To remove the APP sensor/ pedal assembly:

A   Disconnect the electrical connector

B   Remove the three APP sensor/ pedal assembly mounting nuts

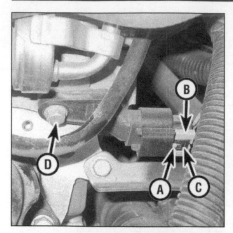

4.3 To remove the Camshaft Position (CMP) sensor from 2007 and later four-cylinder models, depress the tab (A) on the connector lock, slide the lock (B) away from the connector, depress the connector button (C) and pull the connector off, then remove the sensor mounting bolt (D)

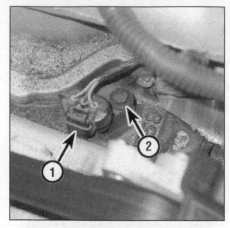

4.7 To remove the Camshaft Position (CMP) sensor, disconnect the electrical connector (1), then remove the sensor mounting bolt (2) (3.5L/3.9L V6 models)

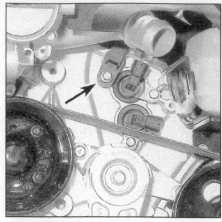

4.13 On 3.6L V6 models, the camshaft position sensors are located on the front of the cylinder heads, just below the valve cover (front bank cylinder head intake sensor shown, rear bank cylinder head sensors similar)

4 Camshaft Position (CMP) sensor - replacement

Four cylinder engines

Note: *All 2007 and later four cylinder engines use an intake camshaft position sensor. Additionally, all 2008 and later four-cylinder engines use an exhaust camshaft position sensor. The sensor(s) are located at the left (driver's) side of the cylinder head with the intake sensor located on the front (radiator side) of the engine. The procedure for both sensors is the same.*

1 Turn the ignition key to OFF.
2 Remove the engine cover and air cleaner duct.
3 Unplug the electrical connector from the CMP sensor (see illustration).
4 Remove the CMP sensor retaining bolt

5.3 To remove the CKP sensor (A) from a four-cylinder engine, disconnect the electrical connector (B) and remove the sensor mounting bolt (C)

and remove the sensor.
5 Installation is the reverse of removal. Lubricate the CMP sensor O-ring with clean engine oil prior to installing the sensor.

3.5L and 3.9L V6 engines

Note: *The CMP sensor is located at the front (drivebelt) end and on top of the block, directly behind the timing chain cover.*

6 Turn the ignition key to off.
7 Disconnect the CMP sensor connector (see illustration)
8 Remove the CMP sensor retaining bolt.
9 Gently twist the CMP sensor to free it, then remove the sensor.
10 Installation is the reverse of removal. Lubricate the CMP sensor O-ring with clean engine oil prior to installing the sensor.

3.6L V6 engine

Note: *This engine uses four CMP sensors. Each cylinder head contains a separate intake and exhaust camshaft position sensor. The sensors are located on the front (drivebelt) end of the cylinder heads.*

11 Turn the ignition key to off.
12 Remove the air cleaner.
13 Disconnect the CMP sensor connector (see illustration).
14 Remove the CMP sensor retaining bolt.
15 Gently twist the CMP sensor to free it, then remove the sensor.
16 Installation is the reverse of removal. Lubricate the CMP sensor O-ring with clean engine oil prior to installing the sensor.

5 Crankshaft Position (CKP) sensor - replacement

Note: *Whenever the crankshaft sensor is replaced, the crankshaft pulley is removed and installed, the crankshaft is replaced and/or the*

engine is removed from the vehicle, the PCM must be reprogrammed using special equipment. A crankshaft sensor variation relearn procedure must be performed using a special TECH-2 scan tool. Have the vehicle PCM reprogrammed by a dealer service department or other qualified automotive repair facility.

Four-cylinder models

Note: *The CKP sensor is located on the front of the engine block, right above the starter motor.*

1 Raise the vehicle and place it securely on jackstands.
2 Remove the starter motor (see Chapter 5).
3 Disconnect the electrical connector from the CKP sensor (see illustration).
4 Unscrew the CKP sensor mounting bolt and remove the CKP sensor.
5 Even if you're planning to reuse the old CKP sensor, be sure to remove the old O-ring (see illustration) and discard it. Always install a new O-ring when installing the CKP sensor.
6 Installation is the reverse of removal. Be sure to tighten the CKP sensor mounting bolt securely.

V6 models

Note: *The CKP sensor is located on the back (firewall side) of the engine block.*

7 Raise the vehicle and place it securely on jackstands.
8 Remove the exhaust manifold heat shield.
9 Disconnect the electrical connector from the CKP sensor.
10 Remove the CKP sensor mounting bolt (see illustration).
11 Remove and discard the old CKP sensor O-ring.
12 Installation is the reverse of removal. Be sure to use a new O-ring and tighten the CKP sensor mounting bolt securely.

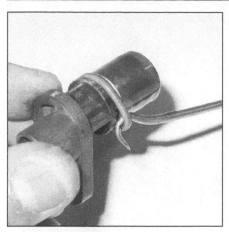

5.5 Be sure to remove the old O-ring from the CKP sensor; always install the sensor with a new O-ring (even if you're installing the old sensor)

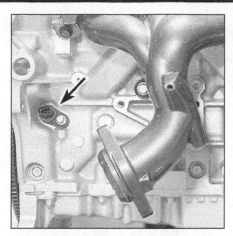

5.10 The CKP sensor is located at the rear of the engine block on the firewall side (3.6L V6 engine shown)

6.2 Disconnect the electrical connector from the ECT sensor (four-cylinder engines)

6 Engine Coolant Temperature (ECT) sensor - replacement

Warning: *Wait until the engine is completely cool before beginning this procedure.*

Four-cylinder models

Note: *2007 and earlier engines, the ECT sensor is located on the thermostat housing, which is at the left rear corner of the cylinder head. On 2008 and later engines, the ECT sensor is located at the left rear of the engine below the CMP exhaust sensor.*

1   Drain the cooling system (see Chapter 1). It's not necessary to fully drain the coolant, but it must be drained to a level that's below the level of the ECT sensor.
2   Disconnect the electrical connector from the ECT sensor (see illustration).

3   Unscrew the ECT sensor (see illustration) and remove it.
4   Wrap the threads of the ECT sensor with Teflon tape (see illustration).
5   Installation is the reverse of removal. Be sure to tighten the ECT sensor to the torque listed in this Chapter's Specifications.
6   Refill the cooling system when you're done (see Chapter 1).

V6 models

Note: *On 3.5L and 3.9L engines, the ECT sensor is located on the left end of the rear cylinder head. On 3.6L engines, the sensor is located in the front cylinder head between the number 2 and 4 exhaust ports.*

7   Drain the cooling system (see Chapter 1). It's not necessary to fully drain the coolant, but it must be drained to a level that's below the level of the ECT sensor.

8   Disconnect the electrical connector from the ECT sensor (see illustration).
9   Unscrew and remove the ECT sensor.
10   Wrap the threads of the new ECT sensor with Teflon tape (see illustration 6.4).
11   Installation is the reverse of removal. Be sure to tighten the ECT sensor to the torque listed in this Chapter's Specifications.
12   Refill the cooling system when you're done (see Chapter 1).

7 Intake Air Temperature (IAT) sensor - replacement

1   The IAT sensor is an integral component of the Mass Air Flow (MAF) sensor (see Section 10).

6.3 Use a deep socket, an extension and a ratchet to unscrew the ECT sensor from a four-cylinder engine

6.4 Before installing the ECT sensor, wrap the threads of the sensor with Teflon tape to prevent leaks

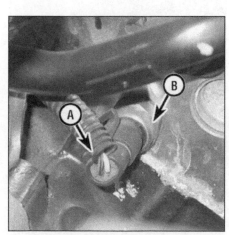

6.8 Disconnect the electrical connector (A) from the ECT sensor (B), then unscrew and remove the sensor (V6 engines)

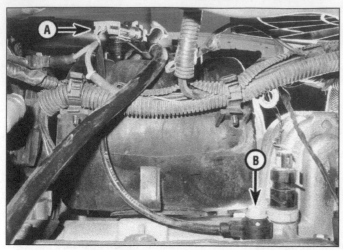

8.3 To remove the knock sensor from a four-cylinder engine, disconnect the electrical connector (A) and remove the sensor retaining bolt (B)

8.8 To detach either of the knock sensors on a V6, disconnect the electrical connector (A), then remove the retaining bolt (B) (knock sensor No. 1, on the back of the block, shown; knock sensor No. 2, on the front of the block, identical)

8   Knock sensor(s) - replacement

Four-cylinder models

Note: *The knock sensor is located on the block, near the starter motor.*
1   Raise the front of the vehicle and place it securely on jackstands.
2   Remove the starter motor (see Chapter 5).
3   Disconnect the knock sensor electrical connector (see illustration).
4   Remove the knock sensor retaining bolt.
5   Remove the knock sensor.
6   Installation is the reverse of removal. Be sure to tighten the knock sensor retaining bolt to the torque listed in this Chapter's Specifications.

V6 models

Note: *There are two knock sensors, one for each side of the block. They're located in the upper middle of each side of the block, right below the exhaust manifolds. You can access them only by raising the vehicle.*

Knock sensor No. 1

Note: *Knock sensor No. 1 is located on the rear side of the block.*
7   Raise the front of the vehicle and place it securely on jackstands.
8   Disconnect the electrical connector from the knock sensor (see illustration).
9   Remove the knock sensor retaining bolt and remove the knock sensor.
10   Installation is the reverse of removal. Be sure to tighten the knock sensor retaining bolt

to the torque listed in this Chapter's Specifications.

Knock sensor No. 2

Note: *Knock sensor No. 2 is located on the front side of the block.*
11   Raise the front of the vehicle and place it securely on jackstands.
12   Disconnect the electrical connector from the knock sensor.
13   Remove the knock sensor retaining bolt and remove the knock sensor.
14   Installation is the reverse of removal. Be sure to tighten the knock sensor retaining bolt to the torque listed in this Chapter's Specifications.

9   Manifold Absolute Pressure (MAP) sensor - replacement

Four-cylinder models

Note: *The MAP sensor is located on the intake manifold, right behind the throttle body.*
1   Remove the throttle body (see Chapter 4).
2   Disconnect the MAP sensor electrical connector (see illustration).
3   Rotate the MAP sensor clockwise until it clears the locator tab on the backside of the sensor (see illustration 9.2), then pull the sensor straight up to remove it.
4   Remove the old sealing grommet (see illustration) from the MAP sensor and discard it. Always use a new grommet when installing the MAP sensor, regardless of whether you're installing the old sensor or a new one.
5   Installation is the reverse of removal.

V6 models

Note: *The MAP sensor is located on top of and at the left rear corner of the intake manifold.*
6   Disconnect the MAP sensor electrical

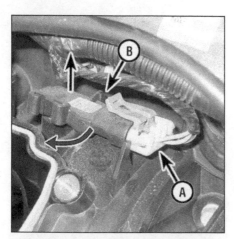

9.2 To remove the MAP sensor from the intake manifold on a four-cylinder engine, disconnect the electrical connector (A), rotate the sensor clockwise to clear the locator tab (B) on the backside of the sensor, then pull the sensor straight up

9.4 Remove and discard the old MAP sensor sealing grommet and install a new grommet before installing the MAP sensor

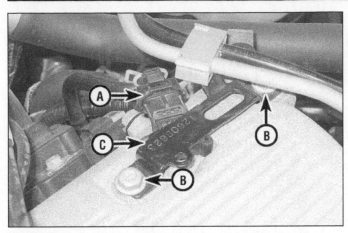

9.6 To remove the MAP sensor from the intake manifold on a V6 engine, disconnect the electrical connector (A), remove the retaining bolt (B), remove the hold-down clamp (C), then pull the sensor straight up

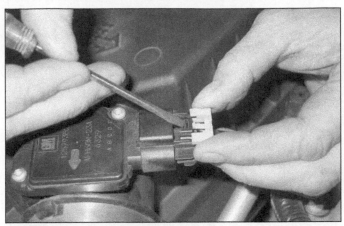

10.1a To disconnect the electrical connector from the MAF sensor on a four-cylinder model, use a small screwdriver to disengage the lock from the connector release tab . . .

connector (see illustration).

7   Remove the MAP sensor retaining bolt and the hold-down clamp, then remove the MAP sensor.

8   Remove the old sealing grommet (see illustration 9.4) from the MAP sensor and discard it. Always use a new grommet when installing the MAP sensor, regardless of whether you're installing the old sensor or a new one.

9   Installation is the reverse of removal. Be sure to tighten the MAP sensor retaining bolt securely.

10   Mass Air Flow (MAF) sensor - replacement

Four-cylinder models

Note: *The MAF sensor is located on the air filter housing.*

1   Disconnect the electrical connector from the MAF sensor (see illustrations).

2   Remove the MAF sensor mounting screws and remove the sensor from the air filter housing (see illustration 10.1b).

3   Installation is the reverse of removal.

V6 models

Note: *The MAF sensor is located between the air filter housing and the air intake duct.*

4   Disconnect the electrical connector from the MAF sensor (see illustration).

5   Loosen the hose clamp screws at each end of the MAF sensor and remove the MAF sensor from between the air filter housing and the air intake duct.

6   Installation is the reverse of removal.

11   Oxygen sensors - general information and replacement

General information

1   An oxygen sensor is a galvanic battery that produces a very small voltage out-

put in response to the amount of oxygen in the exhaust gases. This voltage signal is the "input" side of the feedback loop between the oxygen sensor and the Powertrain Control Module (PCM). Without it, the PCM would be unable to correct the injector on-time (which determines the air/fuel ratio) to maintain the "perfect" (known as stoichiometric) air/fuel ratio of 14.7:1 that the catalyst needs for optimal operation.

2   All vehicles covered by this manual have On-Board Diagnostics II (OBD-II) engine management systems, which means they have the ability to verify the accuracy of the basic feedback loop between the oxygen sensor and the PCM. They accomplish this by using an oxygen sensor ahead of the catalytic converter and another oxygen sensor behind the catalytic converter. By comparing the amount of oxygen in the post-catalyst exhaust gas to the oxygen content of the exhaust gas before it enters the catalyst, the PCM can determine the efficiency of the converter.

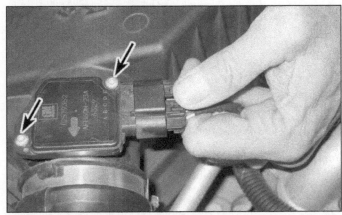

10.1b . . . then depress the release tab and pull off the connector. To detach the sensor from the air filter housing, remove the two sensor mounting screws

10.4 To remove the MAF sensor from a V6 model, disconnect the electrical connector (1), loosen the hose clamp screws (2) and remove the sensor (3) from between the air filter housing and the air intake duct

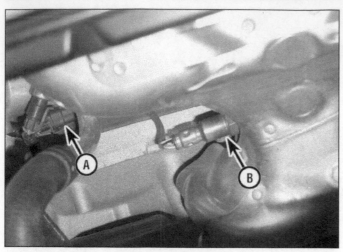

11.7 To remove the upstream oxygen sensor from a four-cylinder engine, disconnect the electrical connector (A), then unscrew the oxygen sensor (B)

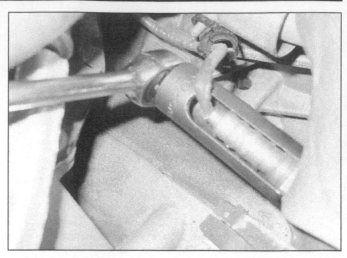

11.8 Use an oxygen sensor socket to unscrew the upstream oxygen sensor (four-cylinder models)

Four-cylinder models

3   All four-cylinder vehicles covered by this manual have two heated oxygen sensors: one upstream sensor (ahead of the catalytic converter) and a downstream oxygen sensor (after the catalyst). The upstream sensor is located on the exhaust manifold, right above the flange. The downstream sensor is located below the catalytic converter. You'll have to raise the vehicle to access this sensor.

V6 models

4   There are four heated oxygen sensors, one upstream and one downstream sensor for each cylinder head, on V6 models. The upstream sensors are located in the exhaust manifolds. The downstream sensors are located on the exhaust pipes, directly below the two small upstream catalysts.

All models

5   The upstream and downstream oxygen sensors on all models are heated to speed up the warm-up time during which the sensors are unable to produce an accurate voltage signal. The circuit for each oxygen sensor

heater is controlled by the PCM, which opens the ground side of the circuit to shut off the heater as soon as the sensor reaches its normal operating temperature.

6   Special care must be taken whenever a sensor is serviced.

a)   *Oxygen sensors have a permanently attached pigtail and an electrical connector that cannot be removed. Damaging or removing the pigtail or electrical connector will render the sensor useless.*

b)   *Keep grease, dirt and other contaminants away from the electrical connector and the louvered end of the sensor.*

c)   *Do not use cleaning solvents of any kind on an oxygen sensor.*

d)   *Oxygen sensors are extremely delicate. Do not drop a sensor, throw it around or handle it roughly.*

e)   *Make sure that the silicone boot on the sensor is installed in the correct position. Otherwise, the boot might melt and it might prevent the sensor from operating correctly.*

Replacement

Note: *Because it is installed in the exhaust manifold or exhaust pipe, both of which contract as they cool down, an oxygen sensor can be very difficult to loosen when the engine is cold. Rather than risk damage to the sensor or its mounting threads, start and run the engine for a minute or two, then shut it off. Be careful not to burn yourself during the following procedure.*

Four-cylinder models

Upstream oxygen sensor

Note: *The upstream oxygen sensor is located on the left side of the exhaust manifold, right above the flange.*

7   Disconnect the upstream oxygen sensor electrical connector (see illustration), which is located near the left rear corner of the engine.

8   Using an oxygen sensor socket (available at most auto parts stores), unscrew the upstream oxygen sensor (see illustration). If the sensor is difficult to loosen, spray some penetrant onto the sensor threads and allow it to soak in for awhile.

9   If you're going to install the old sensor, apply anti-seize compound to the threads of the sensor to facilitate future removal. If you're going to install a new oxygen sensor, it's not necessary to apply anti-seize compound to the threads. The threads on new sensors already have anti-seize compound on them.

10   Installation is otherwise the reverse of removal. Be sure to tighten the sensor to the torque listed in this Chapter's Specifications.

Downstream oxygen sensor

Note: *The downstream oxygen sensor is located on the exhaust pipe after the catalytic converter.*

11   Raise the vehicle and place it securely on jackstands.

12   Locate the downstream oxygen sensor (see illustration) then trace the sensor electrical lead to the connector and disconnect it.

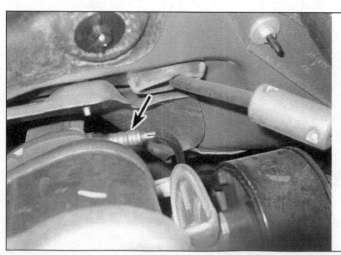

11.12 To remove the downstream oxygen sensor from a four-cylinder engine, trace the electrical lead from the sensor to the electrical connector and disconnect it, then unscrew the sensor with an oxygen sensor socket

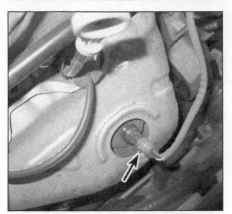

11.16a To remove an upstream oxygen sensor from a V6 engine, trace the electrical lead up to the electrical connector and disconnect it, then unscrew the sensor with an oxygen sensor socket (front upstream oxygen sensor shown, rear upstream sensor similar)

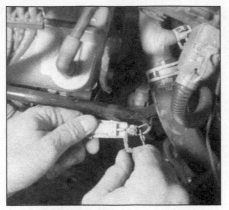

11.16b To disconnect the upstream oxygen sensor connector on a V6 engine, remove this lock first, then unplug the connector

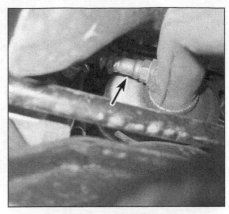

11.21 To remove a downstream oxygen sensor from a V6 engine, trace the electrical lead up to the electrical connector and disconnect it, then unscrew the sensor with an oxygen sensor socket

13   Using an oxygen sensor socket (available at most auto parts stores), unscrew the downstreatm oxygen sensor (see illustration 11.8). If the sensor is difficult to loosen, spray some penetrant onto the sensor threads and allow it to soak in for awhile.

14   If you're going to install the old sensor, apply anti-seize compound to the threads of the sensor to facilitate future removal. If you're going to install a new oxygen sensor, it's not necessary to apply anti-seize compound to the threads. The threads on new sensors already have anti-seize compound on them.

15   Installation is otherwise the reverse of removal. Be sure to tighten the sensor to the torque listed in this Chapter's Specifications.

V6 models

Upstream oxygen sensors

Note: *The upstream oxygen sensors are located on the exhaust manifolds, right above the flanges.*

16   Locate the upstream oxygen sensor (see illustration), then trace the sensor electrical lead up to the electrical connector and disconnect it (see illustration).

17   Using an oxygen sensor socket (available at most automotive retailers), unscrew the upstream oxygen sensor from the exhaust manifold. If the sensor is difficult to loosen, spray some penetrant onto the sensor threads and allow it to soak in for awhile.

18   If you're going to install the old sensor, apply anti-seize compound to the threads of the sensor to facilitate future removal. If you're going to install a new oxygen sensor, it's not necessary to apply anti-seize compound to the threads. The threads on new sensors already have anti-seize compound on them.

19   Installation is otherwise the reverse of removal. Be sure to tighten the sensor to the torque listed in this Chapter's Specifications.

Downstream oxygen sensor

Note: *The downstream oxygen sensors are located on the exhaust pipes, directly behind and below the upstream catalysts.*

20   Raise the vehicle and place it securely on jackstands.

21   Locate the downstream oxygen sensor (see illustration) and trace the sensor electrical lead up to the electrical connector and disconnect it.

22   Using an oxygen sensor socket (available at most automotive retailers), unscrew the downstream oxygen sensor. If the sensor is difficult to loosen, spray some penetrant onto the sensor threads and allow it to soak in for awhile.

23   If you're going to install the old sensor, apply anti-seize compound to the threads of the sensor to facilitate future removal. If you're going to install a new oxygen sensor, it's not necessary to apply anti-seize compound to the threads. The threads on new sensors already have anti-seize compound on them.

24   Installation is otherwise the reverse of removal. Be sure to tighten the sensor to the torque listed in this Chapter's Specifications.

12   Transmission Range (TR) switch - replacement

Removal

Note: *You'll need a special alignment tool (J41545, or a suitable equivalent) to adjust the TR switch.*

Note: *The TR switches on the various transaxles used on these models vary somewhat in appearance depending on the transaxle unit and the model year, but they're all removed and installed the same way.*

1   Set the parking brake, then place the shift lever in the NEUTRAL position.

2   Locate the TR switch on top of the transaxle.

3   Disconnect the shift control cable from the TR switch lever (see illustration).

4   Disconnect the electrical connector from the TR switch (see illustration).

12.3 Use a screwdriver or a trim panel removal tool (shown) to pop the end of the shift control cable loose from the TR switch lever

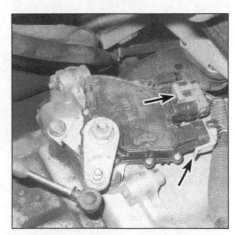

12.4 Release this lock, then disconnect the electrical connector from the TR switch

12.5 Using large water pump pliers to immobilize the lever, loosen this nut and detach the lever from the TR switch

12.6 To detach the TR switch from the transaxle, remove these two mounting bolts

5   Remove the TR switch lever nut (see illustration) and remove the lever.
6   Remove the TR switch mounting bolts (see illustration) and remove the switch.

Installation

Note: *The following procedure applies to a new or old TR switch that's being installed on the transaxle, as well as a TR switch that's already installed, but out of adjustment.*
7   Make sure that the shift lever is still in NEUTRAL.
8   To install the TR switch, align the flats on the transaxle shift shaft with the flats on the TR switch (see illustration), then loosely install the TR switch mounting bolts. Install the special TR switch alignment tool (available through specialty tool dealers and some deal-

ership parts departments) (see illustration) and rotate the TR switch until the tool falls into place. When the TR switch is correctly aligned, tighten the switch mounting bolts to the torque listed in this Chapter's Specifications.
9   Installation is otherwise the reverse of removal. Be sure to tighten the TR switch lever nut to the torque listed in this Chapter's Specifications (and hold it with your water pump pliers while doing so).

13   Transmission speed sensors - replacement

Note: *There are two transmission speed sensors: the Input Shaft Speed (ISS) sensor and*

the Output Shaft Speed (OSS) sensor. The ISS sensor is located inside the transaxle and cannot be replaced at home. It must be replaced by a dealer service department or by a qualified transmission shop. This section covers the OSS sensor, which is located on the right end of the back side of the transaxle, near the right inner CV joint on four-speed transmissions and inside the transmission oil pan, under the valve body on six-speed transmissions. Speed sensor replacement on six speed transmissions should be left to a transmission specialist or qualified repair shop.
1   Raise the front of the vehicle and place it securely on jackstands.
2   Disconnect the electrical connector from the OSS sensor (see illustration) and disen-

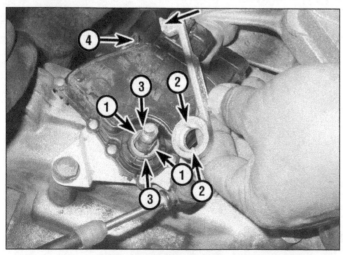

12.8a To install the TR switch, align the flats (1) on the transaxle shift shaft with the flats on the switch, slide the switch onto the shaft and loosely install the switch mounting bolts; to install the alignment tool, align the lugs (2) on the tool with the notches (3) in the switch and align the lug on the other end of the tool with the raised ridge (4) on the switch . . .

12.8b . . . then install the tool and rotate the switch slightly until the alignment tool drops into place (and looks like this), then tighten the TR switch mounting bolts to the torque listed in this Chapter's Specifications

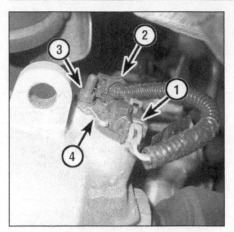

13.2 Disconnect the electrical connector (1) from the OSS sensor, disengage the wiring harness from the clip (2) and set the harness aside, remove the sensor hold-down bolt (3) and the hold-down clamp (4), then pull the sensor out of the transaxle housing

14.2a To unlock the PCM electrical connectors, depress the tab under the connector lock with a small screwdriver . . .

14.2b . . . then flip up the lock and unplug the connector

gage the electrical harness from the clip right above the OSS sensor.
3   Remove the OSS sensor hold-down bolt and hold-down clamp and remove the OSS sensor from the transaxle.
4   Remove the old O-ring from the OSS sensor and discard it. Be sure to use a new O-ring when installing the OSS sensor (even if you're planning to reuse the old OSS sensor).
5   Installation is the reverse of removal. Be sure to tighten the sensor retaining bolt securely.

14   Powertrain Control Module (PCM) - removal and installation

Caution: *To avoid electrostatic discharge damage to the PCM, handle the PCM only by its case. Do not touch the electrical terminals during removal and installation. If available, ground yourself to the vehicle with an anti-static ground strap, available at computer supply stores.*
Note: *The PCM is mounted on the front of the fuse and relay box, which is located at the left front corner of the engine compartment.*
Note: *The procedures in this section apply only to removing and installing the PCM that is already installed in your vehicle. If you need a new PCM, it must be programmed with new software and calibrations. This procedure requires the use of GM's TECH-2 scan tool and GM's latest PCM-programming software, so you WILL NOT BE ABLE TO REPLACE THE PCM AT HOME.*
Note: *The PCM is a highly reliable component and rarely requires replacement. Because the PCM is the most expensive part of the engine management system, you should be absolutely positive that it has failed before replacing*

it. If in doubt, have the system tested by an experienced driveability technician at a dealer service department or other qualified repair shop.
1   Disconnect the cable from the negative terminal of the battery (see Chapter 5, Section 1).
2   Pry loose, then flip up the lock on each PCM electrical connector (see illustrations), then disconnect all of the connectors from the PCM.
3   Release the two plastic retaining clips adjacent to the front of the PCM (see illustration) and remove the PCM from its mounting bracket.
4   Installation is the reverse of removal.

15   Catalytic converters - description, check and replacement

Note: *Because of a Federally-mandated extended warranty which covers emission-related components such as the catalytic converter, check with a dealer service department before replacing the converter at your own expense.*

Description

1   A catalytic converter (or catalyst) is an emission control device in the exhaust system that reduces certain pollutants in the exhaust gas stream. There are two types of converters. An oxidation catalyst reduces hydrocarbons (HC) and carbon monoxide (CO). A reduction catalyst reduces oxides of nitrogen (NOx). Catalysts that can reduce all three pollutants are known as "three-way catalysts." The models covered by this manual are equipped with three-way catalysts.

Check

2   The test equipment for a catalytic converter (a "loaded-mode" dynamometer and a five-gas analyzer) is expensive. If you suspect that the converter on your vehicle is malfunctioning, take it to a dealer or authorized emission inspection facility for diagnosis and repair.
3   Whenever you raise the vehicle to service underbody components, inspect the converter for leaks, corrosion, dents and other damage. Carefully inspect the welds and/or flange bolts and nuts that attach the front and rear ends of the converter to the exhaust system. If you note any damage, replace the converter.

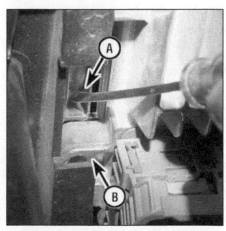

14.3 To detach the PCM from its mounting bracket, use a small screwdriver to depress each clip (A) adjacent to the front of the PCM. Push each clip back far enough to disengage it from the PCM mounting boss (B) (left clip shown, right clip identical)

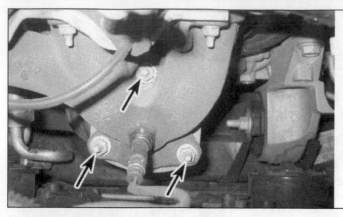

15.8 To disconnect the upper flange of the front catalyst/exhaust pipe assembly from the exhaust manifold flange on a V6 engine, remove these three nuts

4   Although catalytic converters don't break too often, they can become plugged up. The easiest way to check for a restricted converter is to use a vacuum gauge to diagnose the effect of a blocked exhaust on intake vacuum.

a) *Connect a vacuum gauge to an intake manifold vacuum source (see Chapter 2).*

b) *Warm the engine to operating temperature, place the transaxle in Park (automatic models) or Neutral (manual models) and apply the parking brake.*

c) *Note the vacuum reading at idle and write it down.*

d) *Quickly open the throttle to near its wide-open position and then quickly get off the throttle and allow it to close. Note the vacuum reading and write it down.*

e) *Do this test three more times, recording your measurement after each test.*

f) *If your fourth reading is more than one in-Hg lower than the reading that you noted at idle, the exhaust system might be restricted (the catalytic converter could be plugged, OR an exhaust pipe or muffler could be restricted).*

Replacement

Four-cylinder models

5   On four-cylinder models the catalytic converter is an integral component of the exhaust manifold (they're welded together). To replace the catalyst you must replace the exhaust manifold (see *Exhaust manifold/catalytic converter assembly - removal and installation* in Chapter 2A).

V6 models

Catalyst/exhaust pipe assembly (front cylinder bank)

6   Disconnect the electrical connector for the upstream oxygen sensor (see illustration 11.16b).

7   Remove the heat shield from the exhaust manifold (see *Exhaust manifold - removal and installation* in Chapter 2B or Chapter 2C).

8   Remove the three nuts (see illustration) that secure the upper flange of the catalyst/exhaust pipe assembly to the exhaust manifold flange.

9   Raise the front of the vehicle and place it

securely on jackstands.

10   Remove the downstream oxygen sensor from the catalyst/exhaust pipe assembly (see Section 11).

11   Remove the two nuts from the rear flange of the pipe that connects the catalyst/exhaust pipe assembly to the catalyst/exhaust pipe assembly for the rear cylinder bank.

12   Remove the catalyst/exhaust pipe assembly.

13   Remove and discard the old flange gaskets from both mounting flanges.

14   Be sure to use new gaskets at both mounting flanges.

15   Use new nuts at the front flange and new bolts at the rear flange. Coat the threads of the nuts and bolts with anti-seize compound to facilitate future removal. Tighten the nuts that secure the catalyst/exhaust pipe assembly to the exhaust manifold to the torque listed in Chapter 2B Specifications. Tighten the other flange nuts securely.

16   Installation is otherwise the reverse of removal

Catalyst/exhaust pipe assembly (rear cylinder bank)

17   Disconnect the upstream oxygen sensor electrical connector (see illustration 11.16b).

18   Remove the heat shield from the rear exhaust manifold (see *Exhaust manifold - removal and installation* in Chapter 2B or Chapter 2C).

19   Raise the vehicle and place it securely on jackstands.

20   Remove the downstream oxygen sensor from the catalyst/exhaust pipe assembly (see Section 11).

21   Remove the two nuts at the front flange and the two nuts at the rear flange that connect the catalyst/exhaust pipe assembly to the catalyst/exhaust pipe for the front cylinder bank and the rest of the exhaust system.

22   Remove the catalyst/exhaust pipe assembly.

23   Remove and discard the old flange gaskets from all three mounting flanges.

24   Be sure to use new gaskets at all three mounting flanges.

25   Use new nuts at all three flanges. Coat the threads of the nuts with anti-seize compound to facilitate future removal. Tighten the nuts that secure the catalyst/exhaust

pipe assembly to the exhaust manifold to the torque listed in Chapter 2B Specificaitons or Chapter 2C Specifications. Tighten the other flange nuts securely.

26   Installation is otherwise the reverse of removal.

16 Evaporative emissions control (EVAP) system - description and component replacement

Description

1   The Evaporative Emissions Control (EVAP) system prevents fuel system vapors (which contain unburned hydrocarbons) from escaping into the atmosphere. On warm days, vapors trapped inside the fuel tank expand until the pressure reaches a certain threshold, at which point the fuel vapors are routed from the fuel tank through the fuel vapor vent valve and the fuel vapor control valve to the EVAP canister, where they're stored temporarily, until they can be consumed by the engine during normal operation. When the conditions are right (engine warmed up, vehicle up to speed, moderate or heavy load on the engine, etc.) the Powertrain Control Module (PCM) opens the canister purge solenoid, which allows the fuel vapors to be drawn from the canister into the intake manifold, where they mix with the air/fuel mixture before being consumed in the combustion chambers. This system is complex and virtually impossible to troubleshoot without the right tools and training. However, the following description should give you a good idea of how it works:

2   The EVAP canister is located under the vehicle, at the right front corner of the fuel tank. The EVAP canister, which contains activated charcoal, is the repository for storing the fuel vapors. You'll have to raise the vehicle and lower the fuel tank to inspect or replace the canister (or the fuel tank pressure sensor) but the canister is designed to be maintenance-free and should last the life of the vehicle.

3   The fuel tank pressure sensor, which is located on top of the mounting flange for the in-tank fuel pump/fuel level sending unit module, monitors the pressure inside the tank, and transmits its measurement to the PCM during an OBD-II leak test.

4   The EVAP canister vent solenoid, which is mounted above the right side of the fuel tank, is normally open. But it seals off the EVAP system for inspection and maintenance (I/M 240) testing and for OBD-II leak and pressure tests.

5   The EVAP canister purge solenoid, which is under the control of the Powertrain Control Module (PCM), regulates the flow of vapors being purged from the EVAP canister into the intake manifold. The canister purge solenoid is normally closed. It opens only when directed to do so by the PCM, which uses the availability of intake manifold vacuum and data from various information sensor inputs

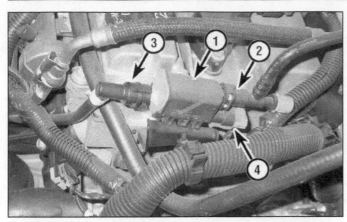

16.9a To remove the canister purge solenoid (1) from a four-cylinder engine, disconnect the inlet (2) and outlet (3) EVAP purge line fittings, disconnect the electrical connector (4) and remove the mounting bracket nut or bolt (not shown)

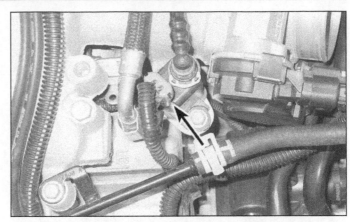

16.9b Canister purge control electrical connector on a 3.5L V6 model

to determine when and how long to open the valve. The interval of time during which the purge valve is opened by the PCM is known as its "duty cycle." On four-cylinder engines the purge valve is located at the left end of the valve cover. On V6 engines the purge valve is located on the left end of the intake manifold, near the throttle body.

General system checks

6   The most common symptom of a faulty EVAP system is a strong fuel odor (particularly during hot weather). If you smell fuel while driving or (more likely) right after you park the vehicle and turn off the engine, check the fuel filler cap first. Make sure that it's screwed onto the fuel filler neck all the way. If the odor persists, inspect all EVAP hose connections, both in the engine compartment and under the vehicle. You'll have to raise the vehicle and place it securely on jackstands to inspect most of the EVAP system, since it's located under the vehicle. Be sure to inspect each hose attached to the canister for damage and leakage along its entire length. Repair or replace as necessary. Inspect the canister for damage and look for fuel leaking from the bottom. If fuel is leaking or the canister is otherwise damaged, replace it.

7   Poor idle, stalling, and poor driveability can be caused by a defective fuel vapor vent valve or canister purge solenoid, a damaged canister, cracked hoses, or hoses connected to the wrong tubes. Fuel loss or fuel odor can be caused by fuel leaking from fuel lines or hoses, a cracked or damaged canister, or a defective vapor valve.

8   To check for excessive fuel vapor pressure in the fuel tank, remove the gas cap and listen for the sound of pressure release. If the fuel tank emits a "whooshing" sound when you open the filler cap, fuel tank vapor pressure is excessive. Inspect the canister vapor hoses and the canister inlet port for blockage or collapsed hoses. Also inspect the vapor vent valve. A complete test can only be done with a proprietary OBD-II scan tool (see Section 2), which will run a series of checks to

detect excessive pressure. You'll have to take the vehicle to a dealer service department to have the EVAP system professionally diagnosed.

Component replacement

EVAP canister purge solenoid

9   Disconnect the canister purge solenoid electrical connector (see illustrations).
10   Disconnect the inlet and outlet EVAP purge line fittings from the canister purge solenoid. Cap the lines to prevent dirt, dust and moisture from entering the EVAP system while the lines are open. If you're unfamiliar with quick-connect fittings, refer to Section 4 in Chapter 4.
11   Remove the canister purge solenoid mounting bracket bolt or nut and remove the purge solenoid and bracket.
12   Separate the canister purge solenoid from its mounting bracket.
13   Installation is the reverse of removal.

EVAP canister vent solenoid

14   Raise the vehicle and place it securely on jackstands.
15   Remove the fuel tank (see Chapter 4).
16   Disconnect the electrical connector from the vent solenoid.
17   Disconnect the EVAP lines from the vent solenoid.
18   Remove the vent solenoid from its mounting bracket.
19   Installation is the reverse of removal.

EVAP canister

20   Raise the vehicle and place it securely on jackstands.
21   Remove the fuel tank (see Chapter 4).
22   Disconnect all electrical connectors from the EVAP canister.
23   Disconnect the EVAP hoses from the EVAP canister.
24   Detach the EVAP canister from its mounting bracket.
25   Installation is otherwise the reverse of removal.

Fuel tank pressure sensor

26   Raise the vehicle and place it securely on jackstands.
27   Remove the fuel tank (see Chapter 4).
28   Disconnect the electrical connector from the fuel tank pressure sensor.
29   Remove the fuel tank pressure sensor from the fuel pump/fuel level sensor module mounting flange.
30   Installation is the reverse of removal.

---

17   Exhaust Gas Recirculation (EGR) system (2007 and earlier V6 models) - description and component replacement

---

Description

1   Oxides of nitrogen, nitrogen oxide, or simply NOx, is a compound that is formed in the combustion chambers when the oxygen and nitrogen in the incoming air mix together. NOx is a natural byproduct of high combustion chamber temperatures (2500 degrees F and higher). When NOx is emitted from the tailpipe, it mixes with reactive organic compounds (ROCs), hydrocarbons (HC) and sunlight to form ozone and photochemical smog. The EGR system reduces NOx by recirculating exhaust gases from the exhaust manifold, through the EGR valve and intake manifold, then back to the combustion chambers, where it mixes with the incoming air/fuel mixture before being consumed. These recirculated exhaust gases "dilute" the incoming air/fuel mixture, which cools the combustion chambers, thereby reducing NOx emissions.
2   An EGR system is only used on 2004 through 2006 3.5L V6 engines. There is no EGR system on four-cylinder engines, on 2007 3.5L V6 engines or on 3.9L V6 engines.
3   The EGR system consists of the Powertrain Control Module (PCM), the EGR valve and various information sensors (ECT, TP, MAP, IAT, RPM and VSS sensors) that the PCM uses to determine when to open the

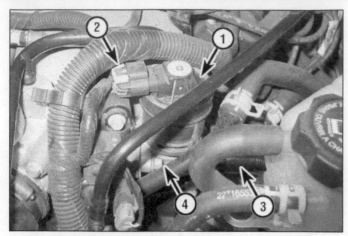

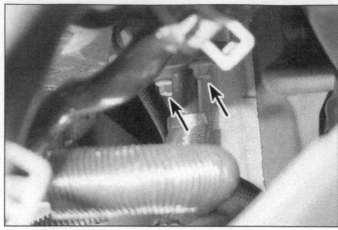

17.6 To remove the EGR valve (1) from the intake manifold on a 2004 through 2006 3.5L V6, disconnect the electrical connector (2), unscrew the tube nut (not shown) for the EGR pipe (3) and remove the EGR valve mounting bolts (4) (other bolt not shown)

17.12 To disconnect the lower end of the EGR pipe from the intake manifold, remove these two bolts

EGR valve. When the PCM closes the power/control circuit for the EGR valve, a solenoid inside the EGR valve is energized. This creates an electromagnetic field, which causes an armature to pull up, lifting the pintle off its seat. The exhaust gas then flows from the exhaust manifold port to the intake manifold.

4   Once activated by the PCM, the EGR valve uses a position feedback circuit to control the position of the pintle valve. The feedback circuit, which functions like a potentiometer, puts out a variable output voltage signal with an operating range between 0.5 and 5.4 volts. This variable output enables the PCM to control the position of the pintle with a high degree of precision. A pintle position sensor monitors the position of the pintle, and the PCM adjusts the current to match the actual pintle position to the optimal pintle position.

5   If there is too much EGR flow at idle, cruise or during cold running conditions, the engine will stop after a cold start, stop at idle after deceleration, surge during cruising speeds or idle roughly. If there is too little EGR flow, combustion chamber temperature can become too high during acceleration or under a heavy load, which can cause spark knock (detonation) and/or engine overheating.

Component replacement
EGR valve
6   Disconnect the electrical connector from the EGR valve (see illustration).
7   Remove the bolt that secures the upper flange of the EGR pipe to the EGR valve.
8   Remove the EGR valve mounting bolts and remove the EGR valve and the old gasket. Discard the old gasket.
9   Clean the gasket mating surfaces on the intake manifold and on the EGR valve mounting flange and install a new gasket.
10   Installation is the reverse of removal. Be sure to tighten the upper EGR pipe flange bolt and the EGR valve mounting bolts securely.

EGR pipe
11   Remove the bolt that secures the upper flange of the EGR pipe to the EGR valve.
12   Remove the two bolts (see illustration) that secure the lower flange of the EGR pipe to the exhaust manifold.
13   Remove the EGR pipe.
14   Clean out the EGR pipe with a wire bottlebrush, then blow it out with compressed air.
15   Coat the threads of the EGR pipe mounting flange bolts with anti-seize compound, then tighten the upper and lower flange bolts securely.

18   Secondary air injection system - description and component replacement

Description
1   An air injection system is used on 2009 3.5L engines. Air injection helps the upstream catalytic converters reach their proper operating temperatures rapidly during engine warm-up. It functions for a maximum of 1 minute after start-up.
2   The system consists of the PCM, an electric air pump, two solenoid/check valves, two injection manifolds and the wiring, relays, hoses and tubing connecting them (see illustration). Each solenoid valve also has an air pressure sensor in it.
3   When the PCM senses that secondary air injection is required during a warm-up period, it energizes the air pump and opens both solenoid valves. The air pump draws fresh air through the engine air filter and pumps it through the solenoid valves into the injection manifolds. These manifolds are located under the exhaust manifolds and direct the air to each exhaust port.
4   The injected air helps complete the burning of the rich start-up fuel mixture. If the fuel were not completely burned, it would have a cooling effect on the catalytic converters and it would take longer for them to be operational.

Component replacement
5   The air injection should be trouble-free for many years because it only works during warm-up. The major components are easily accessible if you do have to service the system.

Air pump
6   The air pump is located directly beneath the inlet of the throttle body. Remove the engine cover and the throttle body intake duct.
7   Disconnect the Y-shape discharge tube from the pump and the manifolds.
8   Remove the pump mounting nuts and lift

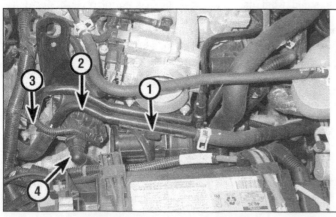

18.2 Air injection system details (2009 and 2010 3.5L engine only)

1   *Air injection pump*
2   *Left-side (front) solenoid/check valve*
3   *Left-side (front) air injection manifold connection tube*
4   *Y-shape pump discharge tube*

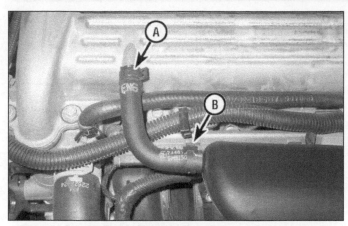

19.2 On four-cylinder engines, the PCV system consists of a single crankcase ventilation hose

A   *Crankcase vapors are drawn through the crankcase through the vent housing (on the underside of the valve cover), then through the ventilation hose . . .*
B   *. . . and into the air intake duct, where they're carried with incoming air through the throttle body and into the intake manifold*

19.3 Typical PCV system on a 3.5L V6 engine:

A   *Fresh air is drawn from the air intake duct through the fresh air inlet hose . . .*
B   *. . . into the crankcase, via a pipe on the rear valve cover*
C   *Crankcase vapors are drawn from the crankcase through a fixed orifice in the valve cover, then through the crankcase ventilation hose . . .*
D   *. . . into the intake manifold, where they're mixed with the incoming air/fuel mixture and consumed*

off the pump.
9   Installation is the reverse of removal.

Solenoid/check valves

10   These valves are mounted adjacent to the air pump, near the throttle body. Remove the engine cover, the intake air duct and any other interfering components.
11   Disconnect the pump discharge tube from the valve.
12   Disconnect the valve's bottom discharge pipe from the manifold by removing the two nuts.
13   Remove the valve mounting bolts and lift off the valve. Remove the gasket and replace it with a new one.
14   Installation is the reverse of removal.

Air injection manifold

15   The air injection manifolds are mounted directly below the exhaust manifolds. Refer to Chapter 2B and remove the exhaust manifold that covers the injection manifold you're removing.
16   Disconnect the tube that connects to the solenoid valve by removing the two nuts. Remove the gasket.
17   Remove the four mounting bolts and lift the injection manifold off. Remove the gasket and replace it with a new one.
18   Installation is the reverse of removal.

19   Positive Crankcase Ventilation (PCV) system - description, check and component replacement

Description

1   The Positive Crankcase Ventilation (PCV) system reduces hydrocarbon emis- sions by scavenging crankcase vapors, which are rich in unburned hydrocarbons.
2   On four-cylinder models, the PCV system consists of a single crankcase ventilation hose (see illustration) between the air intake duct and the valve cover, and a crankcase ventilation housing (or simply, the vent housing) permanently affixed to the underside of the valve cover. Crankcase blow-by vapors are directed through internal passages in the engine block and cylinder head up to the vent housing, from which they're drawn through the hose into the air intake duct, then through the throttle body and manifold and into the combustion chambers where they're consumed along with the air/fuel mixture.
3   On V6 models, the PCV system (see illustration) consists of a fixed orifice (not a PCV valve), which is located in the front valve cover. There are two hoses in this system between the engine vent adapter and the vent housing: one hose carries fresh air from the intake manifold to the vent housing and the other carries a mixture of fresh air and crankcase vapors back to the intake manifold. The vent housing meters the flow of crankcase blow-by gases from the crankcase to the intake manifold in proportion to intake manifold vacuum. At idle, when intake vacuum is high, it restricts the flow, but as intake vacuum decreases, it allows a greater flow. If crankcase pressures become too high, the vent housing allows excess vapors to backflow to the intake manifold.

Check

4   An engine that is operated without a properly functioning crankcase ventilation system can be damaged. So anytime you're servicing the engine, be sure to inspect the PCV system hose(s) for cracks, tears and other damage. Disconnect the hose(s) and check for damage and obstructions. If a hose is clogged, clean it out. If you're unable to clean it satisfactorily, replace it.
5   A plugged PCV hose might cause any or all of the following conditions: A rough idle, stalling or a slow idle speed, oil leaks or sludge in the engine. So if the engine is running roughly, stalling and idling at a lower than normal speed, or is losing oil, or has oil in the throttle body or air intake manifold plenum, or has a build-up of sludge, a PCV system hose might be clogged. Repair or replace the hose(s) as necessary. And clean or replace the vent housing. On four-cylinder engines, remove the valve cover (see Chapter 2A) and inspect the vent housing. Make sure that it's clean by blowing it out with compressed air. On V6 engines, remove the valve cover and inspect the fixed orifice. If it's dirty, clean it out with solvent and compressed air. The orifice cannot be replaced separately. If it's so clogged that you cannot clean it out, replace the valve cover.
6   A leaking PCV hose might cause any or all of the following conditions: a rough idle, stalling or a high idle speed. So if the engine is running roughly, stalling and idling at a higher than normal speed, a PCV system hose might be leaking. Repair or replace the hose(s) as necessary.
7   Here's an easy functional check of the PCV system on a V6 model (this check won't work on the PCV system on a four-cylinder model):

a) *Disconnect the PCV hose from the front valve cover (see illustration 19.3).*
b) *Start the engine and let it warm up to its normal idle.*
c) *Verify that there is vacuum at the PCV hose. If there is no vacuum, look for a plugged hose or manifold port.*

d) *Remove the engine oil dipstick and install a vacuum gauge on the upper end of the dipstick tube.*

e) *Block off the PCV fresh air inlet hose.*

f) *Run the engine at 1500 rpm for 30 seconds, then read the vacuum gauge while the engine is running at 1500 rpm.*

g) *If there's vacuum present, the crankcase ventilation system is operating correctly.*

h) *If there's NO vacuum present, the engine might be drawing in outside air. The PCV system won't function correctly unless the engine is a sealed system. Inspect the valve covers and the oil pan gasket for leaks.*

i) *If the vacuum gauge indicates positive pressure, look for a plugged hose or engine blow-by.*

8   If the PCV system is functioning correctly, but there's evidence of engine oil in the throttle body or air filter housing, it could be caused by excessive crankcase pressure. Have the crankcase pressure tested by a dealer service department or other repair shop.

9   In this type of PCV system, excessive blow-by (caused by worn rings, pistons and/or cylinders, or by constant heavy loads) is discharged into the intake manifold and consumed. If you discover heavy sludge deposits or a dilution of the engine oil, even though the PCV system is functioning correctly, look for other causes (see Troubleshooting and refer to Chapter 2D) and correct them as soon as possible.

Component replacement

Four-cylinder models

PCV hose

10   Loosen the hose clamp (see illustration 19.2) and disconnect the hose from the air intake duct.

11   Loosen the hose clamp and disconnect the hose from the valve cover.

12   Installation is the reverse of removal.

Crankcase vent housing

13   Remove the valve cover (see Chapter 2A).

14   The vent housing is an integral component of the valve cover (it's riveted to the underside of the valve cover). It cannot be replaced separately. If the vent housing is clogged up, clean it with a steel brush and fresh solvent, then blow it out with compressed air.

15   Installation is the reverse of removal.

V6 models

Fresh air inlet hose

16   Disconnect the fresh air inlet hose from the air intake duct (see illustration 19.3).

17   Disconnect the fresh air inlet hose from the rear valve cover.

18   Installation is otherwise the reverse of removal.

Crankcase ventilation hose (PCV hose)

19   Disconnect the crankcase ventilation hose from the front valve cover (see illustration 19.3).

20   Disconnect the crankcase ventilation hose from the intake manifold.

21   Installation is otherwise the reverse of removal.

Fixed orifice

22   Remove the front valve cover (see Chapter 2B or Chapter 2C).

23   The vent housing is an integral component of the valve cover (it's riveted to the underside of the valve cover). It cannot be replaced separately. If the vent housing is clogged up, clean it with a steel brush and fresh solvent, then blow it out with compressed air.

24   Installation is the reverse of removal.

20   Camshaft actuator system - description and component replacement

Description

1   The camshaft actuator system, which is used on 3.9L V6 engines, 2007 3.5L V6 engines and all 2008 and later engines, reduces emissions, improves engine torque, increases fuel mileage and stabilizes engine idle by controlling the amount of intake and exhaust valve overlap.

2   The camshaft actuator system consists of the Powertrain Control Module (PCM), the camshaft actuator solenoid and the cam phaser, which is a variable camshaft timing sprocket. The actuator solenoid opens and closes two passages through which oil flows to the hydraulically operated variable camshaft timing sprocket on the front end of the camshaft. The PCM uses a pulse-width modulated signal to control the camshaft actuator solenoid, which responds by directing oil through one of the two passages feeding oil to the variable cam timing sprocket. When oil is directed by the solenoid through one passage, the variable cam sprocket advances the camshaft timing; when oil is directed through the other passage, the sprocket retards cam timing.

Component replacement

Variable camshaft timing sprocket

3   To replace the variable camshaft timing sprocket, refer to Timing chain and sprockets - removal, inspection and installation in Chapter 2A, Chapter 2B or Chapter 2C.

Camshaft actuator solenoid

Note: *The camshaft actuator solenoid is located on the upper front part of the timing chain cover, directly in front of the camshaft.*

4   Remove the intake manifold cover (see Chapter 2B or Chapter 2C).

5   Remove the air intake duct and air filter housing (see Chapter 4).

6   Remove the engine mount strut bracket (see Chapter 2B or Chapter 2C).

7   Disconnect the electrical connector from the camshaft actuator solenoid.

8   Remove the camshaft actuator solenoid mounting bolts and remove the solenoid.

9   Remove and discard the old O-ring.

10   Installation is the reverse of removal. Be sure to use a new O-ring and tighten the solenoid mounting bolts securely.

21   Intake Manifold Tuning (IMT) valve system - description and component replacement

Description

1   When intake air is drawn into the cylinders at idle or at low engine speeds, less air is needed because the cylinders don't need to be filled so often or so quickly. So at idle and at low engine speeds, the air drawn into an engine with smaller intake runners will have a higher velocity than one with larger intake runners. However, at higher engine speeds, smaller intake runners would prevent the cylinders from filling quickly enough and would therefore limit power. Most intake manifold designs are a compromise between the conflicting demands of low and high engine speeds.

2   The Intake Manifold Tuning (IMT) Valve system helps to maintain a uniformly higher intake air velocity throughout the engine's operating range. Higher intake air velocity promotes better vaporization of the fuel sprayed into the stream of incoming air by the fuel injectors, which means more complete combustion, more power, better fuel economy and less emissions.

3   The IMT Valve system consists of the Powertrain Control Module (PCM), a special intake manifold with an IMT valve inside the manifold, a PCM-controlled solenoid that opens and closes the IMT valve. When the IMT valve is open, the intake manifold is configured as one large plenum. When the IMT valve is closed, the intake manifold is divided into smaller plenums.

4   When the engine is operating at a lower speed but under a higher load, the IMT valve is closed, so incoming air is directed through a longer path. Directing incoming air through a longer path increases torque.

5   When the engine speed and load are both high, the IMT valve is open, which creates a shorter intake path. Directing incoming air through a shorter intake path increases horsepower.

Component replacement

Intake Manifold Tuning (IMT) valve solenoid

Note: *The IMT valve solenoid is located on the intake manifold.*

6   Disconnect the electrical connector from the IMT valve solenoid.

7   Remove the IMT valve solenoid mounting screws and remove the solenoid.

8   Installation is the reverse of removal. Tighten the solenoid screws securely.

Intake manifold

9   Refer to Chapter 2B or Chapter 2C.

Chapter 7 Part A
Manual transaxle

Contents

Specifications

General

Lubricant type .... See Chapter 1

Torque specifications | Ft-lbs (unless otherwise indicated)

Subframe mounting bolts
    Stage 1 73
    Stage 2 .... Tighten an additional 180-degrees
Transaxle-to-engine bolts 44
Transaxle rear brace bolts 22
Transaxle front brace bolts 37

1   General information

1   The vehicles covered by this manual are equipped with either a 6-speed manual or an automatic transaxle. Information on the manual transaxle is included in this Part of Chapter 7A. Information on the automatic transaxle is in Chapter 7B.
2   The 6-speed transaxle consists of a transmission and differential housed in a single all-aluminum assembly. Because of its complexity, the special tools needed to overhaul it, and the difficulty of obtaining replacement parts, overhauling this unit is beyond the scope of the average home mechanic. The information in this Chapter is limited to general diagnosis, external adjustments and removal and installation.
3   Depending on the cost, it may be a good idea to consider replacing the old unit with either a rebuilt or used transaxle instead of a new one. Your local dealer or transmission shop should be able to supply information concerning cost, availability and exchange policy. Regardless of how you decide to remedy a transaxle problem, however, you will save money by removing and installing it yourself.

2   Shift control assembly - removal and installation

1   Remove the shift lever knob, and shift boot. Carefully pry the boot trim ring from the center console, then lift up on the shift knob to remove it and the boot from the shift control lever.
2   Disconnect the shift and select cables from the shift control assembly.
3   Detach the cables from their bracket on the shift control assembly base.
4   Remove the shift control assembly retaining nuts and remove the shift control assembly.
5   Installation is the reverse of removal.

3   Shift cables - removal and installation

1   Remove the air filter duct (see Chapter 4), then disconnect the shift and select cables from the levers on the transaxle.
2   Remove the center console (see Chapter 11).
3   Disconnect the shift and select cables from the shift control assembly. The cables can be pried from the pins on the shift control assembly with a screwdriver.
4   Pry loose the grommet, pull the cables through into the passenger compartment and remove them from the vehicle.
5   Installation is the reverse of removal.

4 Manual transaxle - removal and installation

Removal

1   Disconnect the cable from the negative terminal of the battery (see Chapter 5).

2   Remove the air filter housing (see Chapter 4).

3   Disconnect the shift cables from the levers on the transaxle, then remove the cable bracket.

4   Pull the retaining clip from the end clutch fluid hydraulic line and remove the line from the transaxle.

5   Disconnect the electrical connectors from the Vehicle Speed Sensor and the back-up light switch.

6   Support the radiator and condenser from above with two lengths of rope. Attach an engine support fixture (which is recommended) or an engine hoist or to the engine and raise it sufficiently to just support the weight of the engine.

Note: *The engine must remain supported while the transaxle is out of the vehicle.*

Note: *If you use an engine hoist, position the hoist with its legs inserted under the vehicle from the right (passenger's) side. This will give you room to maneuver the transaxle out with a jack.*

7   Remove the upper transaxle-to-engine studs and nuts.

8   Loosen the driveaxle/hub nuts and the front wheel lug nuts, then raise the vehicle and support it securely on jackstands placed underneath the rocker panel flanges, right behind the wheel openings on each side of the vehicle. Remove the front wheels.

9   Remove both inner fender splash shields.

10   Disconnect the electrical connectors from both front ABS wheel speed sensors harnesses. Detach the harnesses from their clips on the control arms and subframe (also known as the "cradle").

11   Detach both tie-rod ends from the steering knuckle arms (see Chapter 10).

12   Remove the driveaxles (see Chapter 8).

Note: *Instead of separating the struts from the steering knuckles when removing the driveaxles, separate the control arm balljoints from the steering knuckles (see Chapter 10). This will allow you to pull out on the steering knuckles far enough to remove the driveaxles, and has to be done anyway.*

13   Remove the pinch bolt that secures the intermediate steering shaft to the steering gear (see Chapter 10).

14   Remove the power steering gear mounting bolts and hang the gear with a piece of wire.

15   Unbolt the transaxle and engine mounts from the subframe (see Chapter 2C).

16   Position two floor jacks under the subframe and remove the subframe mounting bolts. Slowly lower the jacks, making sure nothing is still connected to the subframe. Remove the subframe out from under the vehicle.

17   Remove the starter motor (see Chapter 5).

18   Unbolt the front and rear transaxle mounts from the transaxle.

19   Support the transaxle with a jack, preferably a jack made for this purpose. Transmission jacks are commonly available at most equipment rental yards. These jacks are equipped with safety chains; use these chains to secure the transaxle to the jack.

20   Remove the remaining transaxle-to-engine bolts.

21   Remove the transaxle from the engine by sliding it toward the left side of the vehicle. It may be necessary to lower the engine to provide clearance for the transaxle to pass below the left side inner body panel.

Installation

22   Installation is the reverse of removal, with attention paid to the following points:

a)   *Apply a film of high-temperature grease to the transaxle input shaft splines.*

b)   *When mating the transaxle to the engine, make sure the transaxle seats against the engine completely before tightening the bolts; if it doesn't, figure out why. Don't use the bolts to draw the transaxle into place, as you could break something.*

c)   *Tighten all transaxle-to-engine bolts to the torque listed in this Chapter's Specifications.*

d)   *Tighten the subframe mounting bolts to the torque listed in this Chapter's Specifications. Also tighten the intermediate shaft pinch bolt, the control arm balljoint-to-steering knuckle nuts and the tie-rod end-to-steering knuckle nuts to the torque values listed in the Chapter 10 Specifications.*

e)   *Tighten the wheel lug nuts to the torque listed in Chapter 1 Specifications.*

f)   *The wheel alignment should be checked and, if necessary, adjusted.*

g)   *Check the transaxle lubricant level and add, as necessary, to bring it to the appropriate level (see Chapter 1).*

5 Manual transaxle overhaul - general information

1   Overhauling a manual transaxle is a difficult job for the do-it-yourselfer. It involves the disassembly and reassembly of many small parts. Numerous clearances must be precisely measured and, if necessary, changed with select fit spacers and snap-rings. As a result, if transaxle problems arise, it can be removed and installed by a competent do-it-yourselfer, but overhaul should be left to a transmission repair shop. Rebuilt transaxles may be available - check with your dealer parts department and auto parts stores. At any rate, the time and money involved in an overhaul is almost sure to exceed the cost of a rebuilt unit.

2   Nevertheless, it's not impossible for an inexperienced mechanic to rebuild a transaxle if the special tools are available and the job is done in a deliberate, step-by-step manner so nothing is overlooked.

3   The tools necessary for an overhaul include internal and external snap-ring pliers, a bearing puller, a slide hammer, a set of pin punches, a dial indicator and possibly a hydraulic press. In addition, a large, sturdy workbench and a vise or transaxle stand will be required.

4   During disassembly of the transaxle, make careful notes of how each piece comes off, where it fits in relation to other pieces and what holds it in place. Note how the parts are installed when you remove them; this will make it much easier to get the transaxle back together.

5   Before taking the transaxle apart for repair, it will help if you have some idea what area of the transaxle is malfunctioning. Certain problems can be closely tied to specific areas in the transaxle, which can make component examination and replacement easier. Refer to the *Troubleshooting* section at the front of this manual for information regarding possible sources of trouble.

Chapter 7 Part B
Automatic transaxle

Contents

Specifications

Torque specifications

	Ft-lbs
Intermediate driveaxle bolts (six-speed models)	44
Transaxle-to-engine bolts	
Malibu	
2010 and earlier models	66
2011 and later models	43
G6 and Aura	
Four-speed models	66
Six-speed models	55
Transaxle brace bolts	37
Driveplate-to-torque converter bolts	46
Fluid pan bolts	See Chapter 1

1   General Information

1   The vehicles covered by this manual are equipped with either a four or six-speed automatic transaxle, or a six-speed manual transaxle. Four models of Hydra-matic transaxles are used: 4T40-E, 4T45-E, 4T65-E and 6T70-E, all of which are electronically controlled.

2   Due to the complexity of the clutches and the hydraulic control system, and because of the special tools and expertise required to perform an automatic transaxle overhaul, it is not included in this manual. Therefore, the procedures in this Chapter are limited to general diagnosis, adjustment and transaxle removal and installation.

3   If the transaxle requires major repair work it should be left to a dealer service department or an automotive or transmission repair shop. You can, however, remove and install the transaxle yourself and save the expense, even if the repair work is done by a transmission specialist.

4   Adjustments that the home mechanic may perform include those involving the shift cable and the Brake Transmission Shift Interlock (BTSI) system.

2   Diagnosis - general

Note: *Automatic transaxle malfunctions may be caused by five general conditions: poor engine performance, improper adjustments, hydraulic malfunctions, mechanical malfunctions or malfunctions in the Powertrain Control Module or its signal network. Diagnosis of these problems should always begin with a check of the easily repaired items: fluid level and condition (see Chapter 1), and shift cable adjustment (see Section 4). Next, perform a road test to determine if the problem has been corrected or if more diagnosis is necessary. Because the transaxle relies on many sensors in the engine control system, and since the transaxle shift points are controlled by the Powertrain Control Module, you'll also want to check to see if any trouble codes have been stored in the PCM (see Chapter 6 for a list of trouble codes and how to extract them). If the problem persists after the preliminary tests and corrections are completed, additional diagnosis should be done by a dealer service department or transmission repair shop. Refer to the Troubleshooting Section at the front of this manual for transaxle problem diagnosis.*

Preliminary checks

1   Drive the vehicle to warm the transaxle to normal operating temperature.

2   Check the fluid level as described in Chapter 1:

   a) *If the fluid level is unusually low, add enough fluid to bring the level within the designated area of the dipstick, then check for external leaks.*

   b) *If the fluid level is abnormally high, drain off the excess, then check the drained fluid for contamination by coolant. The presence of engine coolant in the automatic transmission fluid indicates that a failure has occurred in the internal radiator walls that separate the coolant from the transmission fluid (see Chapter 3).*

   c) *If the fluid is foaming, drain it and refill the transaxle, then check for coolant in the fluid or a high fluid level.*

3   Check the engine idle speed.

Note: *If the engine is malfunctioning, do not proceed with the preliminary checks until it has been repaired and runs normally.*

4   Inspect the shift cable (see Section 4). Make sure that it's properly adjusted and that it operates smoothly.

5   Check the Transaxle Range (TR) sensor adjustment (see Chapter 6). Also check the transmission (TRANS) and transmission control module (TCM) fuses.

Fluid leak diagnosis

6   Most fluid leaks are easy to locate visually. Repair usually consists of replacing a seal or gasket. If a leak is difficult to find, the following procedure may help.

7   Identify the fluid. Make sure it's transmission fluid and not engine oil or brake fluid (automatic transmission fluid is a deep red color).

8   Try to pinpoint the source of the leak. Drive the vehicle several miles, then park it over a large sheet of cardboard. After a minute or two, you should be able to locate the leak by determining the source of the fluid dripping onto the cardboard.

9   Make a careful visual inspection of the suspected component and the area immediately around it. Pay particular attention to gasket mating surfaces. A mirror is often helpful for finding leaks in areas that are hard to see.

10   If the leak still cannot be found, clean the suspected area thoroughly with a degreaser or solvent, then dry it.

11   Drive the vehicle for several miles at normal operating temperature and varying speeds. After driving the vehicle, visually inspect the suspected component again.

12   Once the leak has been located, the cause must be determined before it can be properly repaired. If a gasket is replaced but the sealing flange is bent, the new gasket will not stop the leak. The bent flange must be straightened.

13   Before attempting to repair a leak, check to make sure that the following conditions are corrected or they may cause another leak.

Note: *Some of the following conditions cannot be fixed without highly specialized tools and expertise. Such problems must be referred to a transmission shop or a dealer service department.*

Gasket leaks

14   Check the pan periodically. Make sure the bolts are tight, no bolts are missing, the gasket is in good condition and the pan is flat (dents in the pan may indicate damage to the valve body inside).

15   If the pan gasket is leaking, the fluid level or the fluid pressure may be too high, the vent may be plugged, the pan bolts may be too tight, the pan sealing flange may be warped, the sealing surface of the transaxle housing may be damaged, the gasket may be damaged or the transaxle casting may be cracked or porous. If sealant instead of gasket material has been used to form a seal between the pan and the transaxle housing, it may be the wrong sealant.

Seal leaks

16   If a transaxle seal is leaking, the fluid level or pressure may be too high, the vent may be plugged, the seal bore may be damaged, the seal itself may be damaged or improperly installed, the surface of the shaft protruding through the seal may be damaged or a loose bearing may be causing excessive shaft movement.

17   Make sure the dipstick tube seal is in good condition and the tube is properly seated. Periodically check the area around the speedometer gear or sensor for leakage. If transmission fluid is evident, check the O-ring for damage. Also inspect the driveaxle oil seals for leakage.

Case leaks

18   If the case itself appears to be leaking, the casting is porous and will have to be repaired or replaced.

19   Make sure the oil cooler hose fittings are tight and in good condition.

Fluid comes out vent pipe or fill tube

20   If this condition occurs, the transaxle is overfilled, there is coolant in the fluid, the case is porous, the dipstick is incorrect, the vent is plugged or the drain back holes are plugged.

---

3   Shift lever - removal and installation

Warning: *The models covered by this manual are equipped with a Supplemental Restraint System (SRS), more commonly known as airbags. Always disarm the airbag system before working in the vicinity of any airbag system component to avoid the possibility of accidental deployment of the airbag, which could cause personal injury (see Chapter 12). Do not use a memory saving device to preserve the PCM's memory when working on or near airbag system components.*

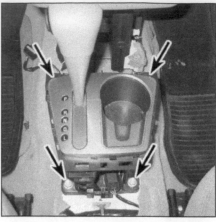

3.7 To detach the shift lever assembly from the floor, remove these four nuts

Removal

1   Disconnect the cable from the negative terminal of the battery (see Chapter 5, Section 1).

2   Remove the center console (see Chapter 11).

3   Disconnect the shift cable from the lever base (see Section 4).

4   Disconnect the BTSI cable from the lever base (see Section 5).

5   Remove the indicator bulb by rotating the bulb socket 1/4 turn and pulling down.

6   Unplug the electrical connector from the BTSI solenoid (see Section 5).

7   Remove the shift lever assembly retaining nuts (see illustration) and remove the shift lever assembly.

Installation

8   Place the shift lever in position on the mounting studs and install the nuts. Tighten the nuts securely.

9   The remainder of installation is the reverse of removal. Adjust the shift cable (see Section 4) and the BTSI cable (see Section 5) when you're done.

---

4   Shift cable - replacement and adjustment

Warning: *The models covered by this manual are equipped with a Supplemental Restraint System (SRS), more commonly known as airbags. Always disarm the airbag system before working in the vicinity of any airbag system component to avoid the possibility of accidental deployment of the airbag, which could cause personal injury (see Chapter 12). Do not use a memory saving device to preserve the PCM's memory when working on or near airbag system components.*

Replacement

1   Disconnect the cable from the negative terminal of the battery (see Chapter 5, Section 1).

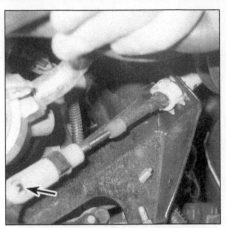

4.3 To disconnect the shift cable from the transaxle manual lever, pry it off the ballstud on the lever with a panel tool or flat-bladed screwdriver

4.4 To disengage the shift cable from the cable bracket on the transaxle, remove this clip then squeeze the two locking tangs and pull the cable housing from the bracket - 2005 Malibu shown

4.6 Location of the shift cable (A) and the BTSI cable (B) on a 2005 Malibu

2   Remove the air filter housing and the air intake duct (see Chapter 4).
3   Working in the engine compartment, disconnect the shift cable from the transaxle manual lever (see illustration).
4   Working in the engine compartment, disengage the shift cable from the cable bracket on the transaxle (see illustration).
Note: *On some models, remove the clip that retains the cable to the bracket. On other models, squeeze the tabs to release the cable from the bracket.*
5   Remove the center console (see Chapter 11).
6   Disconnect the shift cable from the ballstud on the shift lever (see illustration).
7   Detach the cable from the bracket at the front of the shift lever base (see illustration 4.6).
Note: *On some models, remove the clip that retains the cable to the control assembly and on other models, squeeze the tabs to release the cable from the control assembly.*
8   Trace the cable to the cable grommet (the point at which it goes through the firewall). Pry out the grommet and pull the cable through the hole and remove it.
9   Installation is the reverse of removal. When you're done installing the new cable, be sure to adjust it.
Note: *Use new clips when installing a new shift cable.*

Adjustment

All 2004, 2005 models and 2006, 2007 4T45-E models

10   Working at the transaxle, push the shift cable adjustment clip into the released position (see illustration).
Note: *The shift cable adjustment retainer is located on the cable end at the transaxle lever.*
11   Working at the transaxle, move the transaxle lever into the Park position.
12   Working in the passenger compartment,

place the shift lever at the center console into the Park position.
13   With the transaxle lever and the shift lever moved to the Park position, push the shift cable adjustment clip into the locked position.

2006, 2007 4T65-E models

14   Detach the cable from the manual lever on the transaxle. Place the manual lever on the transaxle in the Neutral position; this is accomplished by rotating the lever clockwise from the Park position, through Reverse and into Neutral.
15   Place the shift lever inside the car in Neutral.
16   Pull up on the cable adjuster tab at the cable bracket on the transaxle, then connect the cable to the manual lever on the transaxle. The cable will automatically adjust itself. Push the adjuster tab back into place.

2008 and later models

17   Place the vehicle in park, apply the parking brake and block the wheels to prevent the vehicle from rolling.
18   Working at the transaxle, release the shift cable adjuster clip and slide the two halves of the shift cable together until there is no freeplay felt in the cable.
19   Press the shift cable adjuster clip, allowing the adjuster to lock in place.
20   Pull the shifter cable adjustment halves in opposite directions, assuring the cable and adjustment are secured.
21   Operate the transmission in all gears including park to verify proper cable adjustment.

All models

22   Make sure the engine will start in the Park and Neutral positions only.
23   If the engine can be started in any position other than Park or Neutral, check the adjustment of the Transmission Range Sensor (see Chapter 6), then adjust and check the shift cable again.

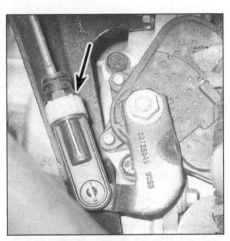

4.10 Pull out the locking clip (arrow), with the transmission and shifter both in Park, push it back in to adjust the cable

5   Brake Transmission Shift Interlock (BTSI) system - description and component replacement

Warning: *The models covered by this manual are equipped with a Supplemental Restraint System (SRS), more commonly known as airbags. Always disarm the airbag system before working in the vicinity of any airbag system component to avoid the possibility of accidental deployment of the airbag, which could cause personal injury (see Chapter 12). Do not use a memory saving device to preserve the PCM's memory when working on or near airbag system components.*

Description

1   The Brake Transmission Shift Interlock (BTSI) system prevents the shift lever from being moved out of Park unless the brake

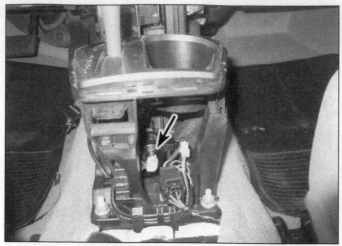

5.4 Location of the Brake Transmission Shift Interlock solenoid

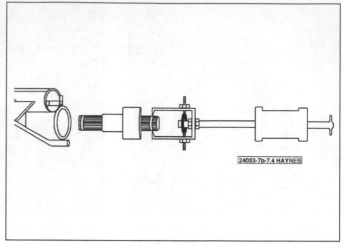

6.4 Remove the stub axle with a slide hammer and adapter (all 2004, 2005 models and 2006 and later 4T45-E models)

pedal is depressed simultaneously. It also prevents the ignition key from being removed from the ignition switch unless the shift lever is in the Park position. When the car is started, the BTSI (Brake Transmission Shift Interlock) solenoid is energized, locking the shift lever in Park; when the brake pedal is depressed, the solenoid is de-energized, unlocking the shift lever so that it can be moved into some other gear.

BTSI solenoid replacement

2   Remove the center console (see Chapter 11).
3   Unplug the electrical connector from the BTSI solenoid.
4   Pry off each end of the BTSI solenoid assembly from the ball sockets (see illustration) and remove it from the shift lever base.
5   Installation is the reverse of removal.

BTSI cable replacement

6   Disconnect the cable from the negative terminal of the battery (see Chapter 5, Section 1).
7   Remove the center console (see Chapter 11).
8   Remove the left side trim panel, the knee bolster and the steering column covers (see Chapter 11).
9   Place the shift lever in the Park position.
10   Turn the ignition key to the Run position.
11   Insert a screwdriver blade into the slot in the ignition switch inhibitor, depress the cable latch and detach the cable from the inhibitor.
12   Detach the cable end from the pin on the shift lever.
13   Depress the tabs on the sides of the cable housing and disengage the housing from the cable bracket.
14   Remove any cable clips and remove the BTSI cable, noting how it's routed (the cable is routed between the HVAC housing and the cowl).
15   Install the new cable, following the same

routing as the original.
16   Position the shift lever in Park.
17   With the ignition key in the Run position, snap the cable into the inhibitor housing until the snap lock is seated.
Note: *The white plastic collar must be flush or recessed approximately 0.04 inch within the ignition BTSI cable housing. If the plastic collar is not in its proper position, the BTSI cable will need to be adjusted.*
18   Attach the cable end to the ballstud on the shift lever.
19   Turn the ignition key to the Lock position.
20   Push in on the locking tab of the cable connector until it clicks back into place.
21   Check the operation of the BTSI cable.
22   If it operates as described above, the BTSI cable system is properly adjusted.
23   If the BTSI system does not operate as described, follow the adjustment procedure. Push the cable connector down and recheck the operation.
24   Install the steering column covers, the knee bolster, and the left-side under-dash panel (see Chapter 11).
25   Install the center console (see Chapter 11).

BTSI cable adjustment

26   Apply the parking brake. Press the brake pedal and move the shift lever to Park. Turn the ignition key to the LOCK position.
27   Remove the center console (see Chapter 11).
28   Use a small screwdriver to release the adjuster lock tang and rotate the threaded cable adjuster until approximately three threads are showing. The adjuster lock tang is located on the cable end at the shift lever.
29   Recheck the operation of the BTSI system and if there is any binding or malfunction, turn the adjuster 1/2-turn at a time until the BTSI system operates properly.
30   Installation is the reverse of the removal.

6   Driveaxle oil seals - replacement

Removal

1   Oil leaks frequently occur due to wear of the driveaxle oil seals. Replacement of these seals is relatively easy, since the repairs can be performed without removing the transaxle from the vehicle.
2   The driveaxle oil seals are located in the sides of the transaxle, where the driveaxles are attached. If leakage at the seal is suspected, raise the vehicle and support it securely on jackstands. If the seal is leaking, fluid will be found on the sides of the transaxle.
3   Remove the driveaxles (see Chapter 8).

All 2004, 2005 models and 2006, 2007 4T45-E models

4   Before the driveaxle seal can be replaced, the stub axle shaft will have to be removed. To do this, remove the snap-ring on the end of the stub shaft, then pull out on the stub axle and rotate it until the inner snap-ring seats in the differential side gear taper. Now attach a slide hammer and puller adapter to the stub axle and pull it from the transaxle (see illustration).
5   Remove the inner snap-ring from the stub axle and discard both snap-rings. Always use new snap rings.
6   The stub axle shaft sleeve and seal must be replaced together. Always use new parts.

Installation
All models

7   Note how deep the seal is installed, then use a screwdriver or prybar to carefully pry the oil seal out of the transaxle bore. If the oil seal cannot be removed with a screwdriver or prybar, a special oil seal removal tool (available at most auto parts stores) will be required.
8   Compare the old seal to the new one to

7.2 The Transmission Control Module (TCM) is located in the left side of the engine compartment, mounted on the unibody structure

7.3 Push back the retaining tang and slide the TCM out of its bracket

8.6 During transaxle removal, the preferred way to support the engine is with a support fixture designed for this purpose; they are often available from rental yards

be sure it's the correct one.

9   Coat the outside and inside diameters of the new seal with a small amount of transmission fluid.

10   Using a seal installation tool, install the new oil seal. Drive it into the bore squarely and make sure it's seated to the original depth.

All 2004, 2005 models and 2006 and later 4T45-E, 6T40 and 6T70 models

11   A special driver will be required to install the new sleeve and it must be installed to the same depth as the original; if you don't have suitable tools to do this, have it done at a dealer service department or other qualified repair shop.

12   Install new snap-rings on the stub shaft, then carefully guide the stub shaft through the seal (don't let the shaft splines contact the seal lips). Tap the stub shaft into place until it is seated.

13   Install the axle seal to the correct depth.

All models

14   Install the driveaxle (see Chapter 8).

15   Installation is the reverse of removal. Check the transaxle fluid level, adding as necessary (see Chapter 1).

7   Transmission Control Module (TCM) - removal and installation

Caution: *The TCM is an Electro-Static Discharge (ESD) sensitive electronic device, meaning a static electricity discharge from your body could possibly damage electrical components. Be sure to properly ground yourself and the TCM before handling it. Avoid touching the electrical terminals of the TCM.*

Note: *The procedures in this section apply only to removing and installing the TCM that is already installed in your vehicle. If you need a new TCM, it must be programmed with new*

software and calibrations. This procedure requires the use of GM's TECH-2 scan tool and GM's latest TCM-programming software, so you WILL NOT BE ABLE TO REPLACE THE TCM AT HOME.

1   Disconnect the cable from the negative terminal of the battery (see Chapter 5, Section 1).

2   Disconnect the TCM electrical connector (see illustration).

3   Detach the TCM from its bracket and remove it from the engine compartment (see illustration).

4   Installation is the reverse of removal.

8   Automatic transaxle - removal and installation

Removal

1   Disconnect the cable from the negative terminal of the battery (see Chapter 5, Section 1).

2   Remove the air filter housing (see Chap-

ter 4). Remove the battery and battery tray.

3   Disconnect the shift cable from the manual lever and the bracket on the transaxle (see Section 4).

4   Support the radiator and condenser from above using wire or large plastic tie-wraps, connected to the condenser tabs.

5   Clearly label, then unplug, all electrical connectors from the transaxle which are accessible from the top. Detach the harnesses from the brackets along the top of the transaxle.

6   Attach an engine support fixture or an engine hoist to the engine and raise it sufficiently to just support the weight of the engine (see illustration).

Note: *The engine must remain supported while the transaxle is out of the vehicle.*

Note: *If you use an engine hoist, position the hoist with its legs inserted under the vehicle from the right (passenger's) side. This will give you room to maneuver the transaxle out with a jack.*

7   Remove the upper transaxle-to-engine bolts (see illustration).

8.7 Remove these three transaxle-to-engine bolts from above

8.13 Location of the transaxle brace mounting bolts

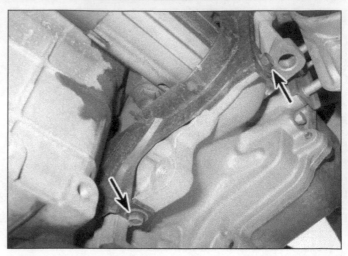

8.14 Location of the starter cover mounting bolts - one bolt hidden from view

8   Loosen the driveaxle/hub nuts and the front wheel lug nuts, then raise the vehicle and support it securely on jackstands. Remove the front wheels.

9   Remove both inner fender splash shields (see Chapter 11).

10   Detach both tie-rod ends from the steering knuckle arms (see Chapter 10). Also unbolt the control arms from the subframe.

11   Remove the driveaxles (see Chapter 8).

Note: *With the control arms detached from the subframe, you will be able to pull out on the steering knuckles far enough to remove the driveaxles (it won't be necessary to separate the control arm balljoint from the steering knuckle as described in Chapter 10).*

12   Disconnect the electrical connectors from both front ABS wheel speed sensors harnesses. Detach the harnesses from their clips on the control arms and subframe.

13   Remove the engine-to-transaxle brace (see illustration) and the transaxle-to-oil pan brace (see Chapter 2A, Chapter 2B or Chapter 2C).

14   Remove the starter cover bolts and the cover (see illustration).

15   Remove the starter motor (see Chapter 5).

16   Mark the relationship of the torque converter to the driveplate and remove the driveplate-to-torque converter bolts (see illustration). You can use a socket and breaker bar on the crankshaft pulley bolt to rotate the engine for access to the bolts, or you can turn the driveplate with a screwdriver by prying against the ring gear teeth.

17   Disconnect and plug the transaxle cooler lines (see illustration).

18   Remove the oil pan-to-bellhousing bracket, if equipped.

19   Disconnect all ground wires and any other electrical connectors accessible from underneath the vehicle.

20   Detach the brake line from the clips along the front of the subframe.

21   Remove the transaxle cooler lines from the clip at the front of the subframe.

22   Remove the pinch bolt that secures the intermediate steering shaft to the steering gear (see Chapter 10).

23   Unscrew the power steering line fittings from the power steering gear, if equipped (see Chapter 10).

24   Remove the bolt that secures the power steering line bracket at the right side of the subframe, if equipped.

25   On 2007 3.5L V6 models, remove the catalytic converter, the exhaust pipe and the exhaust manifold heat shield from the left side of the engine (see Chapter 4).

26   Unbolt all three powertrain mounts from the transaxle (see Chapter 2A, Chapter 2B or Chapter 2C).

27   Remove the subframe (see Chapter 10).

28   Support the transaxle with a jack, preferably a jack made for this purpose. Transmis-

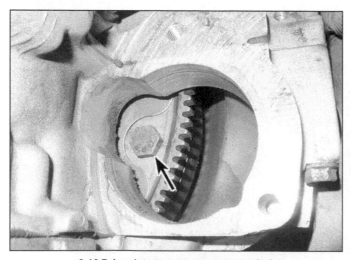

8.16 Driveplate-to-torque converter bolt

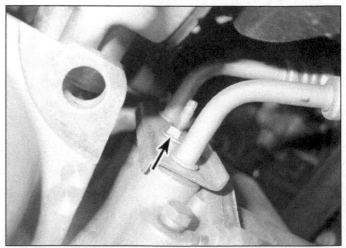

8.17 To detach the transaxle fluid cooler lines, remove the nut between the two lines (4T65-E shown, 6T40 and 6T70 cooling lines are attached at the control valve body cover)

sion jacks are commonly available at most equipment rental yards. These jacks are equipped with safety chains; use these chains to secure the transaxle to the jack.

29   Remove the left transaxle mount bracket-to-transaxle bolts and remove the mount and bracket assembly.

30   Remove the nut and bolt from the heater core hose pipe-to-transaxle bracket.

31   Remove the remaining transaxle-to-engine bolts.

32   Remove the transaxle from the engine by sliding it toward the left side of the vehicle.

Installation

33   Installation is the reverse of removal, with attention paid to the following points:

a) *Before installing the transaxle, make sure the torque converter is completely seated. To do this, push in on the converter while turning it. If it wasn't seated, it will "clunk" into place (it may even "clunk" more than once).*

b) *Apply a film of multi-purpose grease to the nose of the converter.*

c) *When mating the transaxle to the engine, make sure the transaxle seats against the engine completely before tightening the bolts; if it doesn't, figure out why.*

Don't use the bolts to draw the transaxle into place, as you could break something.

d) *Tighten all transaxle-to-engine bolts to the torque listed in this Chapter's Specifications 7B.*

e) *Tighten the torque converter-to-driveplate bolts to the torque listed in this Chapter's Specifications 7B.*

Note: *Install all of the bolts before tightening any of them.*

f) *Tighten the subframe mounting bolts to the torque listed in the Chapter 10 Specifications. Also tighten the intermediate shaft pinch bolt, the control arm balljoint-to-steering knuckle nuts and the tie-rod end-to-steering knuckle nuts to the torque values listed in the Chapter 10 Specifications.*

g) *Tighten the transmission (powertrain) mounting fasteners securely.*

h) *Tighten the wheel lug nuts to the torque listed in the, Section 1.*

i) *The front end alignment should be checked and, if necessary, adjusted.*

j) *Adjust the shift cable (see Chapter 4).*

k) *Check the transaxle fluid level and add fluid, as necessary, to bring it to the appropriate level (see Chapter 1).*

9   Automatic transaxle overhaul - general information

1   In the event of a fault occurring, it will be necessary to establish whether the fault is electrical, mechanical or hydraulic in nature, before repair work can be contemplated. Diagnosis requires detailed knowledge of the transaxle's operation and construction, as well as access to specialized test equipment, and so is deemed to be beyond the scope of this manual. It is therefore essential that problems with the automatic transaxle are referred to a dealer service department or other qualified repair facility for assessment.

2   Note that a faulty transaxle should not be removed before the vehicle has been assessed by a knowledgeable technician equipped with the proper tools, as troubleshooting must be performed with the transaxle installed in the vehicle.

Notes

Chapter 8
Clutch and driveaxles

Contents

Specifications

Clutch fluid type .... Chapter 1

Torque specifications
Ft-lbs (unless otherwise indicated)

Note: *One foot-pound (ft-lb) of torque is equivalent to 12 inch-pounds (in-lbs) of torque. Torque values below approximately 15 ft-lbs are expressed in inch-pounds, since most foot-pound torque wrenches are not accurate at these smaller values.*

Clutch pressure plate-to-flywheel bolts	18
Clutch release cylinder mounting fasteners	89 in-lbs
Driveaxle/hub nut*	159 ft-lbs
Intermediate shaft bracket bolts	44
Wheel lug nuts	See Chapter 1

*Nut must be replaced with a new one

1　General Information

1　The information in this Chapter deals with the components that transmit power to the front wheels, except for the transaxle, which is dealt with in Chapter 7A and Chapter 7B. For the purposes of this Chapter, these components are grouped into two categories: clutch and driveaxles. Separate Sections within this Chapter offer general descriptions and checking procedures for both groups.

2　Since nearly all the procedures covered in this Chapter involve working under the vehicle, make sure it's securely supported on sturdy jackstands or on a hoist where the vehicle can be easily raised and lowered.

2　Clutch - description and check

1　All vehicles with a manual transaxle have a single dry plate, diaphragm spring-type clutch. The clutch disc has a splined hub that allows it to slide along the splines of the transaxle input shaft. The clutch and pressure plate are held in contact by spring pressure exerted by the diaphragm in the pressure plate.

2　The clutch release system is operated by hydraulic pressure. The hydraulic release system consists of the clutch pedal, a master cylinder and a reservoir, a release (or slave) cylinder and the hydraulic line connecting the two components. The release cylinder and release bearing are integral parts of a single assembly, which is installed concentric to the input shaft and is bolted to the transaxle.

3　When the clutch pedal is depressed, a pushrod pushes against brake fluid inside the master cylinder, applying hydraulic pressure to the release cylinder, which pushes the release bearing against the diaphragm fingers of the clutch pressure plate.

4　Terminology can be a problem when discussing the clutch components because common names are in some cases different from those used by the manufacturer. For example, the driven plate is also called the clutch plate or disc, the clutch release bearing is sometimes called a throwout bearing, the release cylinder is sometimes called the slave cylinder.

5　Unless you're replacing components with obvious damage, do these preliminary checks to diagnose clutch problems:

a) *The first check should be of the fluid level in the master cylinder. If the fluid level is low, add fluid as necessary and inspect the hydraulic system for leaks. If the master cylinder reservoir is dry, bleed the system as described in Section 5 and recheck the clutch operation.*

b) *To check clutch spin-down time, run the engine at normal idle speed with the transaxle in Neutral (clutch pedal up - engaged). Disengage the clutch (pedal down), wait several seconds and shift the transaxle into Reverse. No grinding*

noise should be heard. A grinding noise would most likely indicate a bad pressure plate or clutch disc.

c) *To check for complete clutch release, run the engine (with the parking brake applied to prevent vehicle movement) and hold the clutch pedal approximately 1/2-inch from the floor. Shift the transaxle between 1st gear and Reverse several times. If the shift is rough, component failure is indicated.*

d) *Visually inspect the pivot bushing at the top of the clutch pedal to make sure there's no binding or excessive play.*

3　Clutch master cylinder - removal and installation

Removal

1　Under the dash on the driver's side, pull the plastic clip near the top of the clutch pedal that secures the clutch master cylinder push-rod to the pedal. Pull the clutch pedal upward to disengage it.

2　Working inside the engine compartment, remove as much fluid as you can from the master cylinder reservoir with a syringe, such as an old turkey baster.

Warning: *If a baster is used, never again use it for the preparation of food.*

3　Place rags under the fluid fittings and prepare caps or plastic bags to cover the ends of lines or fittings once they are disconnected.

Caution: *Brake fluid will damage paint. Cover all body parts and be careful not to spill fluid during this procedure.*

4　Disconnect the cable from the negative terminal of the battery (see Chapter 5).

5　Remove the fasteners securing the clutch fluid reservoir.

6　Remove the air cleaner duct (see Chapter 4).

7　Remove the line clip at the master cylinder hydraulic line fitting, then separate the hydraulic line from the cylinder.

8　Grab the master cylinder at the engine-side of the firewall and twist it 90 degrees clockwise and remove the master cylinder from the vehicle.

Installation

9　Place the master cylinder pushrod through the firewall and twist the master cylinder 90 degrees counterclockwise to secure it, making sure the pushrod is aligned with the clutch pedal.

10　Push the clutch pedal down to engage the master cylinder pushrod and reinsert the plastic clip to hold it.

11　Connect the hydraulic line fitting to the clutch master cylinder and install the retaining clip at the fitting.

12　Install the clutch fluid reservoir.

13　Fill the reservoir with brake fluid conforming to DOT 3 specifications and bleed the clutch system as outlined in Section 5.

4　Clutch release cylinder - replacement

1　Disconnect the hydraulic line fitting from the release cylinder.

2　Remove the transaxle (see Chapter 7A).

3　Remove the fasteners securing the release cylinder to the transaxle and slide the release cylinder assembly off the transaxle input shaft.

4　Installation is the reverse of removal, noting the following points:

a) *Lubricate the inside diameter of the bearing with high-temperature grease.*

b) *Apply a thread locking compound to the release cylinder mounting fasteners and tighten them to the torque listed in this Chapter's Specifications.*

c) *Install the transaxle (see Chapter 7A).*

d) *Connect the hydraulic fitting.*

e) *Check the fluid level in the brake fluid reservoir, adding brake fluid conforming to DOT 3 specifications until the level is correct.*

f) *Bleed the system as described in Section 5.*

5　Clutch hydraulic system - bleeding

1　Bleed the hydraulic system whenever any part of the system has been removed or the fluid level has fallen so low that air has been drawn into the master cylinder. The bleeding procedure is very similar to bleeding a brake system.

2　Fill the clutch master cylinder reservoir with new brake fluid conforming to DOT 3 specifications.

Caution: *Do not re-use any of the fluid coming from the system during the bleeding operation or use fluid which has been inside an open container for an extended period of time.*

Connect a hose to the bleeder valve fitting at the top of the transaxle. Place the other end of the hose in a container partially filled with clean brake fluid.

3　Have an assistant depress the clutch pedal and hold it. Loosen the bleeder valve, allowing fluid and any air to escape. Tighten the bleeder valve when the flow of fluid (and bubbles) ceases. Once closed, have your assistant release the pedal.

4　Continue this process until all air is evacuated from the system, indicated by a solid stream of fluid being ejected from the bleeder valve each time with no air bubbles. Keep a close watch on the fluid level inside the clutch master cylinder reservoir - if the level drops too far, air will get into the system and you'll have to start all over again.

Note: *Wash the area with water to remove any spilled brake fluid.*

5　Check the brake fluid level again, and add some, if necessary, to bring it to the appropriate level. Check carefully for proper operation before placing the vehicle into normal service.

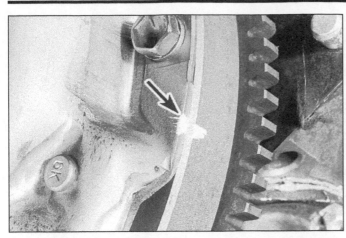

6.4 Mark the relationship of the pressure plate to the flywheel (if you're planning to re-use the old pressure plate)

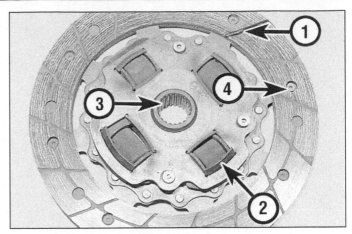

6.8 The clutch disc

1   **Lining** - *this will wear down in use*
2   **Springs or dampers** - *check for cracking and deformation*
3   **Splined hub** - *the splines must not be worn and should slide smoothly on the transmission input shaft splines*
4   **Rivets** - *these secure the lining and will damage the flywheel or pressure plate if allowed to contact the surfaces*

6   Clutch components - removal, inspection and installation

Warning: *Dust produced by clutch wear is hazardous to your health. DO NOT blow it out with compressed air and DO NOT inhale it. DO NOT use gasoline or petroleum-based solvents to remove the dust. Brake system cleaner should be used to flush the dust into a drain pan. After the clutch components are wiped clean with a rag, dispose of the contaminated rags and cleaner in a covered, marked container.*

Removal

1   Access to the clutch components is normally accomplished by removing the transaxle, leaving the engine in the vehicle. If the engine is being removed for major overhaul, check the clutch for wear and replace worn components as necessary. However, the relatively low cost of the clutch components compared to the time and trouble spent gaining access to them warrants their replacement anytime the engine or transaxle is removed, unless they are new or in near-perfect condition. The following procedures are based on the assumption the engine will stay in place.
2   Remove the transaxle from the vehicle (see Chapter 7A). Support the engine while the transaxle is out. Preferably, an engine support fixture or a hoist should be used to support it from above.
3   To support the clutch disc during removal, install a clutch alignment tool through the clutch disc hub. The tool is available at most auto parts stores.
4   Carefully inspect the flywheel and pressure plate for indexing marks. The marks are usually an X, an O or a black mark. If they cannot be found, scribe or paint marks yourself so the pressure plate and the flywheel will be in the same alignment during installation (see illustration).
5   Turning each bolt a little at a time, loosen the pressure plate-to-flywheel bolts. Work in a criss-cross pattern until all spring pressure is relieved evenly. Then hold the pressure plate securely and completely remove the bolts, followed by the pressure plate and clutch disc.

Inspection

6   Ordinarily, when a problem occurs in the clutch, it can be attributed to wear of the clutch driven plate assembly (clutch disc). However, all components should be inspected at this time.
7   Inspect the flywheel for cracks, heat checking, grooves and other obvious defects. If the imperfections are slight, a machine shop can machine the surface flat and smooth, which is highly recommended regardless of the surface appearance. Refer to Chapter 2A, Chapter 2B or Chapter 2C for the flywheel removal and installation procedure.
8   Inspect the lining on the clutch disc. There should be at least 1/16-inch of lining above the rivet heads. Check for loose rivets, distortion, cracks, broken springs/dampers and other obvious damage (see illustration). As mentioned above, ordinarily the clutch disc is routinely replaced, so if in doubt about the condition, replace it with a new one.
9   The clutch release cylinder/release bearing should also be replaced along with the clutch disc (see Section 4).
10 Check the machined surfaces and the diaphragm spring fingers of the pressure plate (see illustrations). If the surface is grooved or otherwise damaged, replace the pressure plate. Also check for obvious damage,

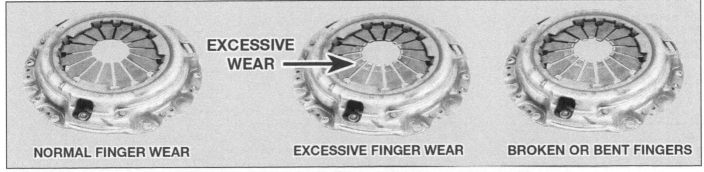

NORMAL FINGER WEAR   EXCESSIVE FINGER WEAR   BROKEN OR BENT FINGERS

6.10a Replace the pressure plate if excessive wear or damage is noted

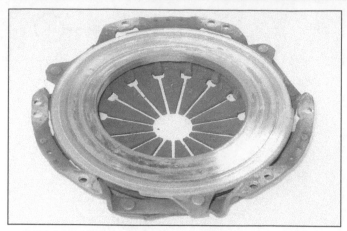

6.10b Inspect the pressure plate surface for excessive score marks, cracks and signs of overheating

6.12 Center the clutch disc in the pressure plate with a clutch alignment tool

distortion, cracking, etc. Light glazing can be removed with emery cloth or sandpaper. If a new pressure plate is required, new and remanufactured units are available.

Installation

11   Before installation, clean the flywheel and pressure plate machined surfaces with brake cleaner, lacquer thinner or acetone. It's important that no oil or grease is on these surfaces or the lining of the clutch disc. Handle the parts only with clean hands.

12   Position the clutch disc and pressure plate against the flywheel with the clutch held in place with an alignment tool (see illustration). Make sure the disc is installed properly (most replacement clutch discs will be marked "flywheel side" or something similar - if not marked, install the clutch disc with the damper springs toward the transaxle).

13   Tighten the pressure plate-to-flywheel bolts only finger tight, working around the pressure plate.

14   Center the clutch disc by ensuring the alignment tool extends through the splined hub and into the pilot bearing in the crankshaft. Wiggle the tool up, down or side-to-side as needed to center the disc. Install and tighten the pressure plate-to-flywheel bolts a little at a time, working in a criss-cross pattern to prevent distorting the cover. After all of the bolts are snug, tighten them to the torque listed in this Chapter's Specifications. Remove the alignment tool.

15   Install the clutch release cylinder (see Section 4).

16   Install the transaxle and all components removed previously.

7   Clutch Pedal Position (CPP) sensor - replacement

Note: *The Clutch Pedal Position (CPP) sensor is an integral component of the clutch pedal. To replace the CPP sensor, you must replace the pedal assembly.*

Note: *A CPP sensor learn procedure must be performed after the replacement of a CPP sensor. The learn procedure must be performed with a factory scan tool by a dealership service department.*

1   Remove the air cleaner duct (see Chapter 4).

2   Working in the engine compartment, remove the fastener securing the clutch pedal to the firewall.

3   In the passenger compartment, remove the driver's side lower trim panel (see Chapter 11).

4   Unplug the electrical connector from the sensor.

5   Pull the plastic clip near the top of the clutch pedal that secures the clutch master cylinder pushrod to the pedal. Pull the clutch pedal upward to disengage it.

6   Remove the remaining fasteners securing the clutch pedal to the firewall. Remove the pedal

7   Installation is the reverse of removal.

Warning: *Have the CPP sensor learn procedure performed by a dealership service department or other repair shop with the proper equipment. Failure to do this might allow the vehicle to start without depressing the clutch pedal.*

8   Driveaxles - general information and inspection

1   Power is transmitted from the transaxle to the wheels through a pair of driveaxles. The inner end of each driveaxle is connected to the transaxle, directly splined to the differential side gears. The outer ends of the driveaxles are splined to the axle hubs and locked in place by a large nut.

2   The inner ends of the driveaxles are equipped with sliding constant velocity joints, which are capable of both angular and axial motion. Each inner joint assembly consists of either a tripod bearing and a joint tulip (housing) in which the joint is free to slide in-and-out as the driveaxle moves up-and-down with

the wheel. The joints can be disassembled and cleaned in the event of a boot failure, but if any parts are damaged, the joints must be replaced as a unit (see Section 10).

3   Each outer joint, which consists of ball bearings running between an inner race and an outer race (housing), is capable of angular but not axial movement.

4   The boots should be inspected periodically for damage and leaking lubricant. Torn CV joint boots must be replaced immediately or the joints can be damaged. Boot replacement involves removal of the driveaxle (see Section 9).

Note: *Some auto parts stores carry "split" type replacement boots, which can be installed without removing the driveaxle from the vehicle. This is a convenient alternative; however, the driveaxle should be removed and the CV joint disassembled and cleaned to ensure the joint is free from contaminants such as moisture and dirt which will accelerate CV joint wear.*

5   The most common symptom of worn or damaged CV joints, besides lubricant leaks, is a clicking noise in turns, a clunk when accelerating after coasting and vibration at highway speeds. To check for wear in the CV joints and driveaxle shafts, grasp each axle (one at a time) and rotate it in both directions while holding the CV joint housings, feeling for play indicating worn splines or sloppy CV joints. Also check the driveaxle shafts for cracks, dents and distortion.

9   Driveaxles - removal and installation

Warning: *The manufacturer recommends replacing the driveaxle/hub nut with a new one whenever they are removed.*

Driveaxle

Removal

1   Set the parking brake. Remove the wheel cover or hubcap.

9.2 Loosen the driveaxle/hub nut with a long breaker bar

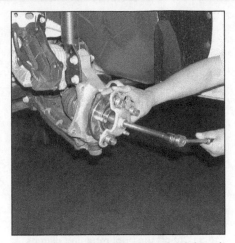

9.7 Attach a puller to the hub and tighten it just enough to break the hub splines loose

9.10 Using a large prybar, carefully pry the inner end of the driveaxle from the transaxle

2   Loosen the driveaxle/hub nut with a large socket and breaker bar, but don't remove it yet (see illustration).

3   Loosen the front wheel lug nuts, raise the vehicle and support it securely on jackstands. Remove the wheel.

4   Remove the driveaxle/hub nut.

5   Disconnect the tie-rod end from the steering knuckle (see Chapter 10).

6   Separate the control arm from the steering knuckle (see Chapter 10).

7   To loosen the driveaxle from the hub splines, tap the end of the driveaxle with a soft-faced hammer. If the driveaxle is stuck in the hub splines and won't move, it may be necessary to push it from the hub with a puller (see illustration).

8   Pull out on the steering knuckle and detach the driveaxle from the hub. Suspend the outer end of the driveaxle on a bungee cord or piece of wire.

9   Before you remove the driveaxle, look for lubricant leakage in the area around the differential seal. If there's evidence of a leak, you'll want to replace the seal after removing the driveaxle (see Chapter 7B).

10   Position a prybar against the inner joint and carefully pry the joint off the transaxle side gear (see illustration). Do not use the driveaxle to pull on the inner joint. Doing so might damage the inner joint components. Remove the driveaxle assembly, being careful not to over-extend the inner joint or damage the axleshaft boots.

11   Should it become necessary to move the vehicle while the driveaxle is out, place a large bolt with two large washers (one on each side of the hub) through the hub and tighten the nut securely.

Installation

12   Installation is the reverse of removal, but with the following additional points:

a) *Remove the old set-ring from the transaxle stub shaft or intermediate shaft (depending on which side you're working*

on) and install a new one (see illustration).

b) *Apply a film of multi-purpose grease around the splines of the joints.*

c) *When installing the driveaxle, hold the driveaxle straight out, then push it in sharply to seat the driveaxle set-ring. To make sure the set-ring is properly seated, attempt to pull the inner CV joint housing out of the transaxle by hand. If the set-ring is properly seated, the inner joint will not move out.*

d) *Clean all foreign matter from the driveaxle outer CV joint threads and coat the splines with multi-purpose grease. Guide the driveaxle into the hub splines and install the new driveaxle/hub nut. Tighten the nut securely but not to the specified torque at this time.*

e) *Using new fasteners, reconnect the control arm and tie-rod end, then tighten the suspension fasteners to the torque listed in the Chapter 10 Specifications.*

f) *Install the wheel and lug nuts, then lower the vehicle.*

g) *Tighten the driveaxle/hub nut to the torque listed in this Chapter's Specifications.*

h) *Tighten the wheel lug nuts to the torque listed in the Chapter 1 Specifications.*

i) *Add transaxle lubricant if it was drained or if any fluid spilled out (see Chapter 1).*

Intermediate shaft

13   Remove the right-side driveaxle (see Steps 1 through 10).

14   Remove the intermediate shaft bracket bolts, then pull the intermediate shaft out of the transaxle.

15   If necessary, replace the oil seal (see Chapter 7B).

16   Installation is the reverse of removal. Install the lower bracket bolt that is closest to the drivebelt end of the engine, but do not tighten it. Install the other two bolts. Tighten the intermediate shaft bracket bolts, begin-

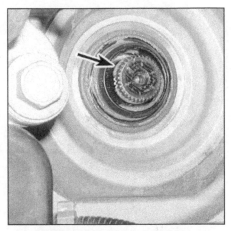

9.12 Remove the old driveaxle set-ring and replace it with a new one

ning with the upper bolt, to the torque listed in this Chapter's Specifications.

17   Install the driveaxle (see Step 12).

10   Driveaxle boot replacement

Note: *If the CV joint boots must be replaced, explore all options before beginning the job. Complete rebuilt driveaxles are available on an exchange basis, which eliminates much time and work. Whichever route you choose to take, check on the cost and availability of parts before disassembling the vehicle.*

1   Remove the driveaxle (see Section 9).

2   Place the driveaxle in a vise lined with rags to avoid damage to the axleshaft. Check the CV joint for excessive play in the radial direction, which indicates worn parts. Check for smooth operation throughout the full range of motion for each CV joint. If a boot is torn, disassemble the joint, clean the components and inspect for damage due to loss of lubrication and possible contamination by foreign matter.

10.3a Cut off the boot retaining clamps
and discard them

10.3b Slide the housing off the
spider assembly

Inner CV joint

3   To replace the inner boot, refer to the accompanying illustrations (see illustrations 10.3a through 10.3t).

10.3c Slide the boot towards the center of
the driveaxle

10.3d Spread the ends of the stop ring
apart and slide it towards the center of
the shaft

10.3e Slide the spider assembly back to
expose the retaining ring and pry off
the ring

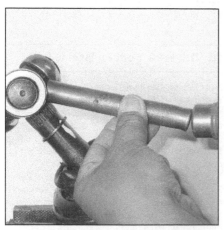

10.3f Carefully tap the spider off the
axleshaft with a brass punch (but don't
hit it so hard that it flies off, or you'll be
picking up needle bearings!)

10.3g When you slide the spider off the
driveaxle, hold the bearings in place with
your hand; even better, use tape or a
cloth wrapped around the spider bearing
assembly to retain them

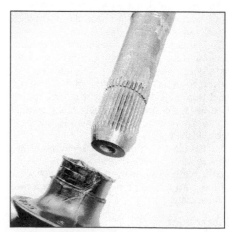

10.3h Slide the boot and the stop ring off
the axleshaft

10.3i Clean all of the old grease out of the housing and spider assembly, then remove each bearing, one at time

10.3j Carefully disassemble each section of the spider assembly, clean the needle bearings with solvent and inspect the rollers, spider cross, bearings and housing for scoring, pitting and other signs of abnormal wear

10.3k Apply a coat of CV joint grease to the inner bearing surfaces to hold the needle bearings in place and slide the bearing over them

10.3l Wrap the axleshaft splines with tape to avoid damaging the boot, then slide the small clamp and boot onto the axleshaft

10.3m Slide the spider stop ring onto the axleshaft, past the groove in which it seats

10.3n Install the spider bearing with the recess in the counterbore facing the end of the driveaxle

10.3o Install the spider retaining ring, then slide the spider assembly against it and install the stop ring in its groove

10.3p Pack the housing with half of the grease furnished with the new boot and place the remainder in the boot

10.3q With the retaining clamps in place (but not tightened), install the tripot housing

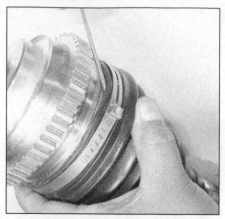

10.3r Seat the boot in the housing and axle seal grooves - a small screwdriver can make the job easier (make sure the boot isn't dimpled, stretched or out of shape)

10.3s Position the CV joint mid-way through its travel

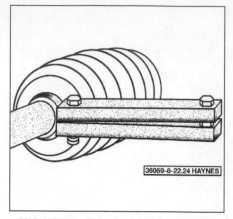

10.3t With the joint at the proper length, equalize the pressure in the boot by inserting a small screwdriver between the boot and the housing, then secure the boot clamps with a clamp crimping tool (available at auto parts stores)

Outer CV joint

4   Refer to the accompanying illustrations and perform the outer CV joint boot replacement procedure (see illustrations 10.4a through 10.4q).

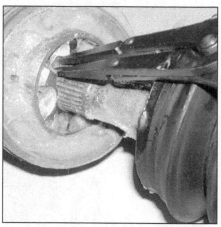

10.4a Cut off the boot retaining clamps, and slide the boot back, spread apart the ends of the internal snap-ring, then slide the CV joint off the shaft

10.4b Press down on the inner race far enough to allow a ball bearing to be removed - if it's difficult to tilt, gently tap the cage and inner race with a brass punch and hammer

10.4c Pry the balls out of the cage, one at a time

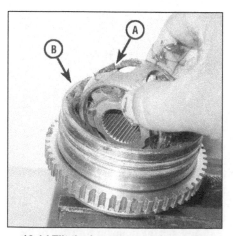

10.4d Tilt the inner race and cage 90-degrees, then align the windows in the cage (A) with the lands of the housing (B) and rotate the inner race up and out of the outer race

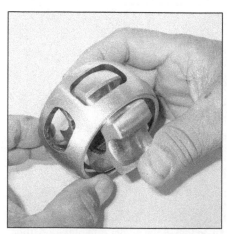

10.4e Align the inner race lands with the cage window and rotate the inner race out of the cage

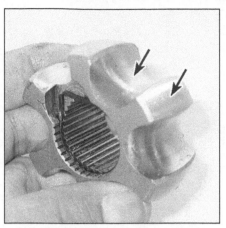

10.4f After cleaning the components with solvent, check the inner race lands and grooves for pitting and score marks

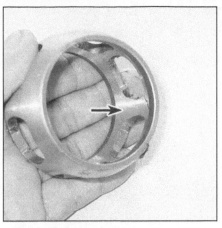

10.4g Check the cage for cracks, pitting and score marks - shiny spots are normal and don't affect operation

10.4h With the race and cage tilted at 90-degrees, lower the assembly into the housing

10.4i Rotate the assembly by gently tapping with a hammer and brass punch, then . . .

10.4j . . . press the balls into the cage windows, repeating until all of the balls are installed

10.4k Use needle-nose pliers to lower a new snap-ring into the groove . . .

10.4l . . . then seat it into the groove with snap-ring pliers

10.4m Apply grease through the splined hole, then insert a wooden dowel (with a diameter slightly less than that of the axle) through the splined hole and push down - the dowel will force the grease into the joint - repeat until the bearing is completely packed

10.4n Install the small clamp and the boot on the driveaxle and apply grease to the inside of the axle boot . . .

10.4o . . . until the level is up to the end of axle

10.4p Position the CV joint assembly on the driveaxle, aligning the splines, then use a soft-face hammer to drive the joint onto the driveaxle until the snap-ring is seated in the groove

10.4q Seat the inner end of the boot in the groove and install the retaining clamp, then do the same on the other end of the boot - tighten boot clamps with the special tool (see illustration 10.3t)

Chapter 9
Brakes

Contents

Specifications

General
Brake fluid type .... See Chapter 1

Disc brakes
Brake pad minimum thickness	See Chapter 1
Disc lateral runout limit	0.002 inch
Disc minimum thickness	Cast into disc
Parallelism (thickness variation) limit	0.001 inch

Drum brakes
Maximum drum diameter	Cast into drum
Shoe lining minimum thickness	See Chapter 1

Torque specifications   Ft-lbs (unless otherwise indicated)

Note: *One foot-pound (ft-lb) of torque is equivalent to 12 inch-pounds (in-lbs) of torque. Torque values below approximately 15 ft-lbs are expressed in inch-pounds, since most foot-pound torque wrenches are not accurate at these smaller values.*

ABS hydraulic control unit mounting bracket bolts	86 in-lbs
Brake hose banjo fitting bolt (drum and disc)	37
Caliper mounting bolts (front and rear)	26
Caliper mounting bracket bolts	
Aura	96
Malibu models	
2007 and earlier	85
2008 and later	96
G6 models	
2006 and earlier	85
2007 and later	96
Master cylinder mounting nuts	18
Master cylinder reservoir mounting bolt (2008 and later models)	71 in-lbs
Proportioning valve bracket mounting nuts	86 in-lbs
Power brake booster mounting nuts	132 in-lbs
Wheel cylinder mounting bolts	89 in-lbs
Wheel lug nuts	See Chapter 1

1   General information

1   The vehicles covered by this manual are equipped with hydraulically operated front and rear brake systems. The front brakes are disc type and the rear brakes are disc or drum type. Both the front and rear brakes are self adjusting. The disc brakes automatically compensate for pad wear, while the drum brakes incorporate an adjustment mechanism that is activated as the parking brake is applied.

Hydraulic system

2   The hydraulic system consists of two separate circuits. The master cylinder has separate reservoir chambers for the two circuits, and, in the event of a leak or failure in one hydraulic circuit, the other circuit will remain operative. A dual proportioning valve in the circuit provides brake balance between the front and rear brakes. Models that are equipped with ABS (see Section 2) do require a proportioning valve.

Power brake booster

3   The power brake booster is mounted on the firewall in the engine compartment. It utilizes engine manifold vacuum and atmospheric pressure to provide assistance to the hydraulically operated brakes resulting in less brake pedal effort.

Parking brake

4   The parking brake system mechanically operates the rear brakes through the use of cables. These models use a self-adjusting parking brake pedal.

Service

5   After completing any operation involving disassembly of any part of the brake system, always test drive the vehicle to check for proper braking performance before resuming normal driving. When testing the brakes, perform the tests on a clean, dry, flat surface. Conditions other than these can lead to inaccurate test results.

6   Test the brakes at various speeds with both light and heavy pedal pressure. The vehicle should stop evenly without pulling to one side or the other. Avoid locking the brakes, because this slides the tires and diminishes braking efficiency and control of the vehicle.

7   Tires, vehicle load and wheel alignment are factors which also affect braking performance.

Precautions

8   There are some general cautions and warnings involving the brake system on this vehicle:

a) *Use only brake fluid conforming to DOT 3 specifications.*

b) *The brake pads and linings contain fibers that are hazardous to your health if inhaled. Whenever you work on brake system components, clean all parts with*

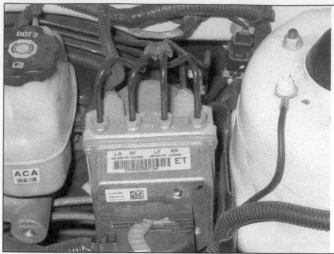

2.2 The ABS hydraulic unit - Delphi unit shown; 2008 and later TRW unit is similar

brake system cleaner. Do not allow the fine dust to become airborne. Also, wear an approved filtering mask.

c) *Safety should be paramount whenever any servicing of the brake components is performed. Do not use parts or fasteners that are not in perfect condition, and be sure that all clearances and torque specifications are adhered to. If you are at all unsure about a certain procedure, seek professional advice. Upon completion of any brake system work, test the brakes carefully in a controlled area before putting the vehicle into normal service. If a problem is suspected in the brake system, don't drive the vehicle until it's fixed.*

2   Anti-lock Brake System (ABS) - general information

General information

1   The Anti-lock Brake System is designed to maintain vehicle steerability, directional stability and optimum deceleration under severe braking conditions on most road surfaces. It does so by monitoring the rotational speed of each wheel and controlling the brake line pressure to each wheel during braking. This prevents the wheels from locking up.

2   The ABS system has three main components - the wheel speed sensors, the electronic control unit (ECU) and the hydraulic unit (see illustration). Four wheel speed sensors - one at each wheel - send a variable voltage signal to the control unit, which monitors these signals, compares them to its program and determines whether a wheel is about to lock up. When a wheel is about to lock up, the control unit signals the hydraulic unit to reduce hydraulic pressure (or not increase it further) at that wheel's brake caliper. Pressure modulation is handled by electrically-operated solenoid valves.

3   If a problem develops within the system,

an "ABS" warning light will glow on the dashboard. Sometimes, a visual inspection of the ABS system can help you locate the problem. Carefully inspect the ABS wiring harness. Pay particularly close attention to the harness and connections near each wheel. Look for signs of chafing and other damage caused by incorrectly routed wires. If a wheel sensor harness is damaged, it must be replaced.

Warning: *Do NOT try to repair an ABS wiring harness. The ABS system is sensitive to even the smallest changes in resistance. Repairing the harness could alter resistance values and cause the system to malfunction. If the ABS wiring harness is damaged in any way, it must be replaced.*

Caution: *Make sure the ignition is turned off before unplugging or reattaching any electrical connections.*

Diagnosis and repair

4   If a dashboard warning light comes on and stays on while the vehicle is in operation, the ABS system requires attention. Although special electronic ABS diagnostic testing tools are necessary to properly diagnose the system, you can perform a few preliminary checks before taking the vehicle to a dealer service department.

a) *Check the brake fluid level in the reservoir.*

b) *Verify that the computer electrical connectors are securely connected.*

c) *Check the electrical connectors at the hydraulic control unit.*

d) *Check the fuses.*

e) *Follow the wiring harness to each wheel and verify that all connections are secure and that the wiring is undamaged.*

5   If the above preliminary checks do not rectify the problem, the vehicle should be diagnosed by a dealer service department or other qualified repair shop. Due to the complex nature of this system, all actual repair work must be done by a qualified automotive technician.

3.5 To make room for the new pads, depress the caliper piston into the bottom of its bore with a large C-clamp

3.6a Always wash the brakes with brake cleaner before disassembling anything

Wheel speed sensor - removal and installation

6   The wheel speed sensors are integrated into the wheel bearing and hub assemblies. Refer to Chapter 10 for replacement of these components.

3   Disc brake pads - replacement

Warning: Disc brake pads must be replaced on both front or rear wheels at the same time - never replace the pads on only one wheel. Also, the dust created by the brake system is harmful to your health. Never blow it out with compressed air and don't inhale any of it. An approved filtering mask should be worn when working on the brakes. Do not, under any circumstances, use petroleum-based solvents to clean brake parts. Use brake system cleaner only!

Note: Disc brake pad replacement for rear disc brakes is similar to the Steps outlined be-

low for the front pads. See Section 4 for details of rear caliper mounting.

1   Remove the cap from the brake fluid reservoir and remove about two-thirds of the fluid.

Caution: Brake fluid will damage paint. Cover all painted surfaces around the work area and be careful not to spill fluid during this procedure. Clean any spilled fluid immediately and rinse the area with lots of water.

2   Loosen the wheel lug nuts, raise the end of the vehicle you're working on and support it securely on jackstands. Block the wheels at the opposite end.

3   Remove the wheels. Work on one brake assembly at a time, using the assembled brake for reference if necessary.

4   Inspect the brake disc carefully as outlined in Section 5. If machining is necessary, follow the information in that Section to remove the disc, at which time the pads can be removed as well.

5   Push the piston back into its bore to provide room for the new brake pads. A C-clamp

can be used to accomplish this (see illustration). As the piston is depressed to the bottom of the caliper bore, the fluid in the master cylinder will rise. Make sure that it doesn't overflow. If necessary, siphon off some of the fluid.

Caution: On rear disc brakes, the caliper contains an integral parking brake mechanism. Depress the piston with the C-clamp just enough to remove the caliper or the mechanism may become damaged.

6   Follow the accompanying photos (illustrations 3.6a through 3.6m) for the actual pad replacement procedure. Be sure to stay in order and read the caption under each illustration. The procedure for the rear disc brake pads is similar to that of the front pads.

7   After the job has been completed, firmly depress the brake pedal a few times to bring the pads into contact with the disc. Check the level of the brake fluid, adding some if necessary. Check the operation of the brakes thoroughly before placing the vehicle into normal service.

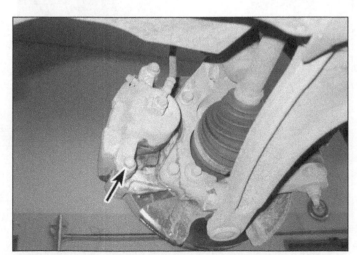

3.6b Remove the lower caliper mounting bolt (on rear calipers, it may be necessary to remove both mounting bolts because of the parking brake cable) . . .

3.6c . . . then, being careful not to damage the seal on the upper caliper guide pin, pivot the caliper up and secure it with a piece of wire

3.6d Remove the outer brake pad from the mounting bracket . . .

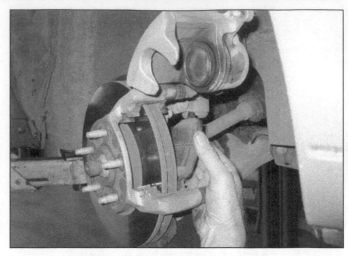

3.6e . . . then remove the inner pad

3.6f Remove both pad support plates. Clean and inspect them for wear or damage and replace them if necessary

3.6g Install the pad support plates

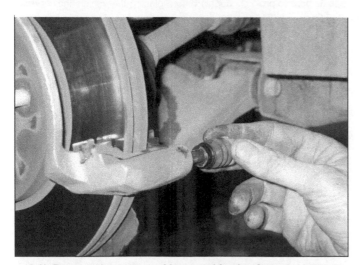

3.6h Remove the upper and lower guide pins from the caliper mounting bracket and clean the old grease off them (the upper guide pin can be removed by pulling the caliper from the mounting bracket with the pin still attached to the caliper)

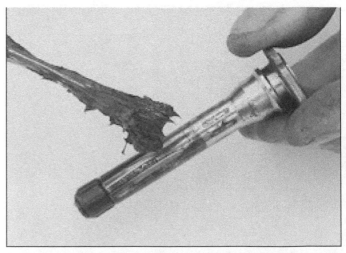

3.6i Inspect the guide pins for wear or damage and replace them if necessary. Lubricate the guide pins with high-temperature grease and install them. Make sure the guide pin seals are installed correctly

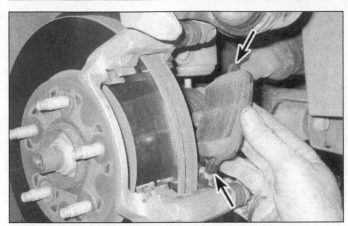

3.6j Make sure that the new pads are just like the old ones and complete with new shims (if equipped). Lubricate the ends of the new pads where they contact the anti-rattle clips with a small amount of high-temperature grease, then install the inner pad

3.6k Install the outer pad

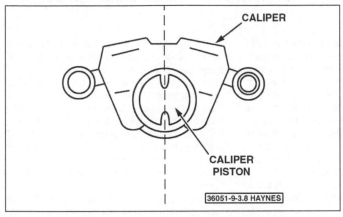

3.6l On rear calipers, rotate the piston with a brake piston tool or needle nose pliers until it's all the way at the bottom and then align the notches in the piston with the openings in the caliper as shown

3.6m Place the caliper over the new pads and back onto the mounting bracket. On rear calipers, carefully align the notch in the piston with the pin on the back of the inner brake pad. Install the caliper mounting bolts and tighten them to the torque listed in this Chapter's Specifications

4   Disc brake caliper - removal and installation

Warning: *Dust created by the brake system is harmful to your health. Never blow it out with compressed air and don't inhale any of it. An approved filtering mask should be worn when working on the brakes. Do not, under any circumstances, use petroleum-based solvents to clean brake parts. Use brake system cleaner only.*

Note: *If replacement is indicated (usually because of fluid leakage), it is recommended that the calipers be replaced, not overhauled. New and factory rebuilt units are available on an exchange basis, which makes this job quite easy. Always replace the calipers in pairs - never replace just one of them.*

Removal

1   Loosen the wheel lug nuts, raise the end of the vehicle you're working on and support it securely on jackstands. Block the wheels at the opposite end. Remove the wheels.

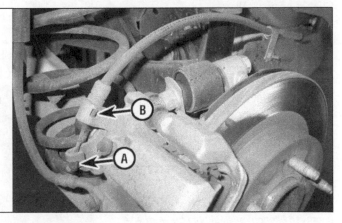

4.2 If you're removing a rear caliper, detach the end of the parking brake cable (A) from the lever, then depress the tangs (B) and detach the cable from the bracket

2   On rear calipers, disconnect the parking brake cable from the caliper (see illustration).
3   Depress the caliper piston with a C-clamp just enough to allow the pad to clear the rotor (see illustration 3.5).
Caution: *On rear disc brakes, the caliper contains an integral parking brake mechanism. Depress the piston with the C-clamp just*

enough to remove the caliper or the caliper may become damaged.
Note: *Anytime the rear calipers are removed, inspect the parking brake side of the caliper for any signs of brake fluid seepage around the shaft and seal for the parking brake lever. The seal is not serviceable, and if there is any leakage here, the caliper must be replaced.*

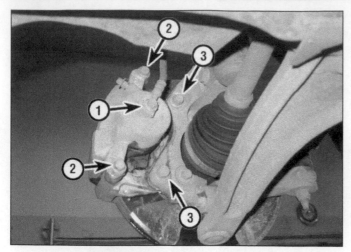

4.4 Brake caliper mounting details:

1	Brake line fitting	3	Caliper mounting
2	Caliper mounting bolts		bracket bolts

5.4 The brake pads on this vehicle were obviously neglected, as they wore down completely and cut deep grooves into the disc - wear this severe means the disc must be replaced

4   Disconnect the brake line from the caliper and plug it to keep contaminants out of the brake system and to prevent losing any more brake fluid than is necessary (see illustration).
Note: *If you're simply removing the caliper for access to other components, don't disconnect the brake line from the caliper.*
5   Remove the caliper mounting bolts.
6   Detach the caliper from its mounting bracket.

Installation

7   Install the caliper by reversing the removal procedure. Remember to replace the copper sealing washers on either side of the brake line fitting with new ones. Tighten the caliper mounting bolts and the brake line banjo fitting bolt to the torque listed in this Chapter's Specifications.
8   Bleed the brake system (see Section 10).
9   Install the wheels and lug nuts and lower the vehicle. Tighten the wheel lug nuts to the torque listed in the Chapter 1 Specifications. Check the operation of the brakes thoroughly before placing the vehicle into normal service.

---

5   Brake disc - inspection, removal and installation

Warning: *Dust created by the brake system is harmful to your health. Never blow it out with compressed air and don't inhale any of it. An approved filtering mask should be worn when working on the brakes. Do not, under any circumstances, use petroleum-based solvents to clean brake parts. Use brake system cleaner only.*

Inspection

Note: *The manufacturer may have installed a thin plate (with a small V-notch pointing to one*

of the wheel studs) between the disc and the hub flange as a method to correct disc runout. Before removing the plate from the hub, make sure that you place a match mark on the stud that the V-notch points to. If the disc is going to be machined or replaced, it's possible that the plate can be removed and discarded. Check the disc runout on the machined or replacement disc with the plate and disc installed. If there is excessive runout, remove the plate and check the disc runout again.
1   Loosen the wheel lug nuts, raise the vehicle and support it securely on jackstands. Remove the wheel.
2   Remove the brake caliper as outlined in Section 4. It's not necessary to disconnect the brake hose for this procedure. After removing the caliper mounting bolts, suspend the caliper out of the way with a piece of wire. Don't let the caliper hang by the hose and don't stretch or twist the hose.
3   Reinstall three lug nuts (inverted) to hold the disc against the hub. It may be necessary to install washers between the disc and the lug nuts to take up space.

4   Visually check the disc surface for score marks, cracks and other damage. Light scratches and shallow grooves are normal after use and may not always be detrimental to brake operation. Deep score marks or cracks may require disc refinishing by an automotive machine shop or disc replacement (see illustration). Be sure to check both sides of the disc. If pulsating has been noticed during application of the brakes, suspect disc runout.
Note: *The most common symptoms of damaged or worn brake discs are pulsation in the brake pedal when the brakes are applied or loud grinding noises caused from severely worn brake pads. If these symptoms are extreme, it is very likely that the disc(s) will need to be replaced.*
5   To check disc runout, place a dial indicator at a point about 1/2-inch from the outer edge of the disc (see illustration). Set the indicator to zero and turn the disc. An indicator reading that exceeds 0.003 of an inch could cause pulsation upon brake application and will require disc refinishing by an automotive

5.5a To check disc runout, mount a dial indicator as shown and rotate the disc

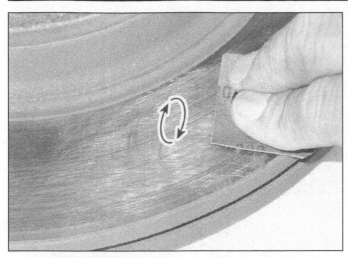

5.5b Using a swirling motion, remove the glaze from the disc surface with sandpaper or emery cloth

5.6a Here is an example of how the minimum thickness specification is cast into a disc (this specification is an example only - check your disc for the actual specification)

machine shop or disc replacement.

Note: *If disc refinishing or replacement is not necessary, you can deglaze the brake pad surface on the disc with emery cloth or sandpaper (use a swirling motion to ensure a non-directional finish)(see illustration).*

6   The disc must not be machined to a thickness less than the specified minimum refinish thickness. The minimum wear (or discard) thickness is cast into either the front or backside of the disc (see illustration). The disc thickness can be checked with a micrometer (see illustration).

Removal and installation

7   Remove the caliper mounting bracket (see illustration 4.4).

8   Mark the disc in relation to the hub so that it can be installed in it's original position on the hub and then remove the disc. If the disc has never been removed, there may be wave washers on the wheel stud securing it to the hub flange. Simply cut them off and discard them. If the disc appears to be stuck to the hub, use a mallet to knock it loose.

Note: *Be sure to look for a thin plate that may be installed between the brake disc and hub flange and refer to the Note at the beginning of this Section.*

9   Clean the hub flange and the inside of the brake disc thoroughly, removing any rust or corrosion, then install the disc onto the hub assembly.

10   Install the caliper mounting bracket and tighten the bolts to the torque listed in this Chapter's Specifications.

11   Place the caliper over the disc and onto the mounting bracket. Install the caliper mounting bolts and tighten them to the torque listed in this Chapter's Specifications.

12   Install the wheel and lower the vehicle to the ground. Torque the wheel lug nuts to the torque listed in Chapter 1 Specifications. Depress the brake pedal a few times to bring the brake pads into contact with the disc.

5.6b Use a micrometer to measure disc thickness

Bleeding of the system will not be necessary unless the brake hose was disconnected from the caliper. Check the operation of the brakes thoroughly before placing the vehicle into normal service.

---

6   Drum brake shoes - replacement

Warning: *Brake shoes must be replaced on both rear wheels at the same time - never replace the shoes on only one side. Also, the dust created by the brake system is harmful to your health. Never blow it out with compressed air and don't inhale any of it. An approved filtering mask should be worn when working on the brakes. Do not, under any circumstances, use petroleum-based solvents to clean brake parts. Use brake system cleaner only!*

Caution: *Whenever the brake shoes are replaced, the return and hold-down springs*

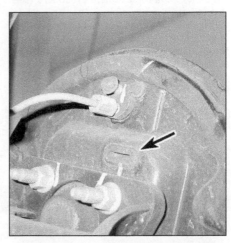

6.2 Remove the rubber plug to get to the adjuster (similar design shown)

should also be replaced. Due to the continuous heating/cooling cycle the springs are subjected to, they lose tension over a period of time and may allow the shoes to drag on the drum and wear at a much faster rate than normal.

1   Loosen the wheel lug nuts, raise the rear of the vehicle and support it securely on jackstands. Block the front wheels to keep the vehicle from rolling and release the parking brake.

2   Remove the rear wheels.

Note: *If the brake drum cannot be removed, make sure the parking brake is completely released. If the drum still cannot be pulled off, the brake shoes will have to be retracted. Remove the rubber plug from the backing plate and insert a screwdriver into the slot (see illustration). Use the screwdriver to turn the adjuster wheel which will move the shoes away from the drum. Refer to illustration 6.4n for a view of the adjuster from the inside. The drum should now come off.*

6.4a Rear drum brake assembly. Be sure to wash the assembly with brake system cleaner before beginning work

1 Shoe retractor spring
2 Trailing brake shoe (parking brake lever behind)
3 Adjuster lever
4 Wheel cylinder
5 Adjuster lever spring
6 Adjuster screw assembly
7 Leading brake shoe

6.4b Remove the spring from the brake adjuster assembly

3   Remove the brake drums.

4   Follow the accompanying illustrations for the brake shoe replacement procedure (see illustrations 6.4a through 6.4n). Be sure to stay in order and read the caption under each illustration.

Note: *All four brake shoes must be replaced at the same time, but to avoid mixing up parts, work on only one brake assembly at a time.*

Note: *The secondary shoe is connected to* a parking brake lever and both parts are replaced together as a single assembly.

5   Before reinstalling the drum, it should be checked for cracks, score marks, deep scratches and hard spots, which will appear as small discolored areas. If the hard spots cannot be removed with fine emery cloth or if any of the other conditions listed above exist, the drum must be taken to an automotive machine shop to have it resurfaced.

Note: *Professionals recommend resurfacing the drums each time a brake job is done. Resurfacing will eliminate the possibility of out-of-round drums. If the drums are worn so much that they can't be resurfaced without exceeding the maximum allowable diameter (stamped into the drum), then new drums will be required. At the very least, if you elect not to have the drums resurfaced, remove the glaze from the surface with emery cloth using a swirling motion.*

6.4c Remove the brake adjuster lever (A) and the adjuster (B)

6.4d With locking pliers, lift the end of the shoe retractor spring away from the shoe, then remove the trailing shoe

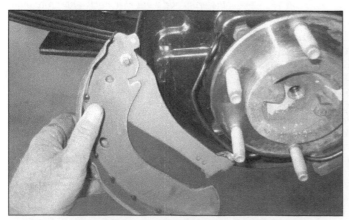

6.4e The trailing shoe is connected to the parking brake lever; lower the shoe and lever away from the backing plate

6.4f Lift the retractor spring and remove the leading shoe from the backing plate

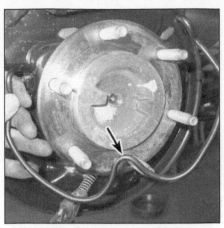

6.4g Remove the shoe retractor spring from the backing plate

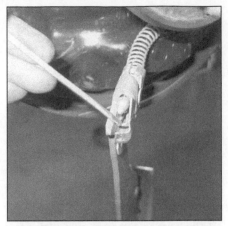

6.4h To remove the trailing shoe/parking brake lever, depress the tab at the cable end and twist the lever off the cable end

6.4i Clean the backing plate with brake cleaner and lightly lube the shoe contact points with brake grease

6.4j Use a socket extension and a hammer to secure the center of the retractor spring in its mount on the backing plate

6.4k Install the new leading brake shoe

6.4l Reconnect the parking brake cable end to the lever and install the new trailing shoe/lever

6.4m Clean and lube the threads of the adjuster and reinstall it between the shoes

6.4n Install the adjuster lever and spring, then adjust the shoes by holding the lever up while turning the star wheel (the drum should just be able to slide over the shoes)

7.5 If you have the proper tools, spread the ends of the shoes away from the wheel cylinder to remove it

7.7 The location of the brake line fitting for the wheel cylinder

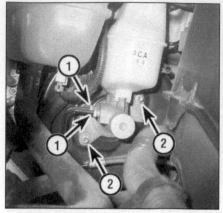

8.5 Brake master cylinder mounting details:

1　*Brake line fittings*
2　*Mounting nuts*

6　Install the brake drums and turn the hubs on each side. The hubs should turn freely. If they do not, remove the drums and turn the adjuster wheel on each side.

7　Install the wheels and lug nuts and then lower the vehicle. Tighten the lug nuts to the torque listed in the Chapter 1 Specifications.

8　Operate the parking brake pedal several times to adjust the rear brakes and then make a number of forward and reverse stops until satisfactory brake pedal action is obtained.

9　Check the operation of the brakes thoroughly before placing the vehicle into normal operation.

7　Wheel cylinder - removal and installation

Note: *If replacement is indicated (usually because of fluid leakage or sticky operation), it is recommended that the wheel cylinders be replaced, not overhauled. Always replace the wheel cylinders in pairs - never replace just one of them.*

Removal

1　Loosen the rear wheel lug nuts, raise the rear of the vehicle and support it securely on jackstands. Block the front wheels to keep the vehicle from rolling.

2　Release the parking brake and remove the rear wheels.

3　Remove the brake drum as described in Section 6.

4　Detach the upper control arm from the rear knuckle (see Chapter 10) and disconnect the rear ABS wheel speed sensor.

Warning: *Do not perform this Step without referring to Chapter 10.*

5　Remove the brake shoes (see Section 6).

Note: *It is not necessary to remove the brake shoes or related hardware if you have a suitable tool to spread the upper part of the brake shoes away from the wheel cylinder (see illustration). This method could save a lot of time*

but it cannot be done without proper tools.

6　Remove all dirt and foreign material from around the wheel cylinder using brake cleaner.

7　On the backside of the backing plate, remove the banjo bolt, sealing washers and brake line fitting from the back of the wheel cylinder (see illustration). Immediately plug the fitting to prevent fluid loss and contamination. A piece of hose inserted through the center of the fitting works well for this. Discard the sealing washers.

8　Remove the wheel cylinder mounting bolts and remove the wheel cylinder, being careful not to spill brake fluid on the shoes.

Installation

9　Place the wheel cylinder in position and install the mounting bolts. Tighten them to the torque listed in this Chapter's Specifications.

10　Using new sealing washers, place the brake line fitting in position and install the banjo bolt. Tighten it to the torque listed in this Chapter's Specifications.

11　Install the brake shoes (if necessary) and position them to mate with the wheel cylinder pistons (see Section 6).

12　Install the brake drum on the hub flange and then install the wheel and lug nuts.

13　Bleed the brake system as outlined in Section 10.

14　Lower the vehicle to the ground and tighten the lug nuts to the torque listed in the Chapter 1 Specifications.

15　Check the operation of the brakes thoroughly before placing the vehicle into normal operation.

8　Master cylinder - removal and installation

Removal

1　The master cylinder is located in the engine compartment, mounted to the power

brake booster.

2　Remove as much fluid as you can from the reservoir with a syringe, such as an old turkey baster.

Warning: *If a baster is used, never again use it for the preparation of food.*

Warning: *Don't depress the brake pedal until the reservoir has been refilled, otherwise air may be introduced into the brake hydraulic system.*

3　Place rags under the fluid fittings and prepare caps or plastic bags to cover the ends of the lines once they are disconnected.

Warning: *Brake fluid will damage paint. Cover all painted surfaces around the work area and be careful not to spill fluid during this procedure.*

4　Depress the brake pedal several times with the engine off to remove vacuum from the power brake booster.

5　Loosen the fittings at the ends of the brake lines where they enter the master cylinder (see illustration). To prevent rounding-off the corners on these nuts, use a flare-nut wrench. Pull the brake lines slightly away from the master cylinder and plug the ends to prevent contamination.

6　Disconnect the electrical connector for the brake fluid level sensor on the reservoir, then remove the nuts attaching the master cylinder to the power booster. Pull the master cylinder off the studs and out of the engine compartment. Again, be careful not to spill the fluid as this is done.

7　If a new master cylinder is being installed, it may be necessary to transfer the old reservoir to the new master cylinder. Tap the retaining pins out with a hammer and punch to remove the old reservoir.

Warning: *Be sure to install new seals when transferring the reservoir.*

Installation

8　Bench bleed the new master cylinder before installing it. Mount the master cylinder in a vise, with the jaws of the vise clamping on the mounting flange.

8.9 The best way to bleed air from the master cylinder before installing it on the vehicle is with a pair of bleeder tubes that direct brake fluid into the reservoir during bleeding

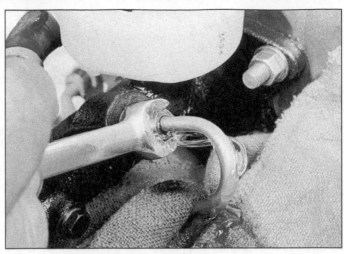

8.18 Have an assistant depress the brake pedal and hold it down, then loosen the fitting nut, allowing the air and fluid to escape; repeat this procedure on both fittings until the fluid is clear of air bubbles

9   Attach a pair of master cylinder bleeder tubes to the outlet ports of the master cylinder (see illustration).

10   Fill the reservoir with brake fluid of the recommended type (see Chapter 1).

11   Slowly push the pistons into the master cylinder (a large Phillips screwdriver can be used for this) - air will be expelled from the pressure chambers and into the reservoir. Because the tubes are submerged in fluid, air can't be drawn back into the master cylinder when you release the pistons.

12   Repeat the procedure until no more air bubbles are present.

13   Remove the bleed tubes, one at a time, and install plugs in the open ports to prevent fluid leakage and air from entering. Install the reservoir cap.

14   Install the master cylinder onto the power brake booster and tighten the mounting nuts only finger tight at this time.

Note: *Inspect the seal on the master cylinder that mates with the power brake booster before installation. If there are signs of damage or wear, replace the seal.*

15   Carefully thread the brake line fittings into the master cylinder by hand until you know they are started straight. Since the master cylinder is still a bit loose, it can be moved slightly in order for the fittings to thread in easily.

Caution: *Do not strip the threads as the fittings are tightened.*

16   Fully tighten the mounting nuts, then the brake line fittings. Tighten the nuts to the torque listed in this Chapter's Specifications.

17   Connect the electrical connector for the brake fluid level sensor on the reservoir.

18   Fill the master cylinder reservoir with fluid, then bleed the master cylinder and the brake system as described in Section 10. To bleed the cylinder on the vehicle, have an assistant depress the brake pedal and hold the pedal to the floor. Loosen the fitting to allow air and fluid to escape, then close the

fitting. Repeat this procedure on both fittings until the fluid is clear of air bubbles (see illustration).

Caution: *Have plenty of rags on hand to catch the fluid - brake fluid will ruin painted surfaces. After the bleeding procedure is completed, rinse the area under the master cylinder with clean water.*

19   Test the operation of the brake system thoroughly before placing the vehicle into normal service.

Warning: *Do not operate the vehicle if you are in doubt about the effectiveness of the brake system. It is possible for air to become trapped in the anti-lock brake system hydraulic control unit, so, if the pedal continues to feel spongy after repeated bleedings or the BRAKE or ANTI-LOCK light stays on, have the vehicle towed to a dealer service department or other qualified shop to be bled with the aid of a scan tool.*

9   Brake hoses and lines - inspection and replacement

1   About every six months, with the vehicle raised and placed securely on jackstands, the flexible hoses which connect the steel brake lines with the front and rear brake assemblies should be inspected for cracks, chafing of the outer cover, leaks, blisters and other damage. These are important and vulnerable parts of the brake system and an inspection should be thorough. A light and mechanics mirror may be needed to see all areas of the flexible hoses. If a hose exhibits any of the above defects, replace it with a new one.

Flexible hoses

2   Clean all dirt away from the ends of the hose. Follow the path of the hose from the metal line to the brake caliper and remove any brackets that may secure sections of the

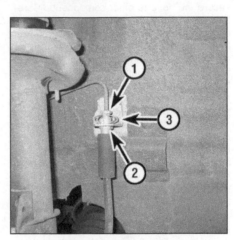

9.3 Brake hose fitting details:

1   *Metal tube nut (use a flare-nut wrench here)*
2   *Brake hose fitting (use a back-up wrench here)*
3   *Retaining clip*

hose to other components.

3   To disconnect a brake hose from the brake line, unscrew the metal tube nut with a flare-nut wrench while holding the hose fitting with wrench. Plug or cap all openings to prevent contamination. Remove the hose from the frame bracket and remove the retaining clip from the hose (see illustration).

4   Disconnect the hose fitting from the caliper, discarding the sealing washers on either side of the fitting.

5   Using new sealing washers, attach the new brake hose to the caliper or wheel cylinder.

6   To reattach a brake hose to the metal line, insert the end of the hose through the frame bracket, make sure the hose is routed properly and isn't twisted. Install the retaining clip on the hose fitting at the frame bracket and then carefully thread the metal tube nut

10.8 When bleeding the brakes, a hose is connected to the bleed screw at the caliper or wheel cylinder and submerged in brake fluid - air will be seen as bubbles in the tube and container (all air must be expelled before moving to the next wheel)

into the hose fitting. Tighten the tube nut securely with a flare-nut wrench.

7   Install any other hose brackets that may secure the hose to other components. Carefully check to make sure the suspension or steering components don't make contact with the hose. Have an assistant push down on the vehicle and also turn the steering wheel lock-to-lock during inspection.

Note: *If your replacement hose is matched to the original one and routed in the same way, you should not experience any problems with the hose making contact with other components.*

8   After installation, check the master cylinder fluid level and add fluid as necessary. Bleed the brake system as outlined in Section 10 and test the brakes carefully before placing the vehicle into normal operation.

Metal brake lines

9   When replacing brake lines, be sure to use the correct parts. Don't use copper tubing for any brake system components. Purchase steel brake lines from a dealer parts department or auto parts store.

10   Prefabricated brake line, with the tube ends already flared and fittings installed, is available at auto parts stores and dealer parts departments. These lines can be bent to the proper shapes using a tubing bender.

11   When installing the new line make sure it's well supported in the brackets and has plenty of clearance between moving or hot components.

12   After installation, check the master cylinder fluid level and add fluid as necessary. Bleed the brake system as outlined in Section 10 and test the brakes thoroughly before placing the vehicle into normal operation.

10   Brake hydraulic system - bleeding

Warning: *If air has found its way into the hydraulic control unit, the system must be bled with the use of a specialized scan tool. If the*

brake pedal feels "spongy" even after bleeding the brakes, or the ABS light or BRAKE warning light on the instrument panel stays on, or if you have any doubts whatsoever about the effectiveness of the brake system, have the vehicle towed to a dealer service department or other repair shop equipped with the necessary tools for bleeding the system.

Warning: *Wear eye protection when bleeding the brake system. If the fluid comes in contact with your eyes, immediately rinse them with water and seek medical attention.*

Note: *Bleeding the brake system is necessary to remove any air that's trapped in the system when it's opened during removal and installation of a hose, line, caliper, wheel cylinder or master cylinder.*

1   It will probably be necessary to bleed the system at all four brakes if air has entered the system due to a low fluid level, or if the brake lines have been disconnected at the master cylinder.

2   If a brake line was disconnected only at a wheel, then only that caliper or wheel cylinder must be bled.

3   If a brake line is disconnected at a fitting located between the master cylinder and any of the brakes, that part of the system served by the disconnected line must be bled, beginning with the fitting closest to the master cylinder and then working downstream, bleeding each fitting of each component as described in Step 18 of Section 8. This includes the proportioning valve or the ABS modulator assembly.

4   Remove vacuum from the brake power booster by applying the brake several times with the engine off.

5   Remove the master cylinder reservoir cap and fill the reservoir with brake fluid. Reinstall the cap.

Note: *Check the fluid level often during the bleeding operation and add fluid as necessary to prevent the fluid level from falling low enough to allow air bubbles into the master cylinder.*

6   Have an assistant on hand, as well as a supply of new brake fluid, an empty clear plastic container, a length of plastic, rubber or

vinyl tubing to fit over the bleeder valve and a wrench to open and close the bleeder valve.

7   Beginning at the right rear wheel, loosen the bleeder screw slightly, then tighten it to a point where it's snug but can still be loosened quickly and easily.

8   Place one end of the tubing over the bleeder screw fitting and submerge the other end in brake fluid in the container (see illustration).

9   Have the assistant slowly depress the brake pedal and hold it in the depressed position.

10   While the pedal is held depressed, open the bleeder screw just enough to allow a flow of fluid to leave the valve. Watch for air bubbles to exit the submerged end of the tube. When the fluid flow slows after a couple of seconds, tighten the screw and have your assistant release the pedal.

11   Repeat Steps 9 and 10 until no more air is seen leaving the tube, then tighten the bleeder screw and proceed to the left front wheel, the left rear wheel and the right front wheel, in that order, and perform the same procedure. Be sure to check the fluid in the master cylinder reservoir frequently.

12   Never use old brake fluid. It contains moisture that can boil, rendering the brake system inoperative.

13   Refill the master cylinder with fluid at the end of the operation.

14   Check the operation of the brakes thoroughly. The pedal should feel solid when depressed, with no sponginess. If necessary, repeat the entire process.

Warning: *Do not operate the vehicle if you are in doubt about the effectiveness of the brake system. It is possible for air to become trapped in the anti-lock brake system hydraulic control unit, so, if the pedal continues to feel spongy after repeated bleedings or the BRAKE or ANTI-LOCK light stays on, have the vehicle towed to a dealer service department or other qualified shop to be bled with the aid of a scan tool.*

11   Power brake booster - removal and installation

Warning: *The ABS hydraulic control unit must be removed in order to remove the power brake booster. Removal of the ABS hydraulic control unit requires opening the hydraulic circuits (brake lines) at the unit, which can introduce air into it. If air has found its way into the hydraulic control unit, the system must be bled with the use of a specialized scan tool. If the brake pedal feels "spongy" even after bleeding the brakes, or the ABS light or BRAKE warning light on the instrument panel stays on, or if you have any doubts whatsoever about the effectiveness of the brake system, have the vehicle towed to a dealer service department or other qualified repair shop equipped with the necessary tools for bleeding the system.*

Note: *It is possible that removal and installation of the power brake booster will result in a*

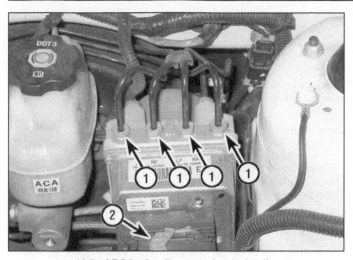

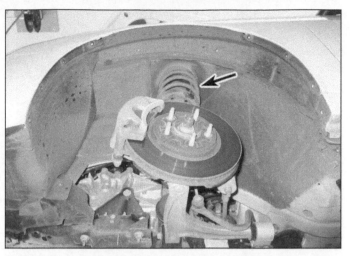

11.7a ABS hydraulic control unit details:

1   Brake lines (label for correct installation)
2   Release lever on the electrical connector

11.7b The mounting fasteners for the ABS hydraulic unit mounting bracket or the proportioning valve can be found near the strut in this area of the wheel well.

CHECK ENGINE warning light and the need for calibration of the brake pedal position sensor. Calibration can only be done with a specialized scan tool at a dealer service department or qualified repair facility.
Note: The power brake booster is not serviceable. If it has failed, replace it with a new or rebuilt unit.

Operating check

1   Depress the brake pedal several times with the engine off and make sure that there is no change in the pedal reserve distance.
2   Depress the pedal and start the engine. If the pedal goes down slightly, operation is normal.

Airtightness check

3   Start the engine and turn it off after one or two minutes. Depress the brake pedal several times slowly. If the pedal goes down farther the first time but gradually rises after the second or third depression, the booster is airtight.
4   Depress the brake pedal while the engine is running, then stop the engine with the pedal depressed. If there is no change in the pedal reserve travel after holding the pedal for 30 seconds, the booster is airtight.

Removal and installation

Note: Removal of the power brake booster requires the removal of the ABS hydraulic control unit (on ABS equipped models) or the proportioning valve (on non-ABS equipped models) in addition to the master cylinder.
5   Turn the ignition key to the OFF position and press the brake pedal several times to remove vacuum in the power brake booster.
Caution: Damage to the ABS control module could occur if the key is not in the OFF position for the following steps.
6   Set the parking brake and loosen the lug nuts on the left-front wheel. Raise the left-front

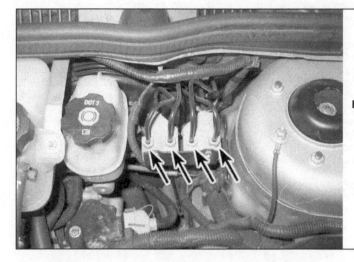

11.8 Label these brake lines in relation to the ports on the proportioning valve for correct installation

part of the vehicle and support it securely on a jackstand. Remove the left-front wheel.
Note: Raise the vehicle just high enough to remove the wheel.
7   On ABS-equipped models, disconnect the electrical connector (see illustration) and remove the fasteners for the ABS hydraulic control unit mounting bracket. On non-ABS models, remove the mounting fasteners for the proportioning valve (see illustration).
Caution: DO NOT turn the ignition key to the ON position until the electrical connector is reconnected to the ABS hydraulic control unit.

Non-ABS equipped models

8   Label the brake lines to the top ports on the proportioning valve, then unscrew the fittings with a flare-nut wrench (see illustration). Cap or plug all openings immediately to prevent contamination.
9   Remove the master cylinder mounting nuts and the electrical connector at the reservoir (see Section 8), then remove it and the

proportioning valve together.
Note: In this step, the brake lines from the master cylinder to the proportioning valve can stay connected.

ABS equipped models

10   Label the brake lines to the top ports on the hydraulic control unit, then unscrew the fittings with a flare-nut wrench (see illustration 11.8). Cap or plug all openings immediately to prevent contamination.
11   Remove the master cylinder mounting nuts and electrical connector at the fluid reservoir (see Section 8), then remove it and the hydraulic control unit together.
Note: In this step, the brake lines from the master cylinder to the hydraulic control unit can stay connected.

All models

12   Disconnect the large vacuum hose by removing the check valve from power brake booster; simply pull it out of the rubber grommet.

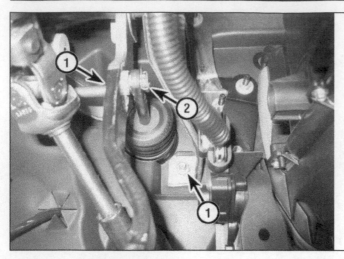

11.14 Power brake booster mounting details:

1   *Mounting nut locations (one hidden, vicinity shown)*
2   *Booster pushrod retaining clip*

13   In the passenger compartment, remove the knee bolster (see Chapter 11). Disconnect the electrical connector for the brake pedal position sensor. This can easily be found by following the brake pedal arm up until you see an small electrical component near the top and left of the arm on the brake pedal bracket.

14   Remove the retaining clip and washers securing the booster pushrod to the brake pedal arm (see illustration).

15   Remove the booster-to-firewall mounting nuts (see illustration 11.14).

16   Tilt the booster slightly to remove the pushrod from the brake pedal arm.

Caution: *Avoid excessive sideways movement and force on the pushrod.*

17   Carefully guide the booster unit out of the engine compartment.

18   To install the booster, place it into posi- tion and attach the booster pushrod to the brake pedal arm. Install the mounting nuts and tighten them to the torque listed in this Chapter's Specifications. Install the pushrod retaining clip and reconnect the brake pedal position sensor electrical connector.

Note: *Don't forget to install one wave washer before attaching the booster pushrod to the brake pedal arm and then the other wave washer before the pushrod retaining clip.*

19   Install the master cylinder with the pro- portioning valve or ABS hydraulic control unit. Reconnect the large vacuum hose by install- ing the check valve into the booster. Tighten all fasteners to the torque listed in this Chap- ter's Specifications

20   Carefully thread the brake line fittings into the proportioning valve or ABS hydraulic control unit and tighten them securely.

Caution: *Do not strip the threads as the fit-* *tings are tightened.*

21   Bleed the brake system (see Sec- tion 10).

22   Test the operation of the brakes thor- oughly before placing the vehicle into normal service. Refer to the Warning at the beginning of this Section.

12   Parking brake - adjustment

1   The parking brake on these models is self-adjusting and there is no need to adjust the parking brake cable. If, however, the park- ing brake pedal travel is excessive or the parking brakes will not hold the vehicle on an incline, the brake shoes (drum brake models) may need to be adjusted or replaced (see Section 6). If your vehicle is equipped with rear disc brakes and the same conditions occur, you may want to check the condition of the rear calipers and brake pads.

13   Brake light switch - replacement

1   These model vehicles do not utilize a typical brake light switch; they are equipped with a Brake Pedal Position Sensor. This sen- sor cannot be removed or installed without the use of a special scan tool that is necessary for its calibration.

Note: *It is very likely that a failure in this sys- tem will set a diagnostic trouble code (DTC) that only a dealership or qualified repair facility can retrieve. Refer to Chapter for more infor- mation regarding DTC's and the related equip- ment needed to access them.*

Chapter 10
Suspension and steering systems

Contents

Specifications

General

Power steering fluid type .... See Chapter 1

Torque specifications   Ft-lbs (unless otherwise indicated)

Front suspension

Strut
  Damper shaft nut.. 52
  Strut upper mounting nut.. 18
  Strut-to-steering knuckle bolts/nuts... 89
Stabilizer bar
  Stabilizer bar link nuts... 48
  Stabilizer bar bracket bolts.. 18
Control arm
  Mounting fasteners (front and rear)
    2009 and earlier models
      Step 1.. 37
      Step 2.. Tighten an additional 90-degrees
    2010 and later models*
      Step 1.. 74
      Step 2.. Tighten an additional 90-degrees
  Balljoint-to-steering knuckle pinch bolt/nut (replace these fasteners with new ones)
    Step 1.. 37
    Step 2
      2009 and earlier models .... Loosen three-quarter turn
      2010 and later models .... Loosen one-third turn
    Step 3
      2005 and earlier models .... Tighten to 37 ft-lbs then tighten an additional 60-degrees
      2006 and later models .... Tighten to 37 ft-lbs then tighten an additional 30-degrees
  Rear bushing-to-control arm bolt.. 32
Hub and bearing assembly bolts 85

*2012 models: replace these fasteners with new ones

Torque specifications (continued)

Ft-lbs (unless otherwise indicated)

Note: *One foot-pound (ft-lb) of torque is equivalent to 12 inch-pounds (in-lbs) of torque. Torque values below approximately 15 ft-lbs are expressed in inch-pounds, since most foot-pound torque wrenches are not accurate at these smaller values.*

Subframe
   Subframe-to-body bolts*
      Step 1 74
      Step 2 .... Tighten an additional 90-degrees
   Subframe support bracket bolts
      Step 1 74
      Step 2 .... Tighten an additional 90-degrees
Driveaxle/hub nut.. See Chapter 8

2012 models: replace with new fasteners

Rear suspension

Upper control arm
   Control arm-to-rear knuckle fasteners
      Step 1 81
      Step 2 .... Tighten an additional 70-degrees
   Control arm-to-suspension support fasteners
      Step 1 44
      Step 2 .... Tighten an additional 60-degrees
Lower control arm
   Control arm-to-rear knuckle fasteners
      Step 1 44
      Step 2 .... Tighten an additional 60-degrees
   Control arm-to-suspension support fasteners 81
Toe link arm
   Toe link arm-to-rear knuckle fasteners
      Step 1 81
      Step 2 .... Tighten an additional 70-degrees
   Toe link arm-to-suspension support fasteners 81
Trailing arm
   Trailing arm-to-rear knuckle fasteners 81
   Trailing arm bracket-to-body bolts
      Step 1 66
      Step 2 .... Tighten an additional 30-degrees
   Trailing arm-to-bracket through-bolt
      Step 1 44
      Step 2 .... Tighten an additional 60-degrees
Hub and bearing assembly nuts 47
Shock absorber
   Upper mounting fasteners
      Sedan models.. 66
      Maxx models... 18
   Lower mounting fasteners (both models)... 133
Stabilizer bar
   Clamp bolts 26
   Bar link-to-knuckle bolts 41

Steering system

Adjustable accelerator pedal assembly mounting nuts .... 89 in-lbs
Power steering pump mounting bolts 18
Steering gear mounting bolts
   Electronically assisted system
      Step 1 52
      Step 2 (2005 and earlier models) .... Tighten an additional 90-degrees)
   Hydraulically assisted system 81
Tie-rod end-to-steering knuckle nut
   2007 and earlier models.. 44
   2008 and later models**
      Step 1 18
      Step 2 .... Tighten an additional 90-degrees
Intermediate shaft-to-steering column pinch-bolt*................................. 36
Steering column mounting fasteners... 18
Steering wheel mounting nut... 24

Fastener must be replaced
**2012 models: replace with new fasteners*

1

General information

1 The front suspension is a MacPherson strut design. The upper end of each strut is attached to the vehicle's body strut support. The lower end of the strut is connected to the upper end of the steering knuckle. The steering knuckle is attached to a balljoint mounted on the outer end of the suspension control arm. A stabilizer bar connected to each strut and mounted to the suspension crossmember reduces body roll during cornering (see illustration).

2 The rear suspension utilizes trailing arms, upper and lower suspension arms, toe link arms, shock absorbers and coil springs. A stabilizer bar is clamped to a suspension support and connected to the rear knuckle by two links (see illustration).

3 The power-assisted rack-and-pinion steering gear is attached to the front suspension subframe. The steering gear actuates the tie-rods, which are attached to the steering knuckles. The steering column is designed to collapse in the event of an accident.

4 Frequently, when working on the sus-pension or steering system components, you may come across fasteners which seem impossible to loosen. These fasteners on the underside of the vehicle are continually subjected to water, road grime, mud, etc., and can become rusted or "frozen" in place, making them extremely difficult to remove. In order to unscrew these stubborn fasteners without damaging them (or other components), be sure to use lots of penetrating oil and allow it to soak in for a while. Using a wire brush to clean exposed threads will also ease removal of the nut or bolt and prevent damage to the threads. Sometimes a sharp blow with a hammer and punch will break the bond between a nut and bolt threads, but care must be taken to prevent the punch from slipping off the fastener and ruining the threads. Heating the stuck fastener and surrounding area with a torch sometimes helps too, but isn't recommended because of the obvious dangers associated with fire. Long breaker bars and extension, or "cheater," pipes will increase leverage, but never use an extension pipe on a ratchet - the ratcheting mechanism could be damaged. Sometimes tightening the nut or bolt first will help to break it loose. Fasteners that require drastic measures to remove should always be replaced with new ones.

5 Most of the procedures in this Chapter involve raising the vehicle and working underneath it; a suitable jack and a good pair of jackstands will be needed. A hydraulic floor jack is the preferred type of jack to lift the vehicle, and it can also be used to support certain components during various operations.

Warning: *Never, under any circumstances, rely on a jack to support the vehicle while working on it.*

Warning: *Whenever any of the suspension or steering fasteners are loosened or removed they must be inspected and, if necessary, replaced with new ones of the same part number or of original equipment quality and design. Torque specifications must be followed for proper reassembly and component retention. Never attempt to heat or straighten any suspension or steering components. Instead, replace any bent or damaged parts with new ones.*

1.1 Front suspension

1 Subframe	4 Tie-rod end	6 Balljoint
2 Control arm	5 Steering knuckle	7 Strut
3 Stabilizer bar		

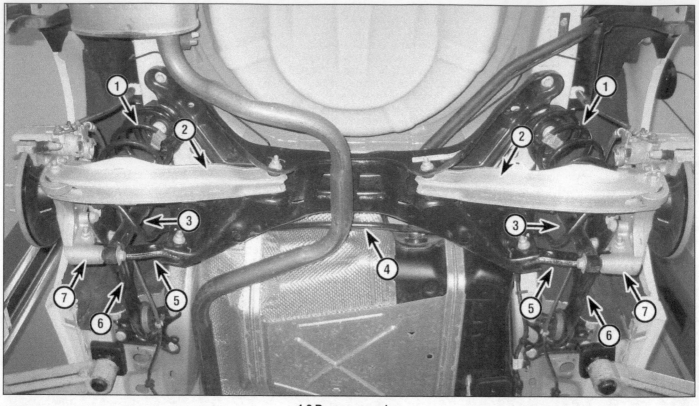

1.2 Rear suspension

1	Coil spring	4	Stabilizer bar	6	Trailing arm
2	Lower control arm	5	Toe link arm	7	Rear knuckle
3	Shock absorber				

---

2   Strut assembly - removal, inspection and installation

Warning: *Always replace the struts and/or coil springs in pairs - never replace just one strut or one coil spring (this could cause dangerous handling peculiarities).*

Removal

1   Loosen the wheel lug nuts, raise the vehicle and support it securely on jackstands. Remove the wheel.

2   Disconnect the stabilizer bar link from the strut (see Section 4), and then remove the brake hose bracket from the strut. Disconnect the ABS harness connector and remove it from its bracket, if equipped. On 2008 and later models, disconnect the stabilizer bar link from the strut.

3   Mark the relationship of the strut to the knuckle (these marks will be used during installation to ensure the camber is returned to its original setting) (see illustration). Remove the strut-to-knuckle nuts and tap the bolts out with a hammer and punch if necessary. Also, remove the bracket for the ABS wheel sensor wiring harness, if equipped.

4   Separate the strut from the steering knuckle. Be careful not to overextend the inner CV joint. Also, don't let the steering

knuckle fall outward, as the brake hose could be damaged. If necessary, support the lower control arm with a jack.

5   Support the strut and spring assembly with one hand and remove the upper mounting nuts (see illustration). Remove the assembly from the fenderwell.

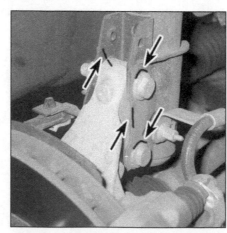

2.3 Thoroughly mark the relationship of the strut to the steering knuckle and mark the fasteners on each side of the strut bracket. Remove the strut-to-steering knuckle fasteners

Inspection

6   Check the strut body for leaking fluid, dents, cracks and other obvious damage that would warrant repair or replacement.

7   Check the coil spring for chips or cracks in the spring coating (this can cause premature spring failure due to corrosion). Inspect

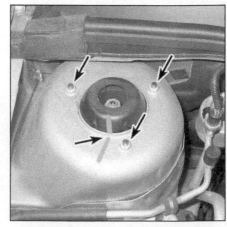

2.5 Mark the strut in relation to the strut tower and then remove the mounting fasteners while supporting the strut

3.3 Install the spring compressor following the tool manufacturer's instructions; compress the spring until all pressure is relieved from the upper spring seat (you can verify this by wiggling the spring)

parts before disassembling your vehicle.

Warning: *Disassembling a strut is potentially dangerous and utmost attention must be directed to the job, or serious injury may result. Use only a high-quality spring compressor and carefully follow the manufacturer's instructions furnished with the tool. After removing the coil spring from the strut assembly, set it aside in a safe, isolated area.*

Disassembly

2   Remove the strut and spring assembly (see Section 2). Mount the strut clevis bracket portion of the strut assembly in a vise.

Caution: *Do not clamp any other portion of the strut assembly in the vise as it will be damaged. Line the vise jaws with wood or rags to prevent damage to the unit and don't tighten the vise excessively.*

3   Following the tool manufacturer's instructions, install the spring compressor (which can be obtained at most auto parts stores or equipment yards on a daily rental basis) on the spring and compress it sufficiently to relieve all pressure from the upper spring seat (see illustration). This can be verified by wiggling the spring.

4   Hold the damper shaft from turning with an appropriate tool, and unscrew the damper shaft nut.

5   Remove the nut and upper mount (see illustration). Lay the parts out in the exact order in which they are removed. Check the rubber portion of the upper mount for cracking and general deterioration. If there is any separation of the rubber, replace it.

6   Remove the upper spring seat from the damper shaft (see illustration). Check the rubber portion of the spring seat for cracking and hardness; replace it if necessary. Inspect the bearing in the spring seat for smooth operation. If it doesn't turn smoothly, replace it.

7   Slide the dust boot and rubber bump stop off the damper shaft. Check the bump stop for cracking and general deterioration. If there is any deterioration of the rubber, replace it.

the spring seat for cuts, hardness and general deterioration.

8   If any undesirable conditions exist, proceed to the strut disassembly procedure (see Section 3).

Installation

9   Guide the strut assembly up into the fenderwell and insert the upper mounting studs through the holes in the strut tower. Once the stud protrudes from the strut tower, install the nuts so the strut won't fall back through. This is most easily accomplished with the help of an assistant, as the strut is quite heavy and awkward.

10   Slide the steering knuckle into the strut flange and insert the two bolts. Install the ABS wheel speed harness bracket, if equipped. Install the nuts, align the marks you made in Step 3, and then tighten the nuts to the torque listed in this Chapter's Specifications.

11   Reattach the brake hose bracket to the strut and reconnect the stabilizer bar link (see Section 4). Reconnect the wheel speed sensor harness connector, if equipped.

12   Install the wheel and lug nuts, then lower

the vehicle and tighten the lug nuts to the torque listed in the Chapter 1 Specifications.

13   Tighten the upper mounting nuts to the torque listed in this Chapter's Specifications.

14   Have the front wheel alignment checked and, if necessary, adjusted.

---

3   Strut/coil spring - replacement

Note: *You'll need a spring compressor for this procedure. Spring compressors can usually be rented at most auto parts stores or equipment yards.*

1   If the struts or coil springs exhibit the telltale signs of wear (leaking fluid, loss of damping capability, chipped, sagging or cracked coil springs) explore all options before beginning any work. The strut/coil spring components are not serviceable and must be replaced if a problem develops. However, strut assemblies, complete with springs, may be available on an exchange basis (which eliminates much time and work). Whichever route you choose to take, check on the cost and availability of

3.5 Remove the upper mount after taking off the damper shaft nut

3.6 Remove the upper spring seat

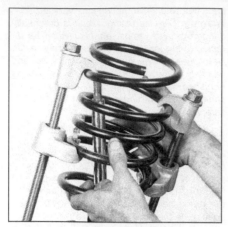

3.8 Carefully remove the compressed spring from the strut

3.10 When installing the spring, make sure the end rests against the raised stop

8   Carefully lift the compressed spring from the assembly (see illustration) and set it in a safe place.

Warning: *When removing the compressed spring, lift it off carefully and set it in a safe place. Keep the ends of the spring away from your body.*

Note: *If you are disassembling both struts, mark the springs LEFT and RIGHT so you don't mix them up (they're different).*

Reassembly

9   Extend the damper rod to its full length and install the rubber bump stop and dust boot.
10   Carefully place the compressed coil spring onto the lower seat of the damper, with the end of the spring resting against the raised stop (see illustration).
11   Install the upper insulator and spring seat.
12   Install the upper mount and mounting nut and then tighten it to the torque listed in this Chapter's Specifications.
13   Remove the spring compressor tool.

14   Install the strut/spring assembly (see Section 2).

---

4   Stabilizer bar, bushings and links (front) - removal and installation

Note: *Stabilizer bar removal involves lowering the subframe. If one becomes damaged it is most likely the result of an accident that was severe enough to damage other major components (such as the subframe itself). Damage this severe will require the services of an auto body shop. For subframe removal, see Section 21.*

1   Loosen the front wheel lug nuts, raise the front of the vehicle, support it securely on jackstands and remove the front wheels.
2   Remove the exhaust pipe from 3.6L V6 models and on other models if needed.
3   Refer to Section 21 and support the subframe with a floor jack.
4   Remove the subframe mounting bolts and lower it in the rear enough to access the

stabilizer bar brackets.
5   Remove the nuts that attach the stabilizer link to the strut/coil spring assembly and to the stabilizer bar and detach the links (see illustration).
6   Remove the bolts from the stabilizer bar bushing retainers (see illustration). Remove the retainers from the bushings, prying them off, if necessary.
7   Inspect the bushings for cracks and tears. If either bushing is broken, damaged, distorted or worn, replace both of them. If the ballstuds on the links are loose or otherwise worn, replace the links.
8   Install the links, tightening the link nuts to the torque listed in this Chapter's Specifications.
9   Install the bushings. Clean the areas on the stabilizer bar where the bushings are located. Lubricate the inside and outside of the new bushings with vegetable oil (used in cooking) to simplify reassembly.

Caution: *Don't use petroleum or mineral-based lubricants or brake fluid - they will lead to deterioration of the bushings. These bushings are split so that you can install them without having to slide them onto the ends of the stabilizer bar. Install the bushings with the slit in each bushing facing towards the rear of the vehicle.*

10   Install the retainers and bolts, tightening the bolts to the torque listed in this Chapter's Specifications.
11   Install the wheels and lug nuts and then lower the vehicle. Tighten the wheel lug nuts to the torque listed in the Chapter 1 Specifications.

---

5   Control arm - removal, inspection and installation

Removal

1   With the vehicle at normal ride height, mark the relationship of the control arm to the

4.5 Remove the nuts on each end of the link to detach it. Hold the ballstud on the link with a tool to keep it from turning while removing the nut

4.6 Remove the stabilizer bar retainer bolts and lift the retainer off the bushing

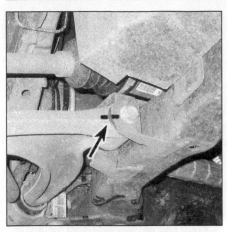

5.1 Mark the relationship of the control arm to the subframe bracket when the vehicle is at normal ride height

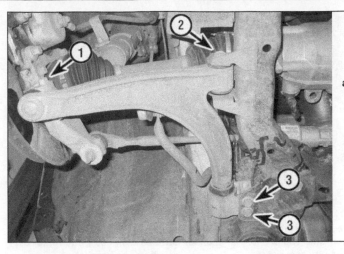

5.5a Lower control arm mounting details:

1   *Steering knuckle/balljoint pinch bolt*
2   *Front pivot bolt*
3   *Rear mounting bolts*

front mounting bracket (where it attaches to the subframe) (see illustration).

Note: *If you are replacing the control arm, transfer this mark to the replacement arm when the old one is removed.*

2   Loosen the wheel lug nuts on the side to be disassembled. Apply the parking brake, raise the front of the vehicle, support it securely on jackstands and remove the wheel.

3   If you're removing the left arm on a 2008 or later model with the 4T45-E transaxle, remove the left transaxle mount (see Chapter 7B).

4   If you're removing the right arm from a 2008 or later model with the 3.5L V6 engine, remove the right engine mount (see Chapter 2B).

5   Remove the pinch bolt securing the balljoint stud to the steering knuckle (see illustration).

Warning: *The manufacturer states that this bolt must be replaced with a new one whenever it is removed. Separate the control arm balljoint from the steering knuckle (see illustration).*

6   Remove the bolts and nuts that attach the control arm to the subframe (see illustration 5.5a).

Note: *On vehicles equipped with a V6 engine, a transaxle or engine mount must first be removed in order to remove the control arm; remove the transaxle mount for removal of the left control arm or remove the engine mount for removal of the right control arm (see Chapter 2B or Chapter 2C).*

7   Remove the control arm.

Inspection

8   Check the control arm for distortion and the bushings for wear, replacing parts as necessary. Do not attempt to straighten a bent control arm. If the bushings are cracked or show signs of wear, take the control arm to an automotive machine shop and have the front pivot bushing replaced. The rear bushing can be removed and replaced with its mounting sleeve by removing the rear bushing bolt.

Installation

9   Installation is the reverse of removal; tighten all of the fasteners to the torque values listed in this Chapter's Specifications, and be sure to use a new balljoint pinch bolt.

Note: *Before tightening the control arm pivot bolt (front), raise the outer end of the control arm with a floor jack and match the reference*

marks made in step one to simulate normal ride height.

10   Install the wheel and lug nuts, lower the vehicle and tighten the lug nuts to the torque listed in the Chapter 1 Specifications.

11   It's a good idea to have the front wheel alignment checked and, if necessary, adjusted after this job has been performed.

---

6   Balljoints - check and replacement

Check

1   Raise the front of the vehicle and support it securely on jackstands. Apply the parking brake and block the rear wheels to keep the vehicle from rolling off the jackstands.

2   Place a large prybar under the balljoint and resting on the wheel, then try to pry the balljoint up while feeling for movement between the balljoint and steering knuckle (see illustration). Now, pry between the control arm and the steering knuckle and try to lever the control arm down while feeling for movement between the balljoint and steering knuckle. If excessive movement is evident (over 0.125 of an inch) in either check, the balljoint is worn. Install a dial indicator and measure the amount of movement to be certain.

3   Have an assistant grasp the tire at the top and bottom and move the top of the tire in-and-out. Touch the balljoint stud nut. If excessive movement is evident (over 0.125 of an inch), the balljoint or knuckle is worn. Install a dial indicator and measure the amount of movement to be certain.

4   Separate the control arm from the steering knuckle (see Section 5). Using your fingers (don't use pliers), try to twist the stud in the socket. If the stud turns, replace the control arm.

Replacement

5   The balljoints on these vehicles are not serviceable. If the balljoint is defective, the control arm must be replaced (see Section 5).

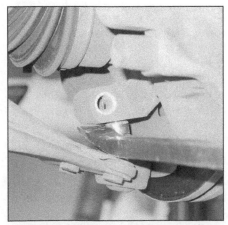

5.5b Use a prybar to separate the balljoint from the steering knuckle. Caution: *Be careful not to damage the balljoint boot*

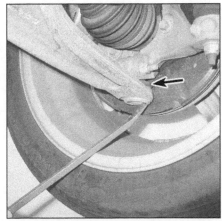

6.2 Pry upward on the balljoint from the wheel to check for balljoint wear

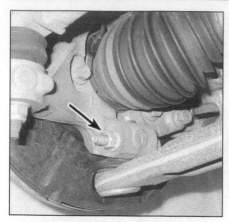

7.5 The steering knuckle is secured to the control arm by this pinch-bolt; replace it with a new one any time you remove it

8.3 Disconnect the ABS wheel speed sensor harness connector and remove it from the bracket

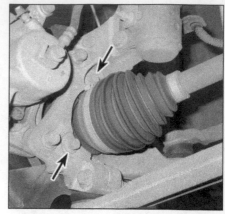

8.5 Hub and bearing assembly mounting bolts (one hidden)

7   Steering knuckle - removal and installation

Warning: *Dust created by the brake system is harmful to your health. Never blow it out with compressed air and don't inhale any of it. Do not, under any circumstances, use petroleum-based solvents to clean brake parts. Use brake system cleaner only.*

Removal

1   Loosen the driveaxle/hub nut (see Chapter 8). Loosen the wheel lug nuts, raise the front of the vehicle and support it securely on jackstands. Remove the wheel.

2   Remove the brake caliper, the caliper mounting bracket and the brake disc from the hub (see Chapter 9).

Caution: *Suspend the caliper to the strut coil spring a using piece of wire. DO NOT let the caliper hang by the brake hose.*

3   Mark the strut to the steering knuckle, then loosen, but do not remove the strut-to-steering knuckle bolts (see illustration 2.3).

4   Remove the nut and separate the tie-rod end from the steering knuckle (see Section 16).

Warning: *The manufacturer states that this nut must be replaced with a new one whenever it is removed.*

5   Remove the balljoint-to-steering knuckle pinch bolt, then separate the balljoint from the steering knuckle (see Section 5) (see illustration).

Warning: *The manufacturer states that this bolt must be replaced with a new one whenever it is removed.*

6   On 2007 and earlier models, remove the driveaxle nut and push the driveaxle from the hub (see Chapter 8). Support the driveaxle with wire. On 2008 and later models, remove the hub and bearing assembly (see Section 8).

7   The strut-to-knuckle bolts can now be removed. On ABS equipped vehicles, note the position of the bracket for the wheel speed sensor harness.

8   Carefully separate the steering knuckle from the strut.

Installation

9   Guide the knuckle into position while carefully inserting the driveaxle into the hub.

10   Push the knuckle into the strut flange and install the bolts and nuts, but don't tighten them yet. On ABS equipped vehicles, place the bracket for the wheel speed sensor harness as the fasteners are installed. Align the steering knuckle using the marks on the strut flange and then tighten the nuts to the torque listed in this Chapter's Specifications

11   Connect the balljoint to the knuckle and tighten the new pinch bolt/nut to the torque listed in this Chapter's Specifications.

12   Attach the tie-rod to the steering knuckle and tighten the new tie-rod nut to the torque listed in this Chapter's Specifications.

13   Place the brake disc on the hub and install the caliper mounting bracket and caliper (see Chapter 9).

14   Install the driveaxle/hub nut and tighten it securely (final tightening will be carried out when the vehicle is lowered).

15   Install the wheel and lug nuts.

16   Lower the vehicle and tighten the lug nuts to the torque listed in the Chapter 1 Specifications. Tighten the driveaxle/hub nut to the torque listed in the Chapter 8 Specifications.

17   Have the front-end alignment checked and, if necessary, adjusted.

8   Hub and bearing assembly (front) - removal and installation

Removal

1   Loosen the driveaxle/hub nut (see Chapter 8).

2   Loosen the wheel lug nuts, raise vehicle and support it securely on jackstands and remove the wheel.

3   Remove the brake caliper, the caliper mounting bracket and the brake disc from the hub (see Chapter 9). On ABS equipped vehicles, disconnect the wheel speed sensor

harness connector and remove it from the bracket attached to the strut (see illustration).

Caution: *Suspend the caliper to the strut coil spring with a piece of wire. DO NOT let the caliper hang by the brake hose.*

4   Remove the driveaxle/hub nut.

5   Remove the hub/bearing assembly mounting bolts from the rear of the steering knuckle (see illustration).

6   Remove the hub/bearing assembly from the steering knuckle.

Note: *If the driveaxle splines stick in the hub, push the driveaxle out of the hub with a two-jaw puller.*

Installation

7   Make sure that the mounting surface inside the steering knuckle and on the drive-axle splines is smooth and free of burrs and nicks prior to installing the hub/bearing assembly.

8   Lubricate the driveaxle splines with multi-purpose grease. Install the hub/bearing assembly onto the driveaxle and into the steering knuckle until it is fully seated on the steering knuckle.

9   Install the hub/bearing assembly-to-steering knuckle bolts. Tighten the bolts to the torque listed in this Chapter's Specifications. On ABS equipped vehicles, connect the wheel sensor harness connector and secure it to the bracket on the strut.

10   Install the driveaxle/hub nut. Do not tighten the nut yet.

11   Install the brake disc, the caliper mounting bracket and the caliper; tighten the fasteners to the torque values listed in the Chapter 9 Specifications.

12   Install the wheel and lug nuts, remove the jackstands, and lower the vehicle.

13   Tighten the driveaxle/hub nut to the torque listed in the Chapter 8 Specifications.

Note: *Have an assistant apply the brakes while tightening the driveaxle/hub nut.*

14   Tighten the lug nuts to the torque listed in the Chapter 1 Specifications.

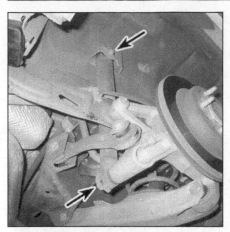

9.3a Rear shock absorber mounting
fasteners (Malibu - base sedan
model shown)

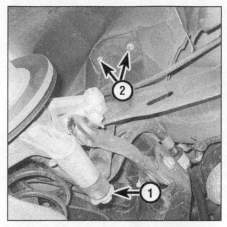

9.3b Lower (1) and upper (2) rear shock
absorber mounting bolts - G6 models

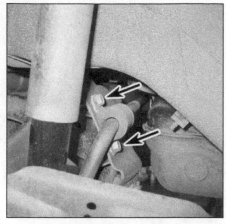

11.4 Rear stabilizer bar clamp
mounting bolts

9   Shock absorber - removal and installation

Warning: *Always replace shock absorbers in pairs - never replace just one of them.*

Removal

1   Loosen the rear wheel lug nuts. Block the front wheels to keep the vehicle from rolling, then raise the rear of the vehicle and support it securely on jackstands. Remove the rear wheels.

2   Place a jack under the rear knuckle, where it connects to the lower control arm, and raise it a few inches to relieve the coil spring tension (see illustration 1.2).

3   Remove the lower shock absorber fastener where it connects to the rear knuckle (see illustrations).

4   Support the shock and remove the upper shock fastener(s) where it connects to the body and then remove the shock.

Installation

5   Installation is the reverse of removal. Tighten the bolts to the torque listed in this Chapter's Specifications.

10   Coil spring - removal and installation

Warning: *Always replace coil springs in pairs - never replace just one of them.*
Warning: *Removingacoil spring is potentially dangerous and utmost attention must be directed to the job or serious injury may result. Use only a high-quality spring compressor and carefully follow the manufacturer's instructions furnished with the tool. After removing the coil spring from the vehicle, set it aside in a safe, isolated area.*
Note: *Spring compressors can usually be rented at most auto parts stores or equipment yards.*

1   Loosen the rear wheel lug nuts. Block the front wheels to keep the vehicle from rolling, then raise the rear of the vehicle and support it securely on jackstands. Remove the rear wheels.

2   Support the lower control arm with a floor jack (see illustration 1.2).

3   Install the spring compressor tool and compress the spring until it is loose against the insulators between the body and lower control arm.

4   Remove the mounting fasteners for the lower control arm and then lower the arm and the coil spring with the floor jack (see Section 12).
Note: *Do not remove these mounting fasteners without referring to Section 12).*
Note: *Do not remove these mounting fasteners without referring to (Section 10).*

5   Remove the coil spring from the control arm.
Note: *Inspect the coil spring insulators for wear or damage and replace as necessary.*

6   If a new spring is to be installed, carefully release the original spring from the compressor tool. If the spring is going to be reused, leave the spring compressed with the tool in place.

7   Installation is the reverse of removal. Tighten the lower control arm mounting fasteners to the torque listed in this Chapter's Specifications.

11   Stabilizer bar, bushings and links (rear) - removal and installation

1   Loosen the rear wheel lug nuts. Block the front wheels to keep the vehicle from rolling, then raise the rear of the vehicle and support it securely on jackstands. Remove the rear wheels.

2007 and earlier models

2   On Malibu MAXX and Pontiac G6 models, remove the muffler and exhaust pipe by

11.6 Remove the stabilizer bar link-to-rear
knuckle bolt

unbolting it from the catalytic converter or resonator. Also refer to Chapter 4 and remove the fuel tank on these models.
Note: *It is best to perform this operation when the tank is nearly empty.*

3   Remove the toe link suspension arm (see Section 12).

4   Unbolt the stabilizer bar bushing clamp bolts (see illustration).

5   Pull the clamps off of the bar and inspect the bushings for cracks, hardness and deterioration. If the bushings are damaged, replace them.

6   Detach the stabilizer bar link-to-knuckle bolts, then remove the bar (see illustration).

2008 and later models

7   Refer to Chapter 4 and remove the rear section of exhaust pipe.

8   Unbolt the trailing arm brackets from the frame.

9   Refer to Section 10 and remove the coil springs.

10   Disconnect the parking brake cables from the trailing arms.

12.3 Upper control arm mounting fasteners

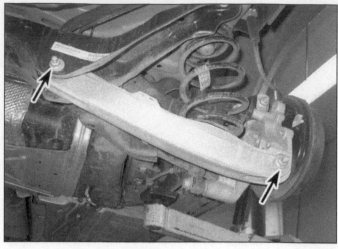

12.7a Lower control arm mounting fasteners

11   Remove the brake calipers (see Chapter 9).

12   Remove the upper shock absorber bolts (see Section 9).

13   Place a floor jack and jackstands under the rear suspension crossmember. Make sure it is securely supported, then remove the crossmember mounting bolts.

14   Carefully lower the crossmember enough to get access to the stabilizer bar bushing clamps.

15   Unbolt the stabilizer bar from the knuckles.

16   Remove the stabilizer bar bushing clamps and disconnect it from the crossmember. Slide the stabilizer bar out.

All models

17   Installation is the reverse of removal, noting the following points:

a) Lubricate the inside and outside of the new bushings with vegetable oil (used in cooking) to simplify reassembly.

Caution: *Don't use petroleum or mineral-based lubricants or brake fluid - they will lead to deterioration of the bushings. These bushings are split so that you can install them with-* out having to slide them onto the ends of the stabilizer bar.

b) On Malibu Maxx sedan and Pontiac G6 models, refer to Chapter 4 for details on the reinstallation of the fuel tank and muffler.

c) Install the link and clamp bolts tightening them to the torque listed in this Chapter's Specifications.

18   Install the wheel and lug nuts, lower the vehicle and tighten the lug nuts to the torque listed in the Chapter 1 Specifications.

12   Suspension arms and knuckle (rear) - removal and installation

1   Loosen the rear wheel lug nuts. Block the front wheels to keep the vehicle from rolling, then raise the rear of the vehicle and support it securely on jackstands. Remove the rear wheels.

Upper control arm

2   On ABS equipped vehicles, disconnect the harness connector for the wheel speed sensor and separate the harness from the

12.7b Mark the control arm adjustment fastener (on both ends) before removing it

control arm.

3   Remove the upper control arm-to-knuckle (outboard) mounting bolt and nut (see illustration). It may be necessary on later models to lower the muffler from the hangers to access the bolt.

Caution: *The head of the bolt must be toward the front when installed on 2008 and later models.*

4   Remove the upper control arm-to-suspension support (inboard) mounting bolt and nut and remove the arm from the vehicle (see illustration 12.3).

5   Inspect the control arm bushings for signs of deterioration. If they are in need of replacement, take the control arm to an automotive machine shop to have the bushings replaced.

6   Installation is the reverse of removal. Tighten the fasteners to the torque listed in this Chapter's Specifications.

Lower control arm

7   Mark the position of the adjustment fastener (on both ends) which mounts the lower control arm to the suspension support (see illustrations).

Caution: *Failure to return the adjustment fastener to its original position during installation will change the 'camber' angle of the rear wheel alignment. This change could adversely affect handling and tire wear (see Section 23).*

8   Remove the coil spring (see Section 10).

9   Remove the lower control arm-to-suspension support mounting nut (see illustration 12.7b).

10   Remove the adjustment fastener (cambolt) and then remove the control arm.

11   Inspect the control arm bushing for signs of deterioration. If it is in need of replacement, take the control arm to an automotive machine shop to have the bushing replaced.

12   Installation is the reverse of removal. Tighten the mounting fasteners to the torque listed in this Chapter's Specifications. On the inboard side of the control arm, tighten the nut while holding the adjustment bolt in position.

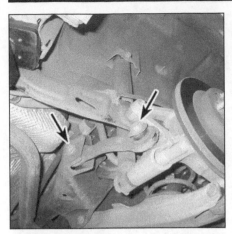

12.13a Toe link arm mounting fasteners

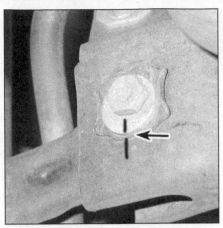

12.13b Mark the toe link adjustment fastener (on both ends) before removing it

12.19 Trailing arm bracket mounting bolts

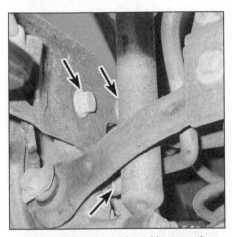

12.21 Trailing arm-to-knuckle mounting bolts (two hidden - vicinity shown)

13.4 The hub nuts are accessed from the back side of the rear knuckle

26   Remove the shock absorber-to-knuckle mounting bolt (see Section 9).
27   Remove the coil spring (see Section 10).
28   Remove the toe link arm (see Step 13).
29   Remove the upper control arm-to-knuckle mounting fasteners (see Step 2).
30   Remove the trailing arm-to-knuckle bolts (see Step 19).
31   Remove the stabilizer bar link-to-knuckle bolt (see Section 11).
32   Remove the knuckle.
33   Installation is the reverse of removal. Tighten all mounting fasteners to the torque listed in this Chapter's Specifications.

All components

34   After the removal and installation of any of the rear suspension components, it's a good idea to have the wheel alignment checked and, if necessary, adjusted.

Toe link arm

13   Mark the position of the adjustment fastener (on both ends) which mounts the toe link arm to the suspension support (see illustrations).
Caution: *Failure to return the adjustment fastener to its original position during installation will change the toe angle of the rear wheel alignment. This change could adversely affect handling and tire wear (see Section 23).*
14   Remove the toe link arm-to-knuckle mounting bolt (see illustration 12.13a).
15   Remove the toe link arm-to-suspension support mounting nut and then the adjustment fastener (cam-bolt) (see illustration 12.13b).
16   Remove the toe link arm.
17   Inspect the control arm bushing for signs of deterioration. If it is in need of replacement, take the control arm to an automotive machine shop to have the bushing replaced.
18   Installation is the reverse of removal. Tighten the mounting fasteners to the torque listed in this Chapter's Specifications. On the inboard side of the toe link arm, tighten the

nut while holding the adjustment bolt in position.

Trailing arm

19   Remove the trailing arm bracket-to-body mounting bolts (see illustration).
20   Remove the parking brake cable bracket from the trailing arm.
21   Remove the trailing arm-to-knuckle mounting bolts and then remove the trailing arm (see illustration).
22   Remove the trailing arm bracket from the trailing arm, if necessary.
23   Inspect the trailing arm bushing for signs of deterioration. If it is in need of replacement, take the trailing arm to an automotive machine shop to have the bushing replaced.
24   Installation is the reverse of removal. Tighten the mounting fasteners to the torque listed in this Chapter's Specifications.

Knuckle

25   Remove the hub and bearing assembly (see Section 13).

13 Hub and bearing assembly (rear) - removal and installation

Warning: *Dust created by the brake system is harmful to your health. Never blow it out with compressed air and don't inhale any of it. Do not, under any circumstances, use petroleum-based solvents to clean brake parts. Use brake system cleaner only.*

1   Loosen the rear wheel lug nuts. Block the front wheels to keep the vehicle from rolling, then raise the rear of the vehicle and support it securely on jackstands. Remove the rear wheels.
2   Remove the brake shoe or disc brake assembly. On ABS equipped vehicles, disconnect the wire harness for the wheel speed sensor (see Chapter 9).
3   Detach the stabilizer bar link from the rear knuckle on 2007 and earlier models (see Section 11).
4   Remove the hub nuts and detach the hub from the knuckle (see illustration).
5   Installation is the reverse of removal.

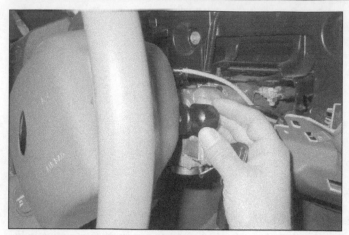

14.4 On 2005 and earlier models, the airbag is secured by spring clips that engage four posts on the airbag (two posts per side); to release them, insert a flat-blade screwdriver into each slot and twist, then repeat on the other side

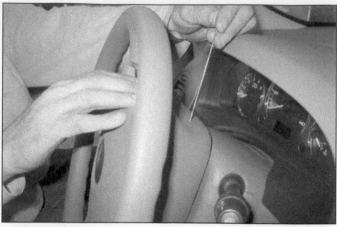

14.5 On 2006 and later models, the airbag is secured by spring clips that engage two posts on the airbag (one post per side); to release them, insert a thin blunt rod into the small hole on one side while gently pulling the airbag away from the steering wheel. Repeat this procedure on the other side

Tighten the hub nuts to the torque listed in this Chapter's Specifications.

6   Install the wheel and lug nuts, lower the vehicle and tighten the lug nuts to the torque listed in the Chapter 1 Specifications.

14   Steering wheel - removal and installation

Warning: *These models are equipped with a Supplemental Restraint System (SRS), more commonly known as airbags. Always disable the airbag system before working in the vicinity of any airbag system component to avoid the possibility of accidental deployment of the airbag(s), which could cause personal injury (see Chapter 12).*

Warning: *Do not use a memory saving device to preserve the PCM or radio memory when working on or near airbag system components.*

Removal

1   Park the vehicle with the wheels pointing straight ahead. Disconnect the cable from the negative terminal of the battery (see Chapter 5, Section 1).

2   Disable the airbag system (see Chapter 12).

3   Remove the upper steering column cover (see Chapter 11).

2005 and earlier models

4   Remove the airbag module by inserting a short flat-bladed screwdriver into one of the corner slots on the back of the steering wheel (there are four slots - two on each side). Turn the blade slightly to release the module retainer while gently pulling the module away (towards you) from the corner where the screwdriver is. Repeat this procedure on the other corner slot on the same side (and back) of the steering wheel. Then, release the other two corners on the other side (and back) of the steering wheel (see illustration).

2006 and later models

5   Remove the airbag module by turning the steering wheel left 90 degrees. Insert a thin straight rod directly into the hole to release the retainer while gently pulling the airbag module towards you (see illustration). Repeat this procedure on the other side of the steering wheel (there is one hole on each side). Return the steering wheel to the original position with the wheels pointing straight ahead.

All models

6   With the airbag module released, disconnect the electrical connectors and remove the airbag module (see illustration). Warning: When carrying the airbag module, keep the driver's side of it away from your body, and when you set it down (in an isolated area), have the driver's side facing up.

7   Note the index mark on the steering wheel hub and the steering shaft and confirm that they are in the 12 o'clock position (see illustration).

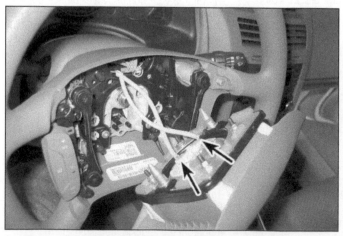

14.6 Airbag module electrical connectors

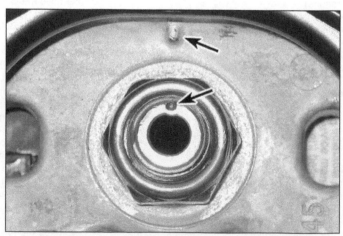

14.7 Steering wheel hub and steering shaft index marks

14.9 Install the two-jaw puller using these two holes in the steering wheel hub

14.10 The electrical connector for the steering wheel mounted switches

8   Remove the steering wheel nut from the steering shaft.

9   Lift the steering wheel off the shaft. If is does not come off easily, remove it by installing a two-jaw puller; placing the puller jaws in the holes on the hub (see illustration).

Caution: *Be careful not to damage the airbag clockspring when installing the two-jaw puller. The jaws of the puller need to be thin enough to fit between the steering wheel hub and the airbag clockspring or the clockspring could be damaged.*

10   Disconnect the electrical connector for the steering wheel mounted control switches (see illustration).

Caution: *While the steering wheel is removed, DO NOT turn the steering shaft. If you do so, the airbag clockspring will need to be re-centered or it could be damaged after the steering wheel is installed and used to turn the front wheels.*

11   If it is necessary to remove the clockspring, disconnect the electrical connectors

attached to it and then remove it from the steering column switch housing (see illustration).

Installation

12   When installing the clockspring, make absolutely sure that the steering shaft is lined up in the 12 o'clock position and that the airbag clockspring is centered by confirming that the small window (above the two small pegs) is yellow (see illustration).

Note: *Some clocksprings may be equipped with a window that shows a view of a thin ribbon to indicate that it's centered as opposed to showing a color. The window may also be to the left of the left peg and not in the center above them. This shouldn't be a problem as long as you have not turned the steering shaft while the wheel was removed. If for some reason the shaft was turned, re-center the steering shaft by placing the front wheels in the straight ahead position and align the index mark on the shaft (and related components) to*

the 12 o'clock position.

Note: *There may be a ring down and around the steering shaft that has square teeth on it (like a tone ring for a sensor). It has one tooth distinctively wider than the others which serves as an index mark for the ring. Before installing the airbag clockspring, make sure that all index marks are in the 12 o'clock position.*

13   Center the clockspring as follows:

a) *Rotate the clockspring clockwise until it stops (don't apply too much force, though).*

b) *Rotate the clockspring counterclockwise about 2-1/2 turns and confirm that the small window on the clockspring is yellow (see illustration 14.12).*

Note: *Some clocksprings may be equipped with a window that shows a view of a thin ribbon to indicate that it's centered as opposed to showing a color. The window may also be to the left of the left peg and not in the center above them.*

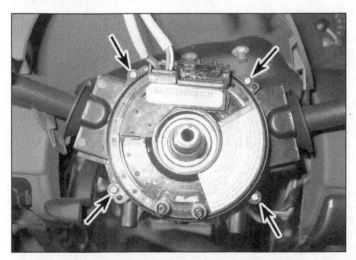

14.11 The airbag clockspring mounting screws

14.12 The two small pegs (A) on the airbag clockspring engage the steering wheel hub. The small window on the clockspring (B) will be yellow when the clockspring is properly centered

15.8 Mark the relationship of the two shafts at the coupler, then remove the coupler pinch bolt

15.12a Location of the upper steering column mounting fasteners

14 The remainder of the installation is the reverse of removal, noting the following points:

a) *Make sure that the airbag clockspring is centered and the steering shaft is in the 12 o'c lock position before installing the steering wheel.*

b) *When installing the steering wheel, align the index mark on the steering wheel hub with the mark on the shaft and make sure that the two small pegs on the airbag clockspring engage with the steering wheel hub properly (see illustrations 14.7 and 14.12).*

c) *Install the steering wheel nut and tighten it to the torque listed in this Chapter's Specifications.*

d) *Install the airbag module on the steering wheel and push it into place until the retaining posts engage with the retaining springs. Enable the airbag system (see Chapter 12).*

15.12b The location of the lower steering column mounting fastener

15 Steering column - removal and installation

Warning: *These models are equipped with airbags. Always disable the airbag system before working in the vicinity of any airbag system component to avoid the possibility of accidental deployment of the airbag(s), which could cause personal injury (see Chapter 12).*

Warning: *Do not use a memory saving device to preserve the PCM's memory when working on or near airbag system components.*

Note: *Models equipped with electronically assisted power steering have a motor and module attached to the steering column. These components can be transferred to a replacement steering column if necessary. The motor and module are removed with the steering column initially and then detached after the steering column is removed.*

Removal

1 Park the vehicle with the wheels pointing straight ahead. Disconnect the cable from the negative terminal of the battery (see Chapter 5, Section 1).

2 Disable the airbag system (see Chapter 12).

3 Remove the steering column covers (see Chapter 11).

4 Remove the steering wheel (see Section 14).

5 Remove the airbag clockspring (see Section 14).

6 Remove the steering column switches and housing (see Chapter 12).

7 Remove the knee bolster from under the steering column (see Chapter 11).

8 Mark the relationship of the steering column shaft to the intermediate shaft coupler, then remove the pinch bolt (see illustration).

9 Disconnect any remaining electrical connectors and remove any wiring harnesses from the steering column that would interfere

with removal.

10 Put the steering wheel in the uppermost position and then lock it there with the tilt lever.

11 On vehicles equipped with adjustable pedals, remove the three assembly bracket nuts for the accelerator pedal and place it aside for access. Do not remove the brake pedal cable or motor from the assembly.

12 Remove the steering column mounting fasteners and then carefully guide the column out from the instrument panel (see illustrations).

Warning: *The steering column is vulnerable to damage and could also expand when it's not installed. Tie a cord around each end of the steering column shaft to keep it from coming apart and handle the column very carefully.*

Installation

13 Guide the column into position, connecting the steering shaft with the intermediate shaft coupler. Be sure to align the marks made in Step 8.

Note: *Leave the cord placed on the steering column until the mounting fasteners are tightened and then remove it.*

14 Install all of the mounting fasteners starting with the lower one first and then the upper two. Tighten them to the torque listed in this Chapter's Specifications beginning with the lower one first, and then the upper two going from left to right.

Note: *This sequence of installing and tightening the steering column mounting fasteners is recommended by the manufacturer.*

15 Install a new pinch bolt and tighten it to the torque listed in this Chapter's Specifications.

16 The remainder of the installation is the reverse of the removal procedure. Refer to Section for the clockspring, steering wheel and airbag module installation details. On vehicles equipped with adjustable pedals, reinstall the accelerator pedal assembly and tighten the fasteners to the torque listed in this Chapter's Specifications.

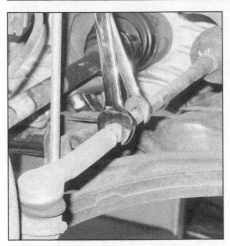

16.2 Using two wrenches, loosen the jam nut

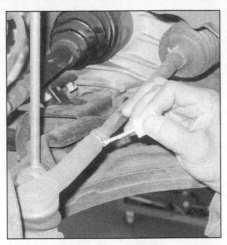

16.3 Mark the position of the tie-rod end in relation to the threads

16.4 Loosen the tie-rod end ballstud from the steering knuckle arm with a puller

16 Tie-rod ends - removal and installation

Removal

1   Loosen the wheel lug nuts, raise the front of the vehicle and support it securely on jackstands. Apply the parking brake and block the rear wheels to keep the vehicle from rolling off the jackstands. Remove the wheel.
2   Loosen the tie-rod end jam nut (see illustration).
3   Mark the relationship of the tie-rod end to the threaded portion of the tie-rod. This will ensure the toe-in setting is restored when reassembled (see illustration).
4   Loosen the tie-rod end ballstud nut a few turns and then install a puller to loosen the ballstud from the steering knuckle arm (see illustration).
5   With the ballstud loose, remove the ballstud nut and then completely separate the tie-rod end from the steering knuckle arm.
Warning: *The manufacturer states that the tie-rod end-to-steering knuckle nut must be replaced with a new one whenever it is removed.*
6   Unscrew the tie-rod end from the tie-rod.

Installation

7   Thread the tie-rod end onto the tie-rod to the marked position and insert the tie-rod end ballstud into the steering knuckle arm. Install the new nut onto the ballstud and tighten it to the torque listed in this Chapter's Specifications.
8   Tighten the jam nut securely and install the wheel. Lower the vehicle and tighten the lug nuts to the torque listed in the Chapter 1 Specifications.
9   Have the front end alignment checked and, if necessary, adjusted.

17 Steering gear boots - removal and installation

1   Loosen the lug nuts, raise the vehicle and support it securely on jackstands. Remove the wheel.
2   Remove the tie-rod end and jam nut (see Section 16).
3   Remove the outer steering gear boot clamp with a pair of pliers (see illustration). Cut off the inner boot clamp with diagonal cutters and slide the boot off the rod.
4   Before installing the new boot, wrap the threads and spines on the end of the steering rod with a layer of tape so the small end of the new boot isn't damaged.
5   Slide the new boot into position on the steering gear until it seats in the groove in the steering rod and install new clamps.
6   Remove the tape and install the tie-rod end (see Section 16).
7   Install the wheel and lug nuts. Lower the vehicle and tighten the lug nuts to the torque listed in the Chapter 1 Specifications.

18 Steering gear - removal and installation

Warning: *Make sure the steering shaft is not turned while the steering gear is removed or you could damage the airbag system clockspring. To prevent the shaft from turning, place the ignition key in the LOCK position or thread the seat belt through the steering wheel and clip it into place.*
Note: *Some models are equipped with an electronically assisted power steering system. On these models, disregard any reference to draining hydraulic fluid or disconnecting any fluid line fittings.*

Removal

1   Disconnect the cable from the negative battery terminal (see Chapter 5, Section 1).
2   Loosen the front wheel lug nuts, raise the front of the vehicle and support it securely on jackstands. Remove both front wheels.
3   Remove the power steering fluid from the power steering reservoir. This can be accomplished with a suction gun or large syringe, or

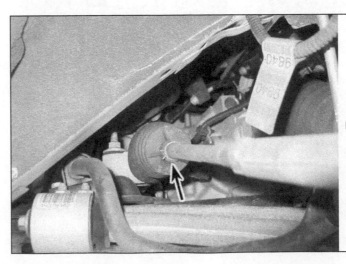

17.3 The location of the steering gear boot outer clamp. The inner clamp is on the other side of the boot

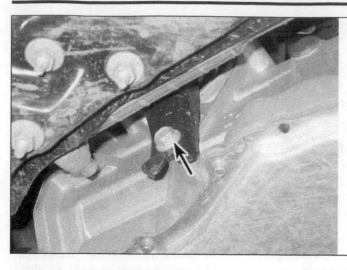

18.4 Remove the transaxle mount-to-transaxle bolt on models with electronically assisted power steering

by disconnecting the fluid hose and draining the fluid into a container.

4   On models equipped with electronically assisted power steering, remove the rear transaxle mount-to-transaxle bolt (see illustration). If you're working on a model with hydraulically assisted power steering, support the transaxle with a floor jack, loosen the transaxle mount through-bolt, unplug and detach the oxygen sensor harness from the mount bracket, then remove the three transaxle mount-to-subframe nuts and the three mount-to-transaxle bolts. Move the transaxle mount out of the way.

5   Mark the relationship of the intermediate shaft U-joint to the steering gear input shaft and then remove the U-joint pinch bolt and detach the U-joint (see illustration). Discard the pinch bolt.

6   Detach the tie-rod ends from the steering knuckles (see Section 16). Remove the steering gear heat shields, if equipped. On V6 engines, remove the transmission mount (and brace, if so equipped) and set it aside. Disconnect the oxygen sensor and unclip the lead from the mount.

7   Place a drain pan under the steering

gear and detach the power steering pressure and return lines. Cap all openings to prevent excessive fluid loss and contamination. Remove the retainer securing the power steering lines to the right-side of the subframe.

8   Remove the steering gear mounting bolts (see illustration).

9   Remove the steering gear through the left wheel opening.

Note: *On 2006 and later vehicle models equipped with an electronically assisted power steering system, rotate the steering gear 90 degrees to ease removal.*

Installation

10   Installation is the reverse of removal, noting the following points:

a) *Tighten the steering gear mounting bolts to the torque listed in this Chapter's Specifications.*

b) *Use a new pinch bolt for the intermediate shaft U-joint and tighten it to the torque listed in this Chapter's Specifications.*

c) *Fill the power steering pump with the recommended fluid (see Chapter 1),*

bleed the system (see Section 20) and recheck the fluid level, if applicable.

d) *Run the engine and check for proper operation and leaks. Shut off the engine and recheck fluid levels, if applicable.*

e) *Have the front end alignment checked and, if necessary, adjusted.*

---

19   Power steering pump - removal and installation

Note: *This Section does not apply to vehicle models equipped with electronically assisted power steering systems.*

Removal

1   Disconnect the cable from the negative battery terminal (see Chapter 5, Section 1).

2   Use a large syringe or suction gun and remove as much fluid out of the power steering fluid reservoir as possible. Place a drain pan under the vehicle to catch any fluid that spills out when the hoses are disconnected.

3   Remove the air filter housing from 2007 and earlier models (see Chapter 4).

4   On 2008 and later 3.5L V6 models, remove the drivebelt idler pulley, the engine lift bracket and the power steering pump pulley.

5   On 2008 and later 3.6L V6 models, remove the right front wheel, the engine mount bracket and the power steering pump pulley.

6   Remove the cover over the intake manifold (see Chapter 2B or Chapter 2C).

7   Remove the drivebelt (see Chapter 1).

8   Disconnect the return hose from the power steering pump. Then, using a flare-nut wrench, unscrew the pressure line fitting from the pump.

9   Remove the mounting bolts securing the pump to the mounting bracket and remove the pump (see illustration).

Note: *Most power steering pump pulleys have holes that allow access to the pump mounting*

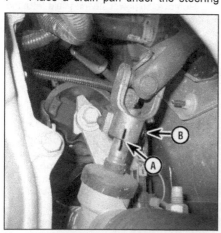

18.5 Mark the relationship of the intermediate shaft U-joint to the steering gear input shaft (A) and remove the U-joint pinch bolt (B)

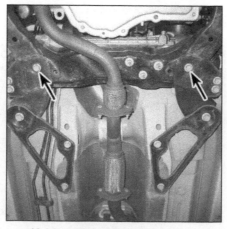

18.8 Location of the steering gear mounting fasteners

19.9 The power steering pump mounting bolts are accessible through the holes in the pulley

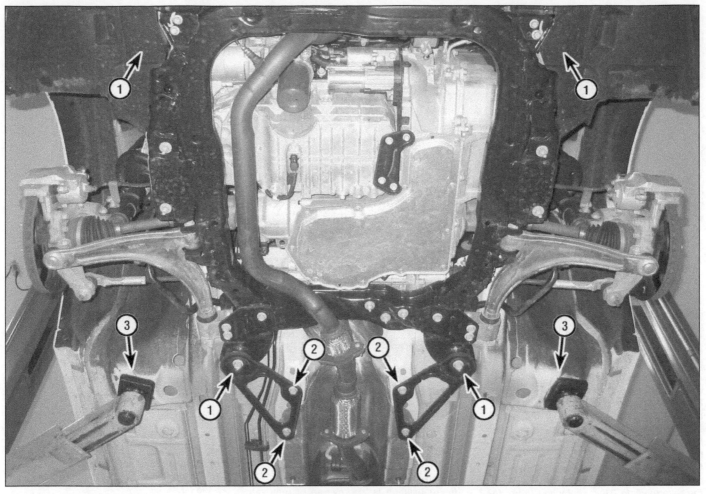

21.2 Subframe mounting details:

1   Main subframe mounting bolts (the front two are hidden beneath the engine splash shields)

2   Subframe support mounting bolts

3   Jacking points away from the subframe

bolts on the other side of the pulley. If the pulley on your model vehicle is not designed with holes, it will be necessary to remove the pulley to gain access to the pump mounting bolts. Pulley removal and installation tools are available at most auto parts stores. When installing the pulley, the hub of the pulley should be flush (even) with the end of the pump shaft.

Installation

Note: *If you're installing a new pump, it may be necessary to transfer the old pulley to the new pump (see above).*

10   Installation is the reverse of removal, noting the following points:

a) *Tighten the mounting bolts to the torque listed in this Chapter's Specifications and tighten line fitting securely.*

b) *Install the drivebelt as outlined in Chapter 1.*

c) *Fill the power steering reservoir with the recommended fluid (see Chapter 1). Bleed the power steering hydraulic system as described in Section 20.*

20   Power steering system - bleeding

Note: *This Section does not apply to vehicle models equipped with electronically assisted power steering systems.*

1   The power steering system must be bled whenever a line is disconnected. Bubbles can be seen in power steering fluid that has air in it and the fluid will often have a tan or milky appearance. Low fluid level can cause air to mix with the fluid, resulting in a noisy pump as well as foaming of the fluid.

2   Open the hood and check the fluid level in the reservoir, adding the specified fluid necessary to bring it up to the proper level (see Chapter 1).

3   Start the engine and slowly turn the steering wheel several times from left-to-right and back again. Do not turn the wheel completely from lock-to-lock. Check the fluid level, topping it up as necessary until it remains steady and no more bubbles are visible.

21   Subframe - removal and installation

1   Disconnect the cable from the negative battery terminal (see Chapter 5, Section 1).

2   Loosen the front wheel lug nuts, raise the front of the vehicle and support it securely on jackstands. Remove both front wheels. On four-cylinder models, remove the plastic fender wells (see Chapter 11). On some four-cylinder models, it will be necessary to remove the front catalytic converter.

Note: *The jackstands must be placed away from the front suspension subframe, not supporting the vehicle by the subframe (see illustration).*

3   Remove the lower radiator mounting brackets and support the radiator and air conditioning condenser by tying them to the upper radiator support - refer to *Cooling fan - check and replacement* in Chapter 3.

4   On V6 engines, remove both the front and rear catalytic converters (see Chapter 4).

5   Remove the fasteners for the engine

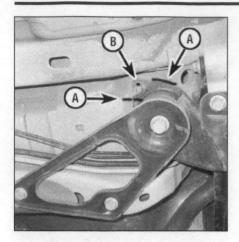

21.14 Mark the relationship of the subframe to the body (A) on both rear corners of the subframe and note the alignment holes (B) - one corner shown

splash shield and the air deflectors on each side and then remove them - refer to *Cooling fan - check and replacement* in Chapter 3.

6   Disconnect the stabilizer bar links from the struts (see Section 4).

7   Detach the tie-rod ends from the steering knuckles (see Section 16).

8   Remove the control arms (see Section 5).

9   Disconnect the intermediate shaft U-joint from the steering gear input shaft (see Section 18).

Caution: *Do not turn the steering wheel or change the direction of the front wheels with the intermediate shaft detached from the steering gear. Moreover, the U-joint and intermediate shaft must be assembled with the steering gear input shaft and steering wheel in the original positions or damage to the airbag clockspring could occur.*

10   Install an engine support fixture to the top of the engine or install an engine hoist to support the engine and transaxle (see Chapter 2A).

11   Remove the front transaxle mount through-bolt (see Chapter 7B).

12   Remove the rear transaxle mount (see Chapter 7B).

13   Remove the fluid line fittings to the steering gear (see Section 18). Also remove the retainer securing the power steering lines to the right-side of the subframe.

Note: *This Step is not applicable to vehicles equipped with electronically assisted power steering.*

14   Carefully mark the position of the subframe in relation to the vehicle chassis and also detach any wiring harnesses or anything else that may be attached to the subframe (see illustration).

15   Using two floor jacks, support the subframe. Position one jack on each side of the subframe; placing them midway between the front and rear mounting points.

16   Remove the bolts securing the subframe and the subframe supports (see illustration 21.2).

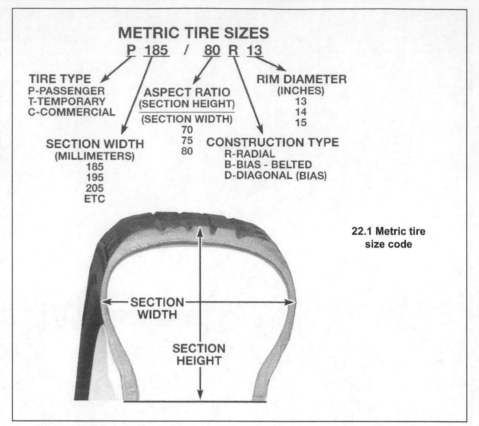

METRIC TIRE SIZES

P 185 / 80 R 13

TIRE TYPE
P-PASSENGER
T-TEMPORARY
C-COMMERCIAL

ASPECT RATIO
(SECTION HEIGHT)
(SECTION WIDTH)
70
75
80

SECTION WIDTH
(MILLIMETERS)
185
195
205
ETC

RIM DIAMETER
(INCHES)
13
14
15

CONSTRUCTION TYPE
R-RADIAL
B-BIAS - BELTED
D-DIAGONAL (BIAS)

SECTION WIDTH

SECTION HEIGHT

22.1 Metric tire size code

17   Lower the jacks until the subframe is sufficiently resting on the ground.

Installation

18   Installation is the reverse of removal, noting the following points:

a) *Raise and support the subframe with the jacks and install the subframe and subframe support mounting bolts until they are all finger-tight.*

b) *Align the subframe using the reference marks made earlier and using a punch or drift in alignment holes provided and then tighten the subframe and subframe support mounting bolts to the torque listed in this Chapter's Specifications.*

c) *Reconnect the negative battery cable (see Chapter 5).*

d) *Add power steering fluid as necessary, then bleed the power steering system (see Section 20).*

e) *Have the front end alignment checked and, if necessary, adjusted.*

22   Wheels and tires - general information

1   All vehicles covered by this manual are equipped with metric-sized fiberglass or steel belted radial tires (see illustration). Use of other size or type of tires may affect the ride and handling of the vehicle. Don't mix different types of tires, such as radials and bias belted, on the same vehicle as handling may

be seriously affected. It's recommended that tires be replaced in pairs on the same axle, but if only one tire is being replaced, be sure it's the same size, structure and tread design as the other.

2   Because tire pressure has a substantial effect on handling and wear, the pressure on all tires should be checked at least once a month or before any extended trips (see Chapter 1).

3   Wheels must be replaced if they are bent, dented, leak air, have elongated bolt holes, are heavily rusted, out of vertical symmetry or if the lug nuts won't stay tight. Wheel repairs that use welding or peening are not recommended.

4   Tire and wheel balance is important in the overall handling, braking and performance of the vehicle. Unbalanced wheels can adversely affect handling and ride characteristics as well as tire life. Whenever a tire is installed on a wheel, the tire and wheel should be balanced by a shop with the proper equipment.

23   Wheel alignment - general information

1   A wheel alignment refers to the adjustments made to the wheels so they are in proper angular relationship to the suspension and the ground. Wheels that are out of proper alignment not only affect vehicle control, but also increase tire wear. The front end angles normally measured are camber, caster and

toe-in (see illustration). Toe-in and camber are adjustable; if the caster is not correct, check for bent components. Rear toe-in and camber are also adjustable.

2　Getting the proper wheel alignment is a very exacting process, one in which complicated and expensive machines are necessary to perform the job properly. Because of this, you should have a technician with the proper equipment perform these tasks. We will, however, use this space to give you a basic idea of what is involved with a wheel alignment so you can better understand the process and deal intelligently with the shop that does the work.

3　Toe-in is the turning in of the wheels. The purpose of a toe specification is to ensure parallel rolling of the wheels. In a vehicle with zero toe-in, the distance between the front edges of the wheels will be the same as the distance between the rear edges of the wheels. The actual amount of toe-in is normally only a fraction of an inch. On the front end, toe-in is controlled by the tie-rod end position on the tie-rod. On the rear end, it's controlled by the position of the suspension toe link on the chassis. Incorrect toe-in will cause the tires to wear improperly by making them scrub against the road surface.

4　Camber is the tilting of the wheels from vertical when viewed from one end of the vehicle. When the wheels tilt out at the top, the camber is said to be positive (+). When the wheels tilt in at the top the camber is negative (-). The amount of tilt is measured in-degrees from vertical and this measurement is called the camber angle. This angle affects the amount of tire tread which contacts the road and compensates for changes in the suspension geometry when the vehicle is cornering or traveling over an undulating surface. Camber can be adjusted on the front end, but will require a modification of the lower strut-to-knuckle bolt hole on the strut. The strut must be disconnected from the steering knuckle, the lower mounting bolt holes elongated (from side-to-side), then reassembled. The relationship of the steering knuckle to the strut can then be altered. On the rear end, camber is controlled by the position of the lower control arm on the subframe.

5　Caster is the tilting of the front steering axis from the vertical. A tilt toward the rear is positive caster and a tilt toward the front is negative caster.

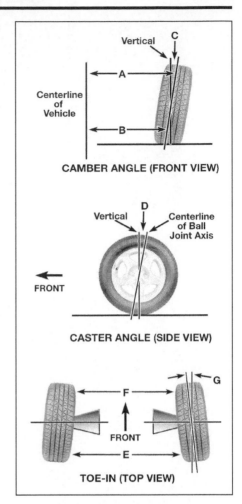

23.1 Camber, caster and toe-in angles

A minus B = C (degrees camber)
D = degrees caster
E minus F = toe-in (measured in inches)
G = toe-in (expressed in degrees)

Notes

Chapter 11
Body

Contents

1 General Information

Warning: *The models covered by this manual are equipped with a Supplemental Restraint System (SRS), more commonly known as airbags. Always disable the airbag system before working in the vicinity of any airbag system components to avoid the possibility of accidental deployment of the airbags, which* could cause personal injury (see Chapter 12).

1 The models covered by this manual feature "unibody" construction, in which the major body components, floor pan and front and rear frame side rails are welded together to create a rigid structure which supports the remaining body components, drivetrain, front and rear suspension and other components.

2 Certain components are particularly vulnerable to accident damage and can be unbolted and repaired or replaced. Among these parts are the body moldings, bumpers, front fenders, doors, the hood and trunk lid. Only general body maintenance practices and body panel repair procedures within the scope of the do-it-yourselfer are included in this Chapter.

3 Although all models are very similar, some procedures may differ somewhat from one body to another.

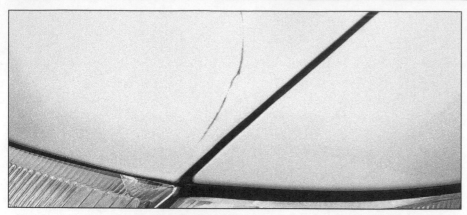

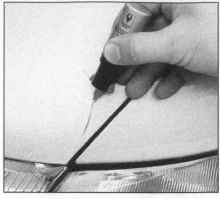

Make sure the damaged area is perfectly clean and rust free. If the touch-up kit has a wire brush, use it to clean the scratch or chip. Or use fine steel wool wrapped around the end of a pencil. Clean the scratched or chipped surface only, not the good paint surrounding it. Rinse the area with water and allow it to dry thoroughly

Thoroughly mix the paint, then apply a small amount with the touch-up kit brush or a very fine artist's brush. Brush in one direction as you fill the scratch area. Do not build up the paint higher than the surrounding paint

2 Repair of minor paint scratches

1   No matter how hard you try to keep your vehicle looking like new, it will inevitably be scratched, chipped or dented at some point. If the metal is actually dented, seek the advice of a professional. But you can fix minor scratches and chips yourself. Buy a touch-up paint kit from a dealer service department or an auto parts store. To ensure that you get the right color, you'll need to have the specific make, model and year of your vehicle and, ideally, the paint code, which is located on a special metal plate under the hood or in the door jamb.

3 Body repair - minor damage

Plastic body panels

1   The following repair procedures are for minor scratches and gouges. Repair of more serious damage should be left to a dealer service department or qualified auto body shop. Below is a list of the equipment and materials necessary to perform the following repair procedures on plastic body panels.

Wax, grease and silicone removing
solvent
Cloth-backed body tape
Sanding discs
Drill motor with three-inch disc holder
Hand sanding block
Rubber squeegees
Sandpaper
Non-porous mixing palette
Wood paddle or putty knife
Curved-tooth body file
Flexible parts repair material

Flexible panels (bumper trim)

2   Remove the damaged panel, if necessary or desirable. In most cases, repairs can be carried out with the panel installed.
3   Clean the area(s) to be repaired with a

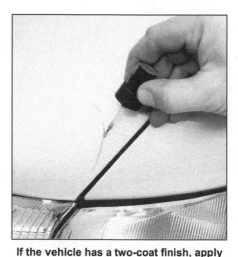

If the vehicle has a two-coat finish, apply the clear coat after the color coat has dried

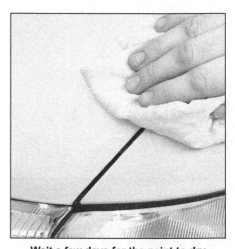

Wait a few days for the paint to dry thoroughly, then rub out the repainted area with a polishing compound to blend the new paint with the surrounding area. When you're happy with your work, wash and polish the area

wax, grease and silicone removing solvent applied with a water-dampened cloth.
4   If the damage is structural, that is, if it extends through the panel, clean the backside of the panel area to be repaired as well. Wipe dry.
5   Sand the rear surface about 1-1/2 inches beyond the break.
6   Cut two pieces of fiberglass cloth large enough to overlap the break by about 1-1/2 inches. Cut only to the required length.
7   Mix the adhesive from the repair kit according to the instructions included with the kit, and apply a layer of the mixture approximately 1/8-inch thick on the backside of the panel. Overlap the break by at least 1-1/2 inches.
8   Apply one piece of fiberglass cloth to the adhesive and cover the cloth with additional adhesive. Apply a second piece of fiberglass cloth to the adhesive and immediately cover the cloth with additional adhesive in sufficient quantity to fill the weave.
9   Allow the repair to cure for 20 to 30 minutes at 60-degrees to 80-degrees F.
10   If necessary, trim the excess repair mate-

rial at the edge.
11   Remove all of the paint film over and around the area(s) to be repaired. The repair material should not overlap the painted surface.
12   With a drill motor and a sanding disc (or a rotary file), cut a "V" along the break line approximately 1/2-inch wide. Remove all dust and loose particles from the repair area.
13   Mix and apply the repair material. Apply a light coat first over the damaged area; then continue applying material until it reaches a level slightly higher than the surrounding finish.
14   Cure the mixture for 20 to 30 minutes at 60-degrees to 80-degrees F.
15   Roughly establish the contour of the area being repaired with a body file. If low areas or pits remain, mix and apply additional adhesive.
16   Block sand the damaged area with sandpaper to establish the actual contour of the surrounding surface.

17   If desired, the repaired area can be temporarily protected with several light coats of primer. Because of the special paints and techniques required for flexible body panels, it is recommended that the vehicle be taken to a paint shop for completion of the body repair.

Steel body panels

Repairing simple dents

Note: *These photos illustrate a method of repairing simple dents. They are intended to supplement* Body repair - minor damage *in this chapter and should not be used as the sole instructions for body repair on these vehicles.*

18   When repairing dents, the first job is to pull the dent out until the affected area is as close as possible to its original shape. There is no point in trying to restore the original shape completely as the metal in the damaged area will have stretched on impact and cannot be restored to its original contours. It is better to bring the level of the dent up to a point that is about 1/8-inch below the level of the surrounding metal. In cases where the dent is very shallow, it is not worth trying to pull it out at all.

19   If the backside of the dent is accessible, it can be hammered out gently from behind using a soft-face hammer. While doing this, hold a block of wood firmly against the opposite side of the metal to absorb the hammer blows and prevent the metal from being stretched.

20   If the dent is in a section of the body which has double layers, or some other factor makes it inaccessible from behind, a different technique is required. Drill several small holes through the metal inside the damaged area, particularly in the deeper sections. Screw long, self-tapping screws into the holes just enough for them to get a good grip in the metal. Now pulling on the protruding heads of the screws with locking pliers can pull out the dent.

21   The next stage of repair is the removal of paint from the damaged area and from an inch or so of the surrounding metal. This is easily done with a wire brush or sanding disk in a drill motor, although it can be done just as effectively by hand with sandpaper. To complete the preparation for filling, score the surface of the bare metal with a screwdriver or the tang of a file or drill small holes in the affected area. This will provide a good grip for the filler material. To complete the repair, see the section on filling and painting.

Repair of rust holes or gashes

22   Remove all paint from the affected area and from an inch or so of the surrounding metal using a sanding disk or wire brush mounted in a drill motor. If these are not available, a few sheets of sandpaper will do the job just as effectively.

23   With the paint removed, you will be able to determine the severity of the corrosion and decide whether to replace the whole panel, if possible, or repair the affected area. New body panels are not as expensive as most people think and it is often quicker to install a new panel than to repair large areas of rust.

24   Remove all trim pieces from the affected area except those which will act as a guide to the original shape of the damaged body, such as headlight shells, etc. Using metal snips or a hacksaw blade, remove all loose metal and any other metal that is badly affected by rust. Hammer the edges of the hole in to create a slight depression for the filler material.

25   Wire-brush the affected area to remove the powdery rust from the surface of the metal. If the back of the rusted area is accessible, treat it with rust inhibiting paint.

26   Before filling is done, block the hole in some way. This can be done with sheet metal riveted or screwed into place, or by stuffing the hole with wire mesh.

27   Once the hole is blocked off, the affected area can be filled and painted. See the following subsection on *Filling and painting*.

Filling and painting

28   Many types of body fillers are available, but generally speaking, body repair kits which contain filler paste and a tube of resin hardener are best for this type of repair work. A wide, flexible plastic or nylon applicator will be necessary for imparting a smooth and contoured finish to the surface of the filler material. Mix up a small amount of filler on a clean piece of wood or cardboard (use the hardener sparingly). Follow the manufacturer's instructions on the package, otherwise the filler will set incorrectly.

29   Using the applicator, apply the filler paste to the prepared area. Draw the applicator across the surface of the filler to achieve the desired contour and to level the filler surface. As soon as a contour that approximates the original one is achieved, stop working the paste. If you continue, the paste will begin to stick to the applicator. Continue to add thin layers of paste at 20-minute intervals until the level of the filler is just above the surrounding metal.

30   Once the filler has hardened, the excess can be removed with a body file. From then on, progressively finer grades of sandpaper should be used, starting with a 180-grit paper and finishing with 600-grit wet-or-dry paper. Always wrap the sandpaper around a flat rubber or wooden block, otherwise the surface of the filler will not be completely flat. During the sanding of the filler surface, the wet-or-dry paper should be periodically rinsed in water. This will ensure that a very smooth finish is produced in the final stage.

31   At this point, the repair area should be surrounded by a ring of bare metal, which in turn should be encircled by the finely feathered edge of good paint. Rinse the repair area with clean water until all of the dust produced by the sanding operation is gone.

32   Spray the entire area with a light coat of primer. This will reveal any imperfections in the surface of the filler. Repair the imperfections with fresh filler paste or glaze filler and once more smooth the surface with sandpaper. Repeat this spray-and-repair procedure until you are satisfied that the surface of the filler and the feathered edge of the paint are perfect. Rinse the area with clean water and allow it to dry completely.

33   The repair area is now ready for painting. Spray painting must be carried out in a warm, dry, windless and dust free atmosphere. These conditions can be created if you have access to a large indoor work area, but if you are forced to work in the open, you will have to pick the day very carefully. If you are working indoors, dousing the floor in the work area with water will help settle the dust that would otherwise be in the air. If the repair area is confined to one body panel, mask off the surrounding panels. This will help minimize the effects of a slight mismatch in paint color. Trim pieces such as chrome strips, door handles, etc., will also need to be masked off or removed. Use masking tape and several thickness of newspaper for the masking operations.

34   Before spraying, shake the paint can thoroughly, then spray a test area until the spray painting technique is mastered. Cover the repair area with a thick coat of primer. The thickness should be built up using several thin layers of primer rather than one thick one. Using 600-grit wet-or-dry sandpaper, rub down the surface of the primer until it is very smooth. While doing this, the work area should be thoroughly rinsed with water and the wet-or-dry sandpaper periodically rinsed as well. Allow the primer to dry before spraying additional coats.

35   Spray on the top coat, again building up the thickness by using several thin layers of paint. Begin spraying in the center of the repair area and then, using a circular motion, work out until the whole repair area and about two inches of the surrounding original paint is covered. Remove all masking material 10 to 15 minutes after spraying on the final coat of paint. Allow the new paint at least two weeks to harden, then use a very fine rubbing compound to blend the edges of the new paint into the existing paint. Finally, apply a coat of wax

---

4   Body repair - major damage

---

1   Major damage must be repaired by an auto body shop specifically equipped to perform body and frame repairs. These shops have the specialized equipment required to do the job properly.

2   If the damage is extensive, the frame must be checked for proper alignment or the vehicle's handling characteristics may be adversely affected and other components may wear at an accelerated rate.

3   Due to the fact that all of the major body components (hood, fenders, etc.) are separate and replaceable units, any seriously damaged components should be replaced rather than repaired. Sometimes the components can be found in a wrecking yard that specializes in used vehicle components, often at considerable savings over the cost of new parts.

These photos illustrate a method of repairing simple dents. They are intended to supplement *Body repair - minor damage* in this Chapter and should not be used as the sole instructions for body repair on these vehicles.

1 If you can't access the backside of the body panel to hammer out the dent, pull it out with a slide-hammer-type dent puller. Tap with a hammer near the edge of the dent to help 'pop' the metal back to its original shape, about 1/8-inch below the surface of the surrounding metal

2 Using coarse-grit sandpaper, remove the paint down to the bare metal. Clean the repair area with wax/silicone remover.

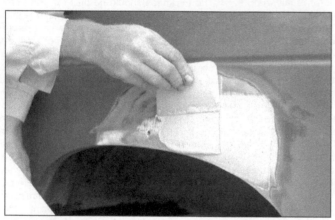

3 Following label instructions, mix up a batch of plastic filler and hardener, then quickly press it into the metal with a plastic applicator. Work the filler until it matches the original contour and is slightly above the surrounding metal

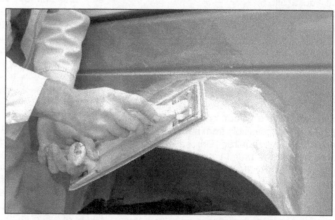

4 Let the filler harden until you can just dent it with your fingernail. File, then sand the filler down until it's smooth and even. Work down to finer grits of sandpaper - always using a board or block - ending up with 360 or 400 grit

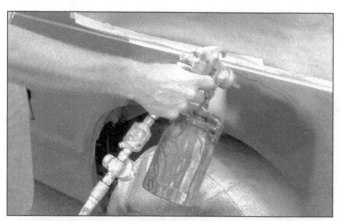

5 When the area is smooth to the touch, clean the area and mask around it. Apply several layers of primer to the area. A professional-type spray gun is being used here, but aerosol spray primer works fine

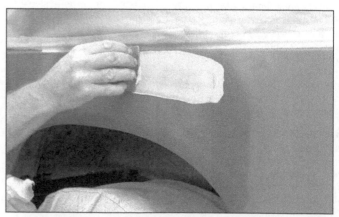

6 Fill imperfections or scratches with glazing compound. Sand with 360 or 400-grit and re-spray. Finish sand the primer with 600 grit, clean thoroughly, then apply the finish coat. Don't attempt to rub out or wax the repair area until the paint has dried completely (at least two weeks)

5 Upholstery, carpets and vinyl trim - maintenance

Upholstery and carpets

1 Every three months remove the floormats and clean the interior of the vehicle (more frequently if necessary). Use a stiff whiskbroom to brush the carpeting and loosen dirt and dust, then vacuum the upholstery and carpets thoroughly, especially along seams and crevices.

2 Dirt and stains can be removed from carpeting with basic household or automotive carpet shampoos available in spray cans. Follow the directions and vacuum again, then use a stiff brush to bring back the "nap" of the carpet.

3 Most interiors have cloth or vinyl upholstery, either of which can be cleaned and maintained with a number of material-specific cleaners or shampoos available in auto supply stores. Follow the directions on the product for usage, and always spot-test any upholstery cleaner on an inconspicuous area (bottom edge of a backseat cushion) to ensure that it doesn't cause a color shift in the material.

4 After cleaning, vinyl upholstery should be treated with a protectant.
Caution: *Do not use protectant on vinyl-covered steering wheels.*
Note: *Make sure the protectant container indicates the product can be used on seats - some products may make a seat too slippery.*

5 Leather upholstery requires special care. It should be cleaned regularly with saddle-soap or leather cleaner. Never use alcohol, gasoline, nail polish remover or thinner to clean leather upholstery.

6 After cleaning, regularly treat leather upholstery with a leather conditioner, rubbed in with a soft cotton cloth. Never use car wax on leather upholstery.

7 In areas where the interior of the vehicle is subject to bright sunlight, cover leather seating areas of the seats with a sheet if the vehicle is to be left out for any length of time.

Vinyl trim

8 Don't clean vinyl trim with detergents, caustic soap or petroleum-based cleaners. Plain soap and water works just fine, with a soft brush to clean dirt that may be ingrained. Wash the vinyl as frequently as the rest of the vehicle.

9 After cleaning, application of a high-quality rubber and vinyl protectant will help prevent oxidation and cracks. The protectant can also be applied to weather-stripping, vacuum lines and rubber hoses, which often fail as a result of chemical degradation, and to the tires.

6 Fastener and trim removal

1 There is a variety of plastic fasteners used to hold trim panels, splash shields and other parts in place in addition to typical screws, nuts and bolts. Once you are familiar with them, they can usually be removed without too much difficulty.

2 The proper tools and approach can prevent added time and expense to a project by minimizing the number of broken fasteners and/or parts.

3 The following illustration shows various types of fasteners that are typically used on most vehicles and how to remove and install them (see illustration). Replacement fasteners are commonly found at most auto parts stores, if necessary.

Fasteners

This tool is designed to remove special fasteners. A small pry tool used for removing nails will also work well in place of this tool

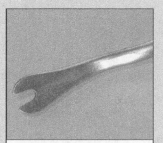

A Phillips head screwdriver can be used to release the center portion, but light pressure must be used because the plastic is easily damaged. Once the center is up, the fastener can easily be pried from its hole

Here is a view with the center portion fully released. Install the fastener as shown, then press the center in to set it

This fastener is used for exterior panels and shields. The center portion must be pried up to release the fastener. Install the fastener with the center up, then press the center in to set it

This type of fastener is used commonly for interior panels. Use a small blunt tool to press the small pin at the center in to release it . . .

. . . the pin will stay with the fastener in the released position

Reset the fastener for installation by moving the pin out. Install the fastener, then press the pin flush with the fastener to set it

This fastener is used for exterior and interior panels. It has no moving parts. Simply pry the fastener from its hole like the claw of a hammer removes a nail. Without a tool that can get under the top of the fastener, it can be very difficult to remove

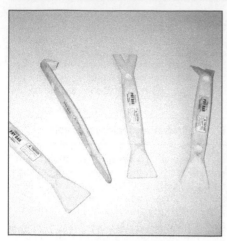

6.4 These small plastic pry tools are ideal for prying off trim panels

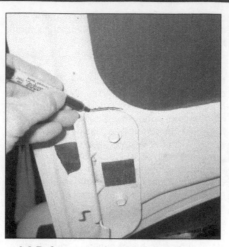

9.2 Before removing the hood, draw a mark around the hinge plate

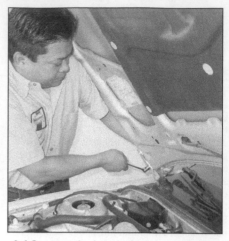

9.4 Support the hood with your shoulder while removing the hood bolts

4   Trim panels are typically made of plastic and their flexibility can help during removal. The key to their removal is to use a tool to pry the panel near its retainers to release it without damaging surrounding areas or breaking-off any retainers. The retainers will usually snap out of their designated slot or hole after force is applied to them. Stiff plastic tools designed for prying on trim panels are available at most auto parts stores (see illustration). Tools that are tapered and wrapped in protective tape, such as a screwdriver or small pry tool, are also very effective when used with care.

7   Hinges and locks - maintenance

1   Once every 3000 miles, or every three months, the hinges and latch assemblies on the doors, hood and trunk should be given a few drops of light oil or lock lubricant. The door latch strikers should also be lubricated with a thin coat of grease to reduce wear and ensure free movement. Lubricate the door and trunk locks with spray-on graphite lubricant.

8   Windshield and fixed glass - replacement

1   Replacement of the windshield and fixed glass requires the use of special fast-setting adhesive/caulk materials and some specialized tools and techniques. These operations should be left to a dealer service department or a shop specializing in glass work.

9   Hood - removal, installation and adjustment

Removal and installation
Note: *The hood is somewhat awkward to remove and install; at least two people should perform this procedure.*
1   Open the hood, then place blankets or pads over the fenders and cowl area of the body then disconnect any cables or wires that will interfere with removal.
Note: *Using blankets or pads will protect the body and paint as the hood is lifted off.*

9.10 Adjust the hood closing height by turning the hood bumpers in or out - Malibu models shown other models similar

2   Make marks or scribe a line around the hood hinge to ensure proper alignment during installation (see illustration).
3   On Aura models, support the hood then use a flat blade screwdriver to release the hood strut retaining tabs, pull the ends of the hood strut off the ball studs and remove the strut from the vehicle.
4   With an assistant supporting one side of the hood and you supporting the other, remove the hinge-to-hood bolts and lift off the hood (see illustration).
5   Installation is the reverse of removal. Align the hinge bolts with the marks made in Step 2.

Adjustment
6   Fore-and-aft and side-to-side adjustment of the hood is done by moving the hinge plate slot after loosening the bolts or nuts.
7   Scribe a line around the entire hinge plate so you can determine the amount of movement (see illustration 9.2).
8   Loosen the bolts or nuts and move the hood into correct alignment. Move it only a little at a time. Tighten the hinge bolts and carefully lower the hood to check the position.
9   If necessary after installation, the entire hood latch assembly can be adjusted up-and-down as well as from side-to-side on the radiator support so the hood closes securely and flush with the fenders.
10   Adjust the hood bumpers on the radiator support so the hood is flush with the fenders when closed (see illustration).
11   If the rear of the hood is too low, insert shims or washers of the correct thickness between the hood and the hinges.
12   The hood latch assembly, as well as the hinges, should be periodically lubricated with white lithium-base grease to prevent sticking and wear.

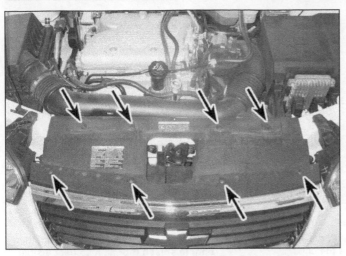

10.1a Remove the fasteners securing the radiator cover -
Malibu models shown, Aura models similar

10.1b Remove the fasteners securing the radiator cover -
G6 models shown

10 Hood release latch and cable - removal and installation

Latch

1   Remove the radiator cover (see illustrations).
2   Scribe a line around the latch to aid alignment when installing, then detach the latch retaining bolts from the radiator support (see illustration) and remove the latch.
3   Disengage the hood release cable from the latch assembly and disconnect any electrical connectors to the latch.
4   Installation is the reverse of the removal procedure.

Cable

5   Disconnect the hood release cable from the latch assembly as described above.
6   On Aura models, remove the battery and battery tray (see Chapter 5).
7   Attach a piece of stiff wire to the end of the cable, trace the cable back to the firewall and detach all cable retaining clips.
8   Working in the passenger compartment, remove the knee bolster (see Section 27).
9   Detach the fasteners securing the hood release lever (see illustration).
10 Pull the old cable into the passenger compartment until you can see the stiff wire that you attached to the cable. A grommet insulates the cable hole in the firewall from the elements. The new cable should have a new grommet, so you can remove and discard the old cable grommet. Make sure the new grommet is already on the new cable (if not, slip the old grommet onto the new cable), then detach the old cable from the wire and attach the new cable to the wire.
11 Working from the engine compartment side of the firewall, pull the wire through the cable hole in the firewall.
12 Installation is otherwise the reverse of the removal. Working from the passenger compartment side, push the grommet into place with your fingers. Make sure it's fully seated in the hole in the firewall.

11 Radiator grille - removal and installation

Malibu and Aura models
2007 and earlier models

1   Remove the headlight housings (see Chapter 12).
2   Remove the radiator cover (see illustration 10.1a or 10.1b).
3   Pull the ends of the grille forward and disengage it from the lower hooks.
4   Installation is the reverse of removal.

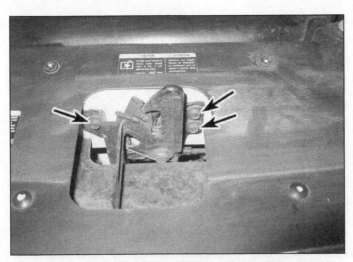

10.2 Scribe a line around the hinge to use as a reference point,
then remove the retaining bolts

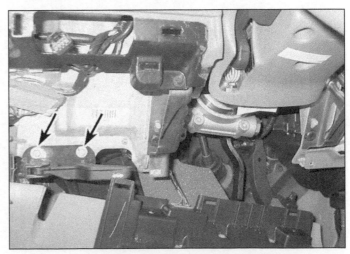

10.9 Remove the hood release lever mounting fasteners and pull
the cable rearward into the passenger compartment

11.8 Use a non-scratching tool to release the grille's clips one at a time

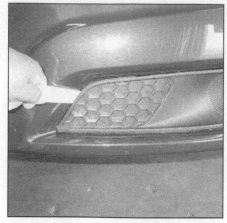

11.11 Carefully pry around the edge of the grille to release the clips

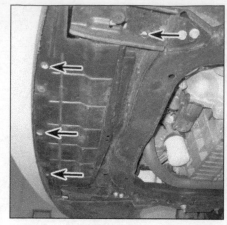

12.4 Remove the fasteners securing the engine splash (not all fasteners visible in photo)

2008 and later models

5   Remove the front bumper covers (see Section 12).

6   Release the attaching tabs to remove the grilles.

7   Installation is the reverse of removal.

G6 models

Upper grilles

8   The upper grilles are retained by clips. Use a non-scratching tool such as a plastic trim removal tool or a screwdriver wrapped with tape to gradually pry the grille out, working slowly around the edge (see illustration).

9   To install it, snap it into place so that it's securely seated.

Lower grille(s)

10   The dual lower grilles are retained with double-sided tape on some models and with clips on others. The single lower grilles are retained by clips. Remove the front bumper cover first (see Section 12).

11   Use a non-scratching tool such as a plastic trim removal tool or a screwdriver wrapped with tape to gradually pry the grille out, working slowly around the edge from the inside (see illustration).

12   To install the double-sided taped grilles, remove the paper from the double-sided tape on the new grille, then press them firmly into place. If you must reinstall the old grilles, purchase new tape first. Grilles with clips can be snapped back into place.

12   Bumper covers - removal and installation

Front

1   Remove the headlight housings on all G6 models and 2007 and earlier Malibu models (see Chapter 12).

2   Remove the radiator cover (see illustra-

tion 10.1a or 10.1b).

3   Raise the vehicle and support it securely on jackstands.

4   Remove the engine splash shield (see illustration).

5   Remove the inner fender splash shields (see illustration).

6   Remove the bumper cover retaining fasteners (see illustration).

7   With the help of an assistant, pull out the bumper to disconnect it from the clips at the fender sides. Pull up on the bumper to disconnect it from the top of the fenders (see illustration).

8   Installation is the reverse of removal.

Rear

Sedan models

9   Raise the vehicle and support it securely on jackstands. On all Aura, G6 and 2008 and later Malibu models, remove the rear inner fender splash shields.

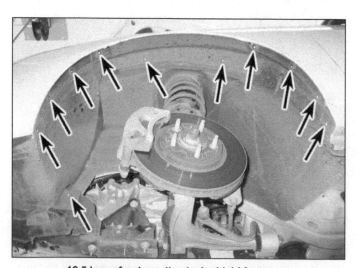

12.5 Inner fenderwell splash shield fasteners

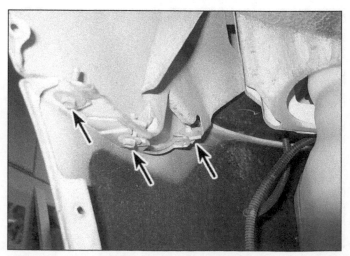

12.6 Peel back the splash shield and remove the bumper cover-to-fender retaining bolts

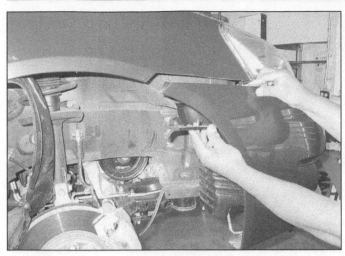

12.7 Pull outward on the ends of the bumper cover to release it from the body

12.11 Bumper cover lower fasteners

10   On Aura models, remove the taillight housings (see Chapter 12).

11   Working under the vehicle, detach the fasteners securing the bottom of the bumper cover (see illustration).

12   Remove the retainers that secure the rear inner wheel wells to the bumper cover. Pull the inner wheel back and remove the push-pin retainers or bolts from the upper edges of the bumper cover (see illustration).

13   Open the trunk or rear liftgate and remove the screws and clips securing the upper edge of the bumper cover (see illustration). Pull the bumper cover out and away from the vehicle. On 2008 and later models, the bumper is attached to the quarter panels with clips that must be released by firm pulling.

14   Remove the screws from the middle of the bumper cover (see illustration), if equipped.

15   Installation is the reverse of removal.

Coupe and convertible models

16   Remove the three fasteners from the rear of each inner fender well that attach the bumper cover.

17   Remove the single screw at each forward upper end of the bumper cover.

18   Open the trunk and pull back the carpet, then remove the two screws along each upper outer edge of the bumper cover.

19   Use a plastic trim tool (or equivalent) to release the five retainers along the upper center part of the bumper cover, then lift it off.

20   Installation is the reverse of removal.

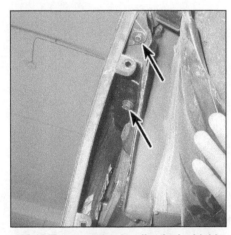

12.12 The rear wheel well splash shield can be folded back for access to the rear bumper cover end bolts

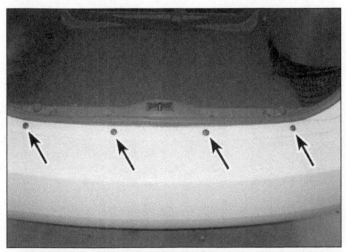

12.13 Remove the screws and clips securing the upper edge of the bumper cover - Malibu model shown, other models similar

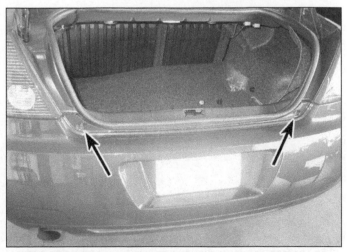

12.14 These screws secure the top of the bumper cover; other models have more fasteners along the edge of the trunk opening - G6 model shown

13.4 Remove the two screws from the front of the rocker panel molding, then carefully pry the clips loose for access to the bolts at the bottom rear corner of the fender

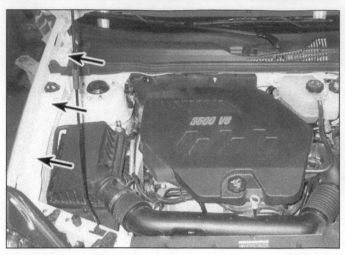

13.5a Remove the fender-to-body upper mounting bolts . . .

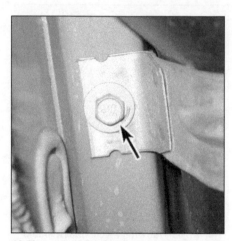

13.5b . . . and the fender to door pillar bolt inside the fenderwell

13 Front fender - removal and installation

Note: *Scribe or mark around the attaching bolts to insure proper alignment if reinstalling the original fender.*

1   Loosen the front wheel lug nuts, raise the vehicle and support it securely on jackstands. Remove the wheel.

2   Remove the inner fender splash shield from the wheel housings (see illustration 12.5).

3   Remove the front bumper cover (see Section 12).

4   Remove the fasteners securing the rocker panel, then remove the rocker panel (see illustration).

5   Remove the fender mounting bolts (see illustrations).

6   Detach the fender. It's a good idea to have an assistant support the fender while it's being moved away from the vehicle to prevent damage to the surrounding body panels.

7   Installation is the reverse of removal.

14 Door trim panels - removal and installation

2007 and earlier Malibu models

1   Remove the power window control switch (see illustration).

2   Remove the door handle trim ring (see illustration).

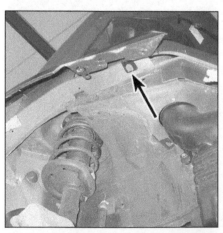

13.5c The front fender mounting screw is visible when the bumper cover has been removed

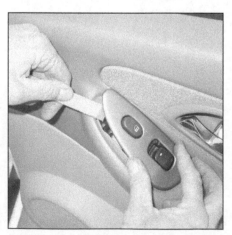

14.1 Using a trim stick, pry out the power window control switch, then unplug the electrical connector

14.2 Carefully pry off the door handle trim ring

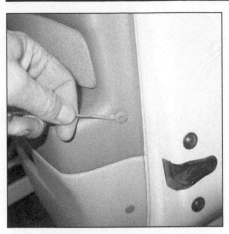

14.3 Remove the push-pin fasteners securing the door panel

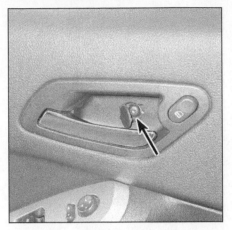

14.9a Use a small screwdriver to pry up the front of this screw cover - it's hinged at the rear

14.9b Carefully pry the latch handle out using a tool that can't scratch the plastic trim

3   Remove the door trim panel retaining fasteners (see illustration).

All Aura models and 2008 and later Malibu models
4   Lift the cover out of the bottom of the door pull handle well, then remove the two screws under it.
5   Pry the switch assembly up at the rear, then pull it to the rear to disengage the hook at the front. Remove the switch assembly and disconnect the wiring.
6   Remove the push-pin retainers from the rear edge of the door panel.
7   Pull the door latch handle, then remove the cap from the screw under it.
8   Remove the two screws under the switch assembly.

G6 models
9   Pry up the door latch handle screw cover at the front (it's hinged at the rear). Remove the screw under the cover, then lift off the latch handle assembly and disconnect it (see illustrations).

10   If you're working on a rear door, pry off the pull handle cover, then remove the screws under it (see illustration). Lift off the rear pull handle.
11   Remove the push-pin fasteners from the rear edge of the door. Press in the center pin of each retainer, then pry the retainer out (see illustration).

All models
12   Remove the door trim panel using a door panel removal tool (see illustration).
13   Once all of the clips are disengaged, carefully detach the trim panel from the door.
14   For access to the handle, latch, lock and window regulator mechanisms, carefully peel back the plastic watershield.
15   Before installing the door trim panel, inspect the condition of all clips and reinstall any clips which may have fallen out.
16   Installation is the reverse of the removal procedure.
Note: *When installing door trim panel retaining clips, make sure the clips are lined up with their mating holes first, then gently tap the clips in with the palm of your hand.*

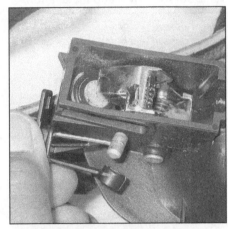

14.9c Disconnect the control cables from the latch handle

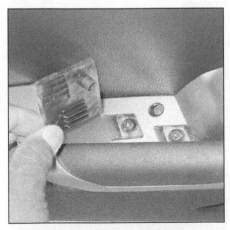

14.10 The front door pull handle screws are concealed by this rubber cover

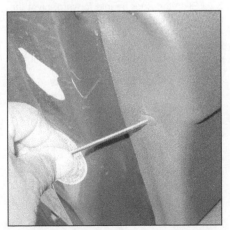

14.11 Push in the center of each push-pin retainer to allow it to be pried loose

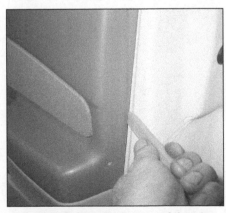

14.12 Start from the bottom of the trim panel and work around the perimeter until all the fasteners have been released from the door

15.6 Remove the door check strap mounting bolt (some models)

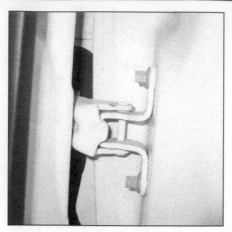

15.8 Remove the door hinge bolts

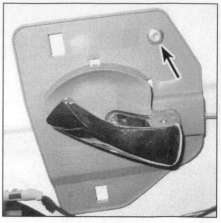

16.2 Remove this rivet securing the door handle to the door

15 Door - removal and installation

Note: *The door is heavy and somewhat awkward to remove and install - at least two people should perform this procedure.*

Removal and installation

1   Open the door all the way and support it on jacks or blocks covered with rags to prevent damaging the paint.
2   Remove the door trim panel and water deflector as described in Section 14.
3   Remove the door speaker (see Chapter 12).
4   Unplug all electrical connections, ground wires and harness retaining clips from the door.
Note: *It is a good idea to label all connections to aid the reassembly process.*
5   Working through the door speaker hole and the door opening, detach the rubber conduit between the body and the door. Then pull the wiring harness through the conduit hole

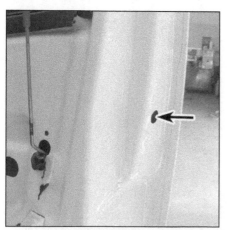

16.12 Working through the access hole, remove the door handle set bolt

and remove from the door.
6   Remove the door check strap mounting bolt (see illustration).
7   Mark around the door hinges with a pen or a scribe to facilitate realignment during reassembly.
8   With an assistant holding the door, remove the hinge-to-door bolts and lift off the door (see illustration).
9   Installation is the reverse of removal.

16 Door handles, key lock cylinder and latch - removal and installation

1   Raise the window, then remove the door trim panel and peel away the watershield (see Section 14).

Inside handle

2007 and earlier Malibu models

2   Remove the rivet securing the handle to the door (see illustration).
3   Slide the door handle forward, pivot the assembly away from the door and unlatch the actuating rod from the door handle lever. Remove the door handle.
4   Installation is the reverse of removal.

2008 and later Malibu models

5   Cut the two integral ribs that retain the bezel to the door trim panel.
6   Remove the light tube from the top of the bezel.
7   Disconnect the electrical connector and remove the handle
8   Snap the new door handle into place. The remaining installation is the reverse of removal.

G6 and Aura models

9   Refer to Section 14 and pry up the door latch handle screw cover at the front edge.

Remove the screw under it (see illustrations 14.9a, 14.9b and 14.9c).
10   Lift off the latch handle assembly and disconnect it.
11   Installation is the reverse of removal.

Outside handle

12   Remove the door handle set bolt (see illustration).
13   Detach the outside cap on the lock cylinder.
14   Pull the outside handle in a reward direction to release it from the door.
15   Remove the door handle gaskets, then remove the remaining door handle set bolt.
16   Working through the access hole, disengage the actuating rods from the outside door handle and from the key lock cylinder.
17   Remove the lock cylinder (see Step 21).
18   Remove the handle from the door.
19   Installation is the reverse of removal.

Key lock cylinder

All G6 models and 2007 earlier models

20   Remove the outside door handle.
21   Remove the lock cylinder retaining clip from the handle, then remove the lock cylinder from the handle
22   Installation is the reverse of removal.

2008 and later Malibu and Aura models

23   Remove the interior trim panel.
24   Remove the outside door handle.
25   Remove the door handle housing bolt, then disconnect the handle and lock rods to remove the housing.
26   Disconnect the electrical connector (if equipped) and remove the retaining clip from the cylinder.
27   Remove the lock cylinder from the housing.
28   Installation is the reverse of removal.

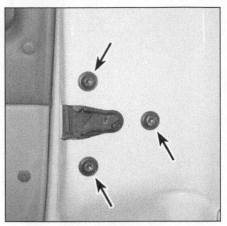

16.31 Remove the latch retaining screws from the end of the door

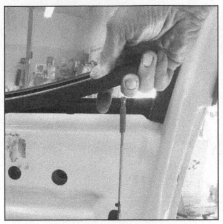

17.3 Remove the window inside sealing strip

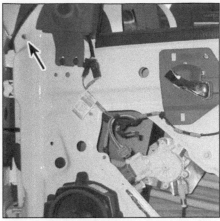

17.4 Remove this fastener to remove the outside sealing strip

17.5 Remove the glass-to-regulator bolts (one of two shown)

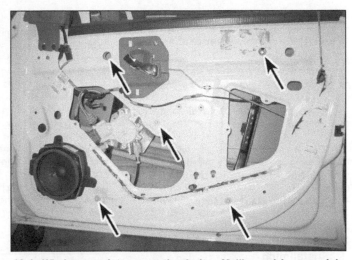

18.4a Window regulator mounting bolts - Malibu and Aura models

Door latch

29 Disengage the inside handle-to-latch actuating rods from the inside handle.

30 Remove the outside door handle (see Steps 12 through 18).

31 Remove the three screws securing the latch to the door (see illustration), then remove the latch assembly from the door.

32 Disconnect the electrical connectors from the latch mechanism.

33 Installation is the reverse of removal.

17 Door window glass - removal and installation

1 Lower the window glass all the way down into the door on 2007 and earlier models, half way down on 2008 and later models.

2 Remove the door trim panel and the plastic watershield (see Section 14).

3 Remove the window sealing strip (see illustration).

4 Remove the sealing strip from the outer door along the window glass (see illustration).

5 Remove the window glass retainer bolts (see illustration).

6 Remove the glass by carefully pulling it up and out.

7 Installation is the reverse of removal.

18 Door window glass regulator and motor - removal and installation

1 Remove the door trim panel and the plastic watershield (see Section 14).

2 Remove the door window glass (see Section 17).

Note: *The window does not have to be completely removed from the door. After removing the bolts that hold the window to the regulator, lift the window to the up position and tape it in place.*

3 Unplug the electrical connector from the window regulator motor.

4 Remove the regulator mounting bolts (see illustrations).

5 Pull the regulator assembly through the service hole in the door frame to remove it.

6 To remove the motor from the regulator assembly, simply remove the three fasteners securing it to the regulator assembly.

7 Installation is the reverse of removal.

18.4b Window regulator mounting bolts - G6 models

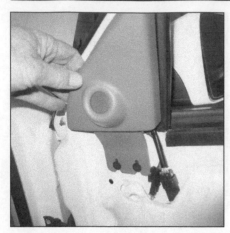

19.2 Pull up on the cover to release it from the door

19.3 Outside mirror mounting fasteners

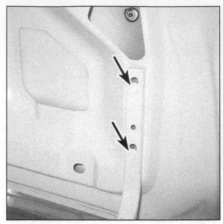

20.3 Mark around the hinge bolts for realignment purposes - then remove the retaining bolts on each side of the trunk lid

19 Outside mirrors - removal and installation

1   Remove the door trim panel (see Section 14).
2   Unplug the mirror electrical connector, then remove the speaker cover (see illustration).
3   Disconnect the electrical connector to the mirror then remove the mirror fasteners (see illustration) and detach the mirror from the vehicle.
4   Installation is the reverse of removal.

20 Trunk lid - removal and installation

Note: *The trunk lid is heavy and somewhat awkward to remove and install - at least two people should perform this procedure.*
1   Open the trunk lid and cover the edges

of the trunk compartment with pads or cloths to protect the painted surfaces when the lid is removed.
2   Disconnect any cables or wire harness connectors attached to the trunk lid that would interfere with removal.
3   Make alignment marks around the hinge mounting bolts with a marking pen (see illustration).
4   While an assistant supports the trunk lid, remove the lid-to-hinge bolts on both sides and lift it off.
5   Installation is the reverse of removal.
Note: *When reinstalling the trunk lid, align the lid-to-hinge bolts with the marks made during removal.*

21 Trunk lid latch and lock cylinder - removal and installation

1   Open the trunk and remove the trunk lid trim panel.

Latch
2   Disengage the rod connecting the lock cylinder to the latch.
3   Remove the trunk lid latch retaining bolts, then remove the latch (see illustration).
4   Installation is the reverse of removal.

Trunk lock cylinder
5   Detach the latch-to-lock cylinder rod from the lock cylinder, then, using a pair of pliers, remove the lock cylinder retaining clip (see illustration).
6   Disconnect the lock cylinder electrical connector.
7   Installation is the reverse of removal.

22 Liftgate - removal, installation and adjustment

Note: *The liftgate is heavy and somewhat awkward to remove and install - at least two people should perform this procedure.*

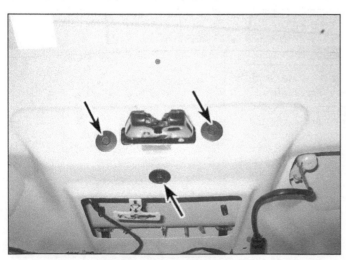

21.3 Trunk latch retaining bolts

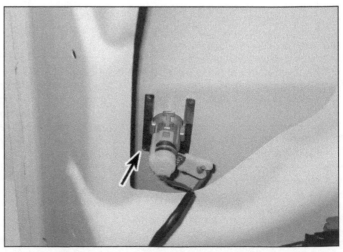

21.5 Remove the retaining clip from the lock cylinder

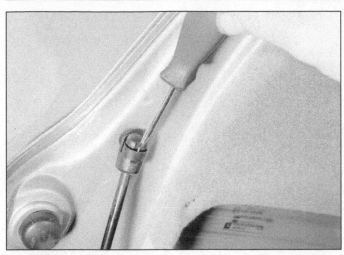

22.3 Pry the clip out and slide the support strut off the ball stud

24.2 Carefully peel off the front panel seal

Removal and installation

1   Using a trim stick, remove the liftgate inside trim panels.
2   Disconnect all electrical connections, ground wires and harness retaining clips from the liftgate.
Note: *It is a good idea to label all connections to aid the reassembly process. Detach the rubber conduit between the body and the liftgate. Then pull the wiring harness through the conduit hole and remove it from the liftgate.*
3   Have an assistant hold the liftgate in the open position. Detach the support struts from the liftgate (see illustration).
4   With an assistant holding the liftgate, remove the hinge mounting fasteners and lift the door off
Note: *Draw a reference line around the hinges before removing the bolts.*
5   Installation is the reverse of removal.

Adjustment

6   Having proper door-to-body alignment is a critical part of a well-functioning liftgate assembly. First, check the liftgate hinge pins for excessive play. Fully open the liftgate and lift up and down on the liftgate door without lifting the body. If a door has 1/16-inch or more excessive play, the hinges should be replaced.
7   Liftgate-to-body alignment adjustments are made by loosening the hinge-to-body bolts and moving the liftgate. Proper body alignment is achieved when the top of the liftgate is parallel with the roof section and the sides of the liftgate are flush with the rear quarter panels and the bottom of the liftgate is aligned with the lower door sill. If these goals can't be reached by adjusting the hinge-to-body bolts, body alignment shims may have to be purchased and inserted behind the hinges to achieve correct alignment.
8   To adjust the door-closed position, scribe a line or mark around the striker plate to provide a reference point, then check that the door latch is contacting the center of the latch striker. If not, adjust the latch striker sideways

position, so that the door panel is flush with the rear quarter panel and provides positive engagement with the latch mechanism.

23 Liftgate latch - removal and installation

Liftgate latch

1   Using a trim stick, remove the liftgate lower trim panel.
2   Remove the liftgate latch electrical connector.
3   Remove the screws securing the latch to the door.
4   Disconnect the latch cable, then remove the latch assembly through the door opening.
5   Installation is the reverse of removal.

24 Cowl cover - removal and installation

1   Remove the windshield wiper arms (see Chapter 12).

2   Carefully peel off the front panel seal (see illustration).
3   Remove the push pin fasteners securing the cowl cover (see illustration).
4   Remove the cowl cover from the vehicle.
5   Installation is the reverse of removal.

25 Center console - removal and installation

Warning: *Models covered by this manual are equipped with a Supplemental Restraint System (SRS), more commonly known as airbags. Always disable the airbag system before working in the vicinity of any airbag system component to avoid the possibility of accidental deployment of the airbag, which could cause personal injury (see Chapter 12).*
Warning: *Block the wheels to prevent the vehicle from rolling when the shifter is moved from the park position.*
1   Turn the ignition key to the On position, depress the brake pedal, then place the shifter in the drive position.
2   Disconnect the cable from the negative battery terminal (see Chapter 5, Section 1).

24.3 Remove the fasteners along the top of the cowl cover

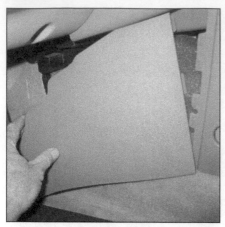

25.6 Carefully detach the center console side panels

25.7a Remove the fastener covers at the front . . .

25.7b . . . and at the rear of the center console, then the retaining screws

25.8 Pry out the console pen tray

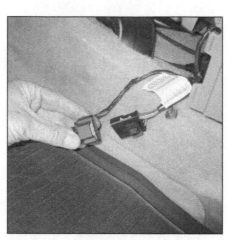

25.9 Disconnect any electrical connectors interfering with removal

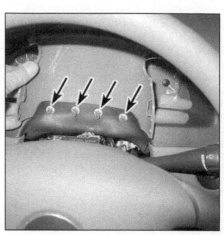

26.2 With the upper cover tilted up, release the lock rings (All G6 models and 2004 through 2007 Aura and Malibu models)

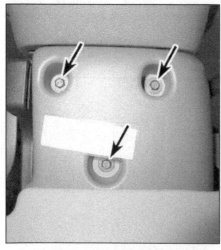

26.3 Remove the screws from the lower cover

2008 and later Malibu models

3   Pull down the gear shift boot, then loosen the shift knob screw. Lift off the knob.
4   Close the front console door, then use a plastic trim tool or a screwdriver wrapped with tape to carefully release the clips of the upper console trim plate.
5   Lift the trim plate up and disconnect the wiring.

All models

6   Remove the center console side panels (see illustration).
7   Remove the console retaining screws (see illustrations).
8   Remove the console pen tray (see illustration).
9   Carefully lift the console up and over the shifter, disconnect any electrical connectors, then remove it from the vehicle (see illustration).
10   Installation is the reverse of removal.
11   Move the shift lever back into the Park position, turn the ignition key to Off, then

reconnect the cable to the negative terminal of the battery.

26   Steering column covers - removal and installation

Warning: *Models covered by this manual are equipped with a Supplemental Restraint System (SRS), more commonly known as airbags. Always disable the airbag system before working in the vicinity of any airbag system component to avoid the possibility of accidental deployment of the airbag, which could cause personal injury (see Chapter 12).*

1   Disconnect the cable from the negative terminal of the battery, see Chapter 5, Section 1.
2   Remove the upper cover (see illustration).
3   Remove the steering column lower cover screws (see illustration), lower the steering column tilt lever, then detach the cover from the steering column.
4   Installation is the reverse of removal.

27.3 After removing the fasteners securing the instrument cluster bezel, use a trim stick to release any clips

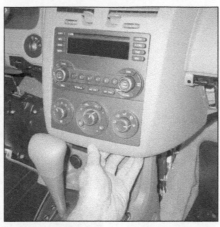

27.8 Pull on the bottom of the center trim panel to detach the clips, then remove the panel

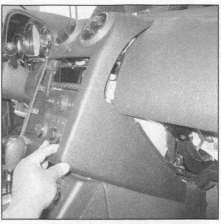

27.12 Pull on the sides of the center trim panel to detach the clips, then remove the panel

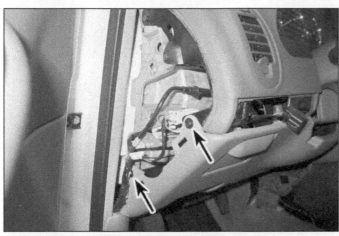

27.19a Remove the fasteners from the side . . .

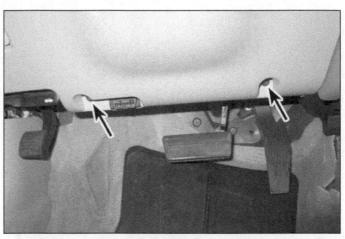

27.19b . . . and along the bottom of the knee bolster - Malibu models shown, other models similar

27 Dashboard trim panels - removal and installation

Warning: *The models covered by this manual are equipped with a Supplemental Restraint System (SRS), more commonly known as airbags. Always disarm the airbag system before working in the vicinity of any airbag system component to avoid the possibility of accidental deployment of the airbag, which could cause personal injury (see Chapter 12). Do not use a memory saving device to preserve the PCM's memory when working on or near airbag system components.*

Instrument cluster bezel

1   Disconnect the cable from the negative battery terminal (see Chapter 5, Section 1).
2   Detach the upper steering column cover (see Section 26). On 2008 and later Malibu and Aura models, remove the driver's knee bolster.
Note: *The steering column upper cover and instrument cluster bezel are attached to each*

other and can only be separated once both pieces have been removed, so you'll have to remove the upper steering column cover at the same time or try to release the retainers.
3   Remove the fasteners securing the instrument cluster bezel, then remove the bezel from the instrument panel (see illustration).
4   Installation is the reverse of the removal procedure.

Center trim panel

Malibu and Aura models
5   Remove the center console side panels (see illustration 25.6).
6   On 2007 and earlier models, remove the glove box (see Steps 23 through 25).
7   Remove the knee bolster (see Steps 16 through 22).
8   Detach the clips securing the center trim panel, then remove the panel from the instrument panel (see illustration).
9   Installation is the reverse of the removal procedure.

G6 models
10   Turn the ignition ON, depress the brake pedal and place the transmission in Neutral.
11   Pull rearward at the bottom of the bezel to release it from its clips.
12   Work upward, prying the panel's clips apart as you go (see illustration). Use a plastic trim tool or a screwdriver wrapped with tape to avoid scratching the plastic.
13   Disconnect the hazard flasher as you remove the bezel.
14   Set the trim panel face up in a safe place so it doesn't become scratched.
15   Installation is the reverse of removal.

Knee bolster
16   Disconnect the cable from the negative battery terminal (see Chapter 5, Section 1).
17   Remove the instrument panel side cover from the left side (see illustration 27.27).
18   Remove the adjustable pedal switch (if equipped), fog lamp switch and instrument panel dimmer switch (see Chapter 12).
19   Remove the fasteners securing the knee bolster (see illustrations).

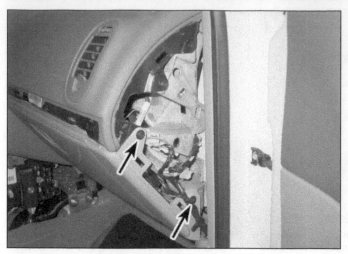

27.25a Remove the fasteners from the side . . .

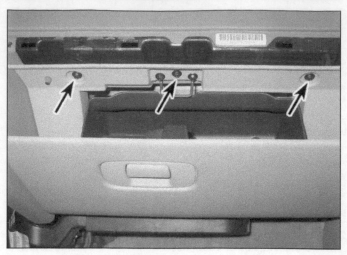

27.25b . . . and along the top of the glove box

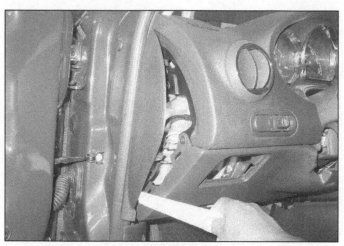

27.27 Using a trim stick, carefully pry around the panel to release the clips

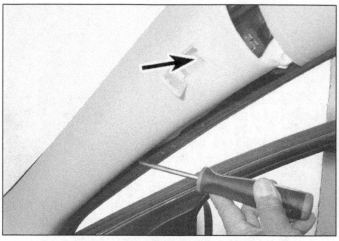

27.29 Open the cover, then remove the A-pillar screw (late models only)

20   Separate the left side sound insulator from the lower dash.

21   Pull the knee bolster toward the front seat to release the clips from the instrument panel. Remove the knee bolster.

22   Installation is the reverse of removal. Make sure the clips are engaged properly before pushing the knee bolster firmly into place.

Glove box

23   Remove the instrument panel side cover from the right side (see illustration 27.27).

24   Separate the right side sound insulator from the lower dash. On 2008 and later models, remove the center console (see Section 25).

25   Open the glove box door and remove the glove box mounting screws (see illustrations). Remove the glove box.

26   Installation is the reverse of removal.

Side covers

27   Using a trim stick, carefully pry around the panels to release the clips (see illustration).

28   Installation is the reverse of removal.

Upper trim panel

29   On models so equipped, pry open the cover and remove the screw, then, using a trim stick, carefully remove the A-pillar trim moldings (see illustration).

30   Using a trim stick, carefully pry around the panel to release the clips, then remove the panel.

31   Installation is the reverse of removal.

28   Instrument panel - removal and installation

Warning: *The models covered by this manual are equipped with a Supplemental Restraint System (SRS), more commonly known as airbags. Always disarm the airbag system before working in the vicinity of any airbag system component to avoid the possibility of accidental deployment of the airbag, which could cause personal injury (see Chapter 12). Do not use a memory saving device to preserve*

the PCM's memory when working on or near airbag system components.

Warning: *The air conditioning system is under high pressure. DO NOT loosen any fittings or remove any components until after the system has been discharged. Air conditioning refrigerant must be properly discharged into an EPA-approved container at a dealer service department or an automotive air conditioning repair facility. Always wear eye protection when disconnecting air conditioning system fittings.*

Note: *This is a difficult procedure for the home mechanic. There are many hidden fasteners, difficult angles to work in and many electrical connectors to tag and disconnect/connect. We recommend that this procedure be done at a dealership or qualified shop.*

Note: *During removal of the instrument panel, make careful notes of how each piece comes off, where it fits in relation to other pieces and what holds it in place. If you note how each part is installed before removing it, getting the instrument panel back together again will be much easier.*

28.12 Remove these nuts securing the brake pedal arm bracket to the firewall

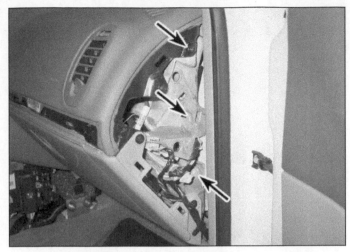

28.21 Remove the fasteners securing the sides of the instrument panel

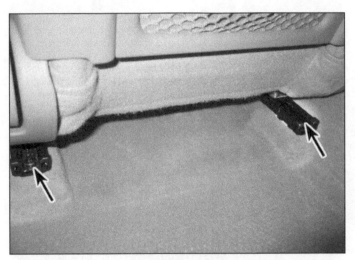

29.2 With the seat positioned all the way forward, remove the retaining bolts

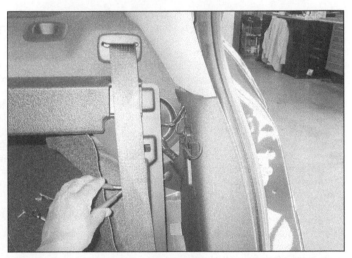

29.14 The rear seat side bolsters are secured with clips; remove the bolsters for access to the rear seat mounting fasteners

Note: *It is not necessary, but it is suggested to remove both front seats to allow additional working space and lessen the chance of damage to the seats during this procedure.*

1   Have the air conditioning system discharged by a dealer service department or an automotive air conditioning shop before proceeding (see Warning above).

2   Disconnect the cable from the negative battery terminal (see Chapter 5, Section 1).

3   Drain the cooling system (see Chapter 1).

4   Disconnect the heater hoses from the heater core fittings at the firewall (see Chapter 3).

5   Disconnect the refrigerant lines to the evaporator core at the firewall (see Chapter 3). Plug all open lines and fittings to prevent contamination of the air conditioning system.

6   Unbolt the heater module from the front

of the dashboard plate.

7   Remove the center console (see Section 25).

8   Remove all of the dashboard trim panels as described in Section 27.

9   Disconnect the instrument panel electrical connectors.

Note: *A number of electrical connectors must be disconnected in order to remove the instrument panel. Most are designed so that they will only fit on the matching connector (male or female), but if there is any doubt, mark the connectors with masking tape and a marking pen before disconnecting them.*

2007 and earlier Malibu models

10   Remove the Accelerator Pedal Position (APP) sensor (see Chapter 6).

11   Disconnect the electrical connector for the brake pedal position sensor, then remove

the retaining clip and washers securing the booster pushrod to the brake pedal arm.

12   Remove the nuts securing the brake pedal arm bracket to the firewall (see illustration), then remove the fasteners securing the bracket to the instrument panel. Remove the brake pedal.

All models

13   Working at the instrument panel center support, remove the fasteners securing the Body Control Module (BCM), then disconnect the electrical connector and remove the module.

14   Remove the fasteners securing the instrument panel center support.

15   Remove the shift lever (see Chapter 7B).

16   Using a trim stick, carefully remove the A-pillar trim moldings.

29.17a Rear seat frame rear fasteners

29.17b Removing the rear seat frame bolts

17   Remove the instrument panel upper trim panel (see Section 27).
18   Remove steering column covers (see Section 26). On G6 and 2007 and earlier models, lower the steering column (see Chapter 10). On 2008 and later Malibu and Aura models, remove the steering column (see Chapter 10).
19   On G6 models, remove the brake pedal assembly.
20   Remove the body control module from the dashboard center support.
21   Remove the fasteners securing the left and right side of the instrument panel (see illustration).
22   Remove the floor heat ducts.
23   Pull the instrument panel towards the rear of the vehicle and detach any electrical connectors interfering with removal.
24   Once all the electrical connectors are detached, lift the instrument panel then pull it away from the windshield and take it out through the driver's door opening.
Note: *This is a two-person job.*
25   Installation is the reverse of removal.

29   Seats - removal and installation

Front seat

Warning: *The models covered by this manual are equipped with a Supplemental Restraint System (SRS), more commonly known as airbags. Always disarm the airbag system before working in the vicinity of any airbag system component to avoid the possibility of accidental deployment of the airbag, which could cause personal injury (see Chapter 12). Do not use a memory saving device to preserve the PCM's memory when working on or near airbag system components.*
1   Disconnect the cable from the negative

terminal of the battery (see Chapter 5, Section 1).
2   Position the seat all the way forward to access the retaining bolts. Remove the trim cover from the side of the seat, then remove the seat belt anchor bolt. Detach any bolt trim covers and remove the retaining bolts (see illustration).
3   Tilt the seat upward to access the underneath, then disconnect any electrical connectors and lift the seat from the vehicle.
4   Installation is the reverse of removal.

Rear seat

Malibu and Aura models

Rear seat back
5   Detach the bolt trim covers and remove the seat cushion retaining bolts. Then fold the seat back down, and remove the seat belt anchor bolt.
6   Detach the retaining bolts at the lower edge of the seat back.
7   Lift up on the seat and remove it from the vehicle.
8   Installation is the reverse of removal.

Rear seat cushion
9   Pull the bottom seat cushion upward to release it from the retainers on the frame.
10   Unhook the rear of the cushion and remove the cushion.
11   Installation is the reverse of removal.

G6 models
12   Open the trunk and pull the latches to allow the rear seat to fold down.
13   Working in the passenger's compartment, remove the four screws from the top of the rear seat back frame.
14   Pull the seat back out for access to the retaining clips at the rear of the seat side bolsters. Remove the clips and the side bolsters (see illustration).

15   Remove the bolts from the outboard seat pivots, then remove the seat back sections from the frame.
16   Pull the bottom seat cushion upward to release it from the retainers on the frame.
17   The rear seat frame can be removed after unbolting it if necessary (see illustrations). Remove the plastic covers from the front lower bolts first.
18   Installation is the reverse of removal.

30   Parcel shelf trim panel - removal and installation

Warning: *The models covered by this manual are equipped with a Supplemental Restraint System (SRS), more commonly known as airbags. Always disarm the airbag system before working in the vicinity of any airbag system component to avoid the possibility of accidental deployment of the airbag, which could cause personal injury (see Chapter 12). Do not use a memory saving device to preserve the PCM's memory when working on or near airbag system components.*
1   Using a trim stick, carefully remove the C-pillar trim molding starting from the top and disconnect the tether.
Warning: *Obtain a new tether clip for installation.*
2   Remove the rear seat backs (see Section 29).
3   Locate the trim panel plastic fasteners at the front corners of the panel, then pry the fasteners out using a trim tool.
4   Remove the rear seat belt anchor bolts, if neseccary.
5   Lift the panel up and pull it outwards disconnecting any electrical connectors and remove the panel.
6   Installation is the reverse of removal.

Chapter 12
Chassis electrical system

Contents

1   General Information

1   The electrical system is a 12-volt, negative ground type. Power for the lights and all electrical accessories is supplied by a lead/acid battery, which is charged by the alternator.

2   This Chapter covers repair and service procedures for the various electrical components not associated with the engine. Information on the battery, alternator and starter motor can be found in Chapter 5.

3   It should be noted that when portions of the electrical system are serviced, the negative battery cable should be disconnected from the battery to prevent electrical shorts and/or fires.

2   Electrical troubleshooting - general information

1   A typical electrical circuit consists of an electrical component, any switches, relays, motors, fuses, fusible links or circuit breakers related to that component and the wiring and connectors that link the component to both the battery and the chassis. To help you pinpoint an electrical circuit problem, wiring diagrams are included at the end of this Chapter.

2   Before tackling any troublesome electrical circuit, first study the appropriate wiring diagrams to get a complete understanding of what makes up that individual circuit. Trouble spots, for instance, can often be narrowed down by noting if other components related to the circuit are operating properly. If several components or circuits fail at one time, chances are the problem is in a fuse or ground connection, because several circuits are often routed through the same fuse and ground connections.

3   Electrical problems usually stem from simple causes, such as loose or corroded connections, a blown fuse, a melted fusible link or a failed relay. Visually inspect the condition of all fuses, wires and connections in a problem circuit before troubleshooting the circuit.

4   If test equipment and instruments are going to be utilized, use the diagrams to plan ahead of time where you will make the necessary connections in order to accurately pinpoint the trouble spot.

5   The basic tools needed for electrical troubleshooting include a circuit tester or voltmeter (a 12-volt bulb with a set of test leads can also be used), a continuity tester, which includes a bulb, battery and set of test leads, and a jumper wire, preferably with a circuit breaker incorporated, which can be used to bypass electrical components (see illustrations). Before attempting to locate a problem with test instruments, use the wiring diagram(s) to decide where to make the connections.

Voltage checks

6   Voltage checks should be performed if a circuit is not functioning properly. Connect one lead of a circuit tester to either the negative battery terminal or a known good ground. Connect the other lead to a connector in the circuit being tested, preferably nearest to the battery or fuse (see illustration). If the bulb of the tester lights, voltage is present, which means that the part of the circuit between the connector and the battery is problem free. Continue checking the rest of the circuit in the same fashion. When you reach a point at which no voltage is present, the problem lies between that point and the last test point with voltage. Most of the time the problem can be traced to a loose connection. Note: Keep in mind that some circuits receive voltage only when the ignition key is in the Accessory or Run position.

Finding a short

7   One method of finding shorts in a circuit is to remove the fuse and connect a test light or voltmeter in place of the fuse terminals. There should be no voltage present in the circuit. Move the wiring harness from side-to-side while watching the test light. If the bulb goes on, there is a short to ground somewhere in that area, probably where the insulation has rubbed through. The same test can be performed on each component in the circuit, even a switch.

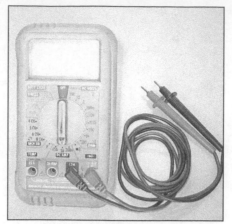

2.5a The most useful tool for electrical troubleshooting is a digital multimeter that can check volts, amps, and test continuity

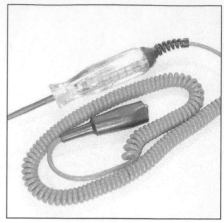

2.5b A simple test light is a very handy tool for testing voltage

Ground check

8   Perform a ground test to check whether a component is properly grounded. Disconnect the battery and connect one lead of a continuity tester or multimeter (set to the ohms scale), to a known good ground. Connect the other lead to the wire or ground connection being tested. If the resistance is low (less than 5 ohms), the ground is good. If the bulb on a self-powered test light does not go on, the ground is not good.

Continuity check

9   A continuity check is done to determine if there are any breaks in a circuit - if it is passing electricity properly. With the circuit off (no power in the circuit), a self-powered continuity tester or multimeter can be used to check the circuit. Connect the test leads to both ends of the circuit (or to the power end and a good ground), and if the test light comes on the circuit is passing current properly (see illustration). If the resistance is low

(less than 5 ohms), there is continuity; if the reading is 10,000 ohms or higher, there is a break somewhere in the circuit. The same procedure can be used to test a switch, by connecting the continuity tester to the switch terminals. With the switch turned On, the test light should come on (or low resistance should be indicated on a meter).

Finding an open circuit

10   When diagnosing for possible open circuits, it is often difficult to locate them by sight because the connectors hide oxidation or terminal misalignment. Merely wiggling a connector on a sensor or in the wiring harness may correct the open circuit condition. Remember this when an open circuit is indicated when troubleshooting a circuit. Intermittent problems may also be caused by oxidized or loose connections.

11   Electrical troubleshooting is simple if you keep in mind that all electrical circuits are basically electricity running from the battery, through the wires, switches, relays, fuses

2.6 In use, a basic test light's lead is clipped to a known good ground, then the pointed probe can test connectors, wires or electrical sockets - if the bulb lights, the circuit being tested has battery voltage

2.9 With a multimeter set to the ohm scale, resistance can be checked across two terminals - when checking for continuity, a low reading indicates continuity, a high reading or infinity indicates high resistance or lack of continuity

and fusible links to each electrical component (light bulb, motor, etc.) and to ground, from which it is passed back to the battery. Any electrical problem is an interruption in the flow of electricity to and from the battery.

Connectors

12   Most electrical connections on these vehicles are made with multi-wire plastic connectors. The mating halves of many connectors are secured with locking clips molded into the plastic connector shells. The mating halves of large connectors, such as some of those under the instrument panel, are held together by a bolt through the center of the connector.

13   To separate a connector with locking clips, use a small screwdriver to pry the clips apart carefully, then separate the connector halves. Pull only on the shell, never pull on the wiring harness as you may damage the individual wires and terminals inside the connectors. Look at the connector closely before trying to separate the halves. Often the locking clips are engaged in a way that is not immediately clear. Additionally, many connectors have more than one set of clips.

14   Each pair of connector terminals has a male half and a female half. When you look at the end view of a connector in a diagram, be sure to understand whether the view shows the harness side or the component side of the connector. Connector halves are mirror images of each other, and a terminal that is shown on the right side end-view of one half will be on the left side end view of the other half.

Electrical connectors

Most electrical connectors have a single release tab that you depress to release the connector

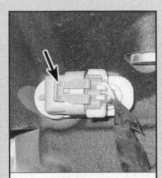

Some electrical connectors have a retaining tab which must be pried up to free the connector

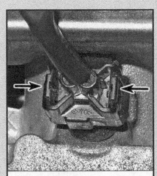

Some connectors have two release tabs that you must squeeze to release the connector

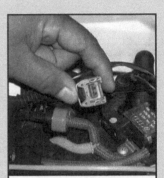

Some connectors use wire retainers that you squeeze to release the connector

Critical connectors often employ a sliding lock (1) that you must pull out before you can depress the release tab (2)

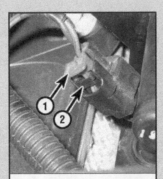

Here's another sliding-lock style connector, with the lock (1) and the release tab (2) on the side of the connector

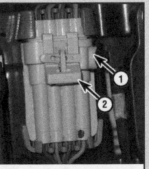

On some connectors the lock (1) must be pulled out to the side and removed before you can lift the release tab (2)

Some critical connectors, like the multi-pin connectors at the Powertrain Control Module employ pivoting locks that must be flipped open

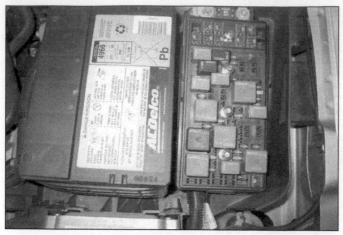

3.1a The engine compartment fuse/relay box is mounted on the left side of the engine compartment, next to the battery. All fuses and relays are listed by location and function on the underside of the fuse/relay box cover

3.1b The fuse/relay box inside the vehicle is located on the right side, and at the forward end of the console, behind a small access door (in this shot the right console trim panel has been removed for clarity)

3   Fuses and fusible links - general information

Fuses

1   The electrical circuits of the vehicle are protected by a combination of fuses, circuit breakers and fusible links. Fuse and relay boxes are located in the engine compartment, under the instrument panel and in the trunk (see illustrations). Each of the fuses is designed to protect a specific circuit, and the various circuits are identified on the fuse panel cover. If the fuse panel cover is difficult to read, or missing, you can also refer to your owner's manual, which includes a complete guide to all fuses and relays in both fuse/relay boxes.

2   Miniaturized fuses are employed in the fuse blocks. If an electrical component fails, always check the fuse first. The best way to check a fuse is with a test light. Check for power at the exposed terminal tips of each

fuse. If power is present on one side of the fuses but not the other, the fuse is blown. A blown fuse can also be confirmed by visually inspecting it (see illustration).

3   Be sure to replace blown fuses with the correct type. Fuses of different ratings are physically interchangeable, but only fuses of the proper rating should be used. Replacing a fuse with one of a higher or lower value than specified is not recommended. Each electrical circuit needs a specific amount of protection. The amperage value of each fuse is molded into the fuse body.

4   If the replacement fuse immediately fails, don't replace it again until the cause of the problem is isolated and corrected. In most cases, this will be a short circuit in the wiring caused by a broken or deteriorated wire.

Fusible links

5   Some circuits are protected by fusible links. The links are used in circuits which are not ordinarily fused, or which carry high cur-

rent, such as the circuit between the alternator and the starter motor. Fusible links, which are usually four wire gauges smaller in size than the circuit that they protect, are designed to melt if the circuit is subjected to more current than it was designed to carry. If you have to replace a blown fusible link, make sure that you replace it with one of the same specification. If the replacement fusible link blows in the same circuit, make sure that you troubleshoot the circuit in which the fusible link melted BEFORE installing another fusible link.

4   Circuit breakers - general information

1   Circuit breakers protect certain circuits, such as the power windows or heated seats. Depending on the vehicle's accessories, there may be one or two circuit breakers, located in the fuse/relay box in the engine compartment.

3.1c To access the fuse/relay box that's located in the left side of the trunk, open this access door in the left trunk trim . . .

3.1d . . . then open the fuse/relay box cover

3.2 When a fuse blows, the element between the terminals melts

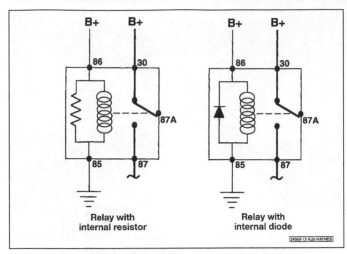

5.2a Typical ISO relay designs, terminal numbering and circuit connections

5.2b Most relays are marked on the outside to easily identify the control circuits and the power circuits - four terminal type shown

2   Because the circuit breakers reset automatically, an electrical overload in a circuit-breaker-protected system will cause the circuit to fail momentarily, then come back on. If the circuit does not come back on, check it immediately.

3   For a basic check, pull the circuit breaker up out of its socket on the fuse panel, but just far enough to probe with a voltmeter. The breaker should still contact the sockets.

4   With the voltmeter negative lead on a good chassis ground, touch each end prong of the circuit breaker with the positive meter probe. There should be battery voltage at each end. If there is battery voltage only at one end, the circuit breaker must be replaced.

5   Some circuit breakers must be reset manually.

5   Relays - general information and testing

General information

1   Several electrical accessories in the vehicle, such as the fuel injection system, horns, starter, and fog lamps use relays to transmit the electrical signal to the component. Relays use a low-current circuit (the control circuit) to open and close a high-current circuit (the power circuit). If the relay is defective, that component will not operate properly. Most relays are mounted in the engine compartment fuse/relay box, with some specialized relays located above the interior fuse box in the dash (see illustrations 3.1a, 3.1b and 3.1c). If a faulty relay is suspected, it can be removed and tested using the procedure below or by a dealer service department or a repair shop. Defective relays must be replaced as a unit.

Testing

2   Most of the relays used in these vehicles are of a type often called "ISO" relays, which

refers to the International Standards Organization. The terminals of ISO relays are numbered to indicate their usual circuit connections and functions. There are two basic layouts of terminals on the relays used in these vehicles (see illustrations).

3   Refer to the wiring diagram for the circuit to determine the proper connections for the relay you're testing. If you can't determine the correct connection from the wiring diagrams, however, you may be able to determine the test connections from the information that follows.

4   Two of the terminals are the relay control circuit and connect to the relay coil. The other relay terminals are the power circuit. When the relay is energized, the coil creates a magnetic field that closes the larger contacts of the power circuit to provide power to the circuit loads.

5   Terminals 85 and 86 are normally the control circuit. If the relay contains a diode, terminal 86 must be connected to battery positive (B+) voltage and terminal 85 to ground. If the relay contains a resistor, terminals 85 and 86 can be connected in either direction with respect to B+ and ground.

6   Terminal 30 is normally connected to the battery voltage (B+) source for the circuit loads. Terminal 87 is connected to the circuit leading to the component being powered. If the relay has several alternate terminals for load or ground connections, they usually are numbered 87A, 87B, 87C, and so on.

7   Use an ohmmeter to check continuity through the relay control coil.

a) Connect the meter according to the polarity shown in illustration 5.2a for one check; then reverse the ohmmeter leads and check continuity in the other direction.

b) If the relay contains a resistor, resistance will be indicated on the meter, and should be the same value with the ohmmeter in either direction.

c) If the relay contains a diode, resistance should be higher with the ohmmeter in the forward polarity direction than with the meter leads reversed.

d) If the ohmmeter shows infinite resistance in both directions, replace the relay.

8   Remove the relay from the vehicle and use the ohmmeter to check for continuity between the relay power circuit terminals. There should be no continuity between terminal 30 and 87 with the relay de-energized.

9   Connect a fused jumper wire to terminal 86 and the positive battery terminal. Connect another jumper wire between terminal 85 and ground. When the connections are made, the relay should click.

10   With the jumper wires connected, check for continuity between the power circuit terminals. Now, there should be continuity between terminals 30 and 87.

11   If the relay fails any of the above tests, replace it.

6   Steering column switches - replacement

Warning: The models covered by this manual are equipped with a Supplemental Restraint System (SRS), more commonly known as airbags. Always disarm the airbag system before working in the vicinity of any airbag system component to avoid the possibility of accidental deployment of the airbag, which could cause personal injury (see Section 24).

Turn signal/multi-function switch

1   Remove the steering column covers (see Chapter 11).

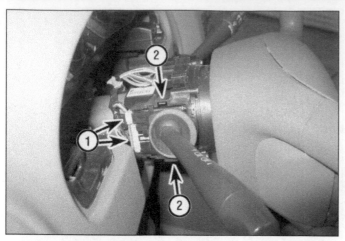

6.2 To detach the turn signal/multi-function switch from the steering column, disconnect the electrical connectors (1), then depress these two locking tabs (2) and pull the switch out of the switch mount housing

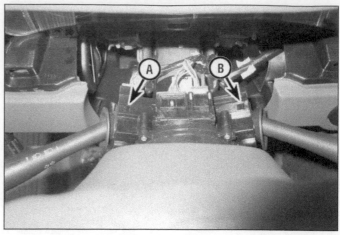

6.3a Upper locking tabs for the turn signal/multi-function switch (A) and the windshield wiper/washer switch (B)

6.3b Lower locking tabs for the turn signal/multi-function switch (A) and the windshield wiper/washer switch (B)

2   Disconnect the electrical connector from the turn signal/multi-function switch (see illustration).

3   Depress the upper and lower locking tabs (see illustrations) and remove the turn signal/multi-function switch from the steering column.

4   Installation is the reverse of removal.

Windshield wiper/washer switch

5   Remove the steering column covers (see Chapter 11).

6   Disconnect the electrical connector from the windshield wiper/washer switch (see illustration).

7   Depress the upper and lower locking tabs (see illustrations 6.3a and 6.3b) and remove the windshield wiper/washer switch from the steering column.

8   Installation is the reverse of removal.

Steering column switch assembly

Note: *If you're simply removing the steering column switches in order to remove the steering column, rather than actually replacing one of the switch units, you can remove the entire switch mount housing and both steering column switches as a single assembly.*

9   Remove the steering wheel and the clockspring (see Chapter 10).

10   Remove the steering column covers (see Chapter 11).

11   Disconnect the electrical connectors from the steering column switches and from the switch mount housing (see illustration).

12   Slide the steering column switches off the steering column.

13   Installation is the reverse of removal.

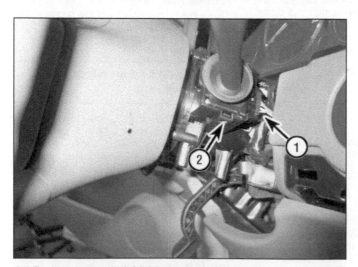

6.6 To detach the windshield wiper/washer switch from the switch mount housing, disconnect the electrical connector (1), then depress the upper and lower locking tabs (2) and pull the switch out of the housing (upper locking tab not visible in this photo)

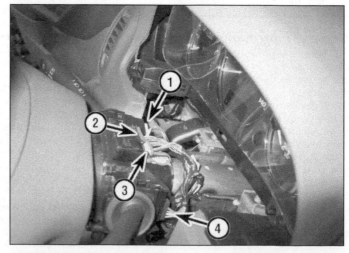

6.11 Before removing the switch mount housing from the steering column, disconnect the electrical connectors from the turn signal/multi-function switch (1), from the switch mount housing (2 and 3) and from the windshield wiper/washer switch (4)

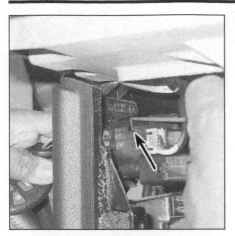

7.3 On 2007 and earlier models, depress this button to pull out the lock cylinder

7.8 Ignition switch mounting nut (1), theft deterrent module electrical connector (2), ignition switch electrical connector (3) and BTSI cable (4)

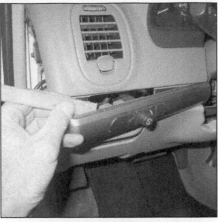

8.1a On Malibu and Aura models, to remove the trim panel that houses the adjustable pedal button/instrument panel dimmer switch/fog light switch assembly, carefully pry it loose from the instrument panel with a trim panel removal tool or with a small, flat piece of wood or plastic . . .

7 Key lock cylinder and ignition switch - replacement

Warning: *The models covered by this manual are equipped with a Supplemental Restraint System (SRS), more commonly known as airbags. Always disarm the airbag system before working in the vicinity of any airbag system component to avoid the possibility of accidental deployment of the airbag, which could cause personal injury (see Section 24).*

Key lock cylinder

Caution: *The following procedure is included only as a prelude to removing the ignition switch (see below). Do NOT try toreplacethe key lock cylinder at home. If you replace the key lock cylinder you must have the transponder in the new key programmed by a dealer service department before the engine will start. Even if you were to use the old key in a new key lock cylinder you would still have to have it re-programmed at the dealer before the engine would start.*

1   Remove the knee bolster (see Chapter 11).
2   Insert the ignition key in the key lock cylinder and turn the key to the ON position.
Note: *On these vehicle, the ON or RUN position, which is located between the accessory (ACC) position and the start position (a clockwise rotating circle), is not indicated by the word "ON" but by a vertical line.*
3   On 2007 and earlier models, insert a 1/8-inch punch through the hole in the casting that holds the key lock cylinder, depress the release tab and pull out the lock cylinder (see illustration). On 2008 and later models, use a flat-blade screwdriver to push in the release tab.
4   Installation is the reverse of removal.

Ignition switch

5   Remove the instrument cluster (see Section 9).

6   Remove the knee bolster and the instrument panel center trim panel (see Chapter 11).
7   Disconnect the electrical connectors from the ignition switch.
8   Disconnect the electrical connector from the theft deterrent control module (see illustration).
9   Remove the ignition switch key lock cylinder (see Steps 2 and 3).
10   Disconnect the Brake Transmission Shift Interlock (BTSI) cable from the ignition switch (see Chapter 7B).
11   Remove the ignition switch mounting screws and remove the ignition switch assembly.
12   Remove the ignition switch mounting bracket screws.
13   Remove the theft deterrent control module.
14   Remove the ignition switch mounting bracket.
15   Installation is the reverse of removal.

8 Dashboard switches - replacement

Warning: *The models covered by this manual are equipped with a Supplemental Restraint System (SRS), more commonly known as airbags. Always disarm the airbag system before working in the vicinity of any airbag system component to avoid the possibility of accidental deployment of the airbag, which could cause personal injury (see Section 24).*

Adjustable pedal button/ instrument panel dimmer switch/fog light switch

Note: *The adjustable pedal button/instrument panel dimmer switch/fog light switch assembly is located on a small trim panel to the left of the steering column. Not all vehicles are equipped with all three of these switches, but this procedure applies to all of them.*

1   On 2007 and earlier models, use a trim panel removal tool or a thin piece of wood or plastic to carefully pry the panel that houses the adjustable pedal button/instrument panel dimmer switch/fog light switch assembly out of the instrument panel and disconnect the electrical connector(s) (see illustrations). On all Aura models and all 2008 and later models, remove the driver's knee bolster trim panel (see Chapter 12)
2   To remove the adjustable pedal button, instrument panel dimmer switch or fog light switch from the trim panel, squeeze the retaining tabs and push it out of the panel.
3   To install a switch, push it into the trim panel until it snaps into place.

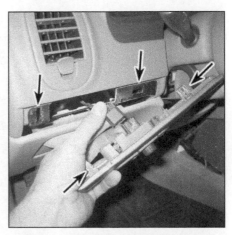

8.1b . . . and disconnect the electrical connector. When installing the panel, make sure that the mounting clips are aligned with their corresponding slots in the instrument panel, then push the trim panel firmly into place until the clips pop into place

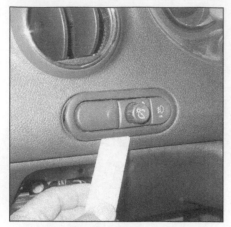

8.1c On G6 models, to remove the trim panel that houses the adjustable pedal button/instrument panel dimmer switch/ fog light switch assembly, carefully pry it loose from the instrument panel with a trim panel removal tool, then disconnect the electrical connector to the switch

8.6 To remove the hazard flasher switch, unclip it from the center trim panel, then disconnect the electrical connector - Malibu and Aura models

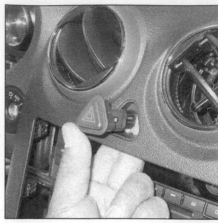

8.7 To remove the hazard flasher switch, pop it out of the center trim panel, then disconnect the electrical connector - G6 models

4   To install the trim panel, reconnect the electrical connector(s), then align the mounting clips on the panel with their corresponding slots in the instrument panel (see illustration 8.1b) and push the panel firmly onto the instrument panel until the clips snap into place.

Hazard flasher switch

5   Remove the center trim panel (see Chapter 11).
6   On Malibu and Aura models, remove the hazard flasher switch from the back of the center trim panel, then disconnect the electrical connector from the switch (see illustration).
7   On G6 models, push the hazard flasher

switch from the back of the center trim panel to the front, then disconnect the electrical connector from the switch (see illustration).
8   Installation is the reverse of removal.

9   Instrument cluster - removal and installation

Warning: *The models covered by this manual are equipped with a Supplemental Restraint System (SRS), more commonly known as airbags. Always disarm the airbag system before working in the vicinity of any airbag system component to avoid the possibility of accidental deployment of the airbag, which could cause personal injury (see Section 24).*

1   Remove the instrument panel cluster trim plate bezel (see Chapter 11).
2   Remove the instrument cluster retaining screws (see illustration), then pull out the cluster, disconnect the electrical connectors and remove the cluster.
3   Installation is otherwise the reverse of removal.

10   Windshield wiper motor - replacement

1   Pry off the windshield wiper trim caps (see illustration).
2   Remove the windshield wiper retaining nuts and washers. Mark the position of each

9.2 To remove the instrument cluster, remove these two screws - Malibu models shown, all other models similar

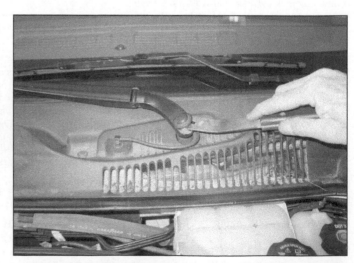

10.1 To remove the windshield wiper arms, carefully pry off each trim cap with a trim removal tool or a small screwdriver

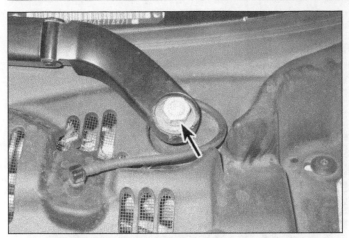

10.2 Remove each wiper arm retaining nut, then mark the relationship of each arm to its shaft BEFORE removing the arm

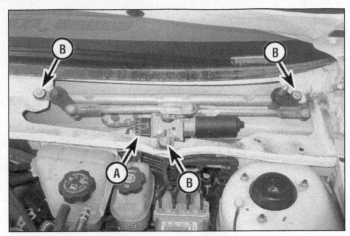

10.4 Wiper motor details

A   *Electrical connector*   B   *Mounting bolts*

wiper arm in relation to its shaft (see illustration), then remove the wiper arms.

3   Remove the hood seal and the cowl cover (see Chapter 11).

4   Disconnect the electrical connector from the wiper motor, then remove the three mounting bolts that secure the windshield wiper motor and linkage assembly to the cowl (see illustration).

5   Pry the linkage off the wiper motor crank arm.

6   Remove the nut that attaches the actuator arm to the motor shaft.

7   Mark the relationship of the actuator arm to the motor shaft, then remove the actuator arm from the shaft.

8   Remove the motor mounting screws and remove the motor.

9   Installation is the reverse of removal. Be sure to align the marks you made on the actuator arm and the motor shaft, and on the windshield wiper arms and the wiper arm shafts.

11   Radio and speakers - removal and installation

Warning: *The models covered by this manual are equipped with a Supplemental Restraint System (SRS), more commonly known as airbags. Always disarm the airbag system before working in the vicinity of any airbag system component to avoid the possibility of accidental deployment of the airbag, which could cause personal injury (see Section 24).*

Radio

1   Remove the center trim panel (see Chapter 11). On Aura and 2008 and later models, remove the HVAC module.

2   Remove the radio mounting screws (see illustration) and pull the radio out from the dash.

3   Disconnect the antenna cable, the electrical connectors and the ground strap from

the back of the radio and remove the radio from the dash.

4   When installing the radio, make sure that the locator pin on the backside of the radio is aligned with its corresponding hole in the radio support bracket.

5   Installation is otherwise the reverse of removal.

Speakers

Front door tweeters

All G6 and 2007 and earlier Malibu models

Note: *The tweeters are located in the upper extensions (the small triangular trim pieces, above the door trim panels, that cover up the spaces in front of the door windows).*

6   Remove the front door trim panel (see Chapter 11).

7   Remove the upper extension (see illustration).

11.2 To detach the radio from the instrument panel, remove these four screws, then pull it out and disconnect the electrical connector and the antenna cable from the backside. When you install the radio, make sure that the locator pin on the back of the radio is aligned with its corresponding hole in the radio support bracket - Malibu models shown, other models similar

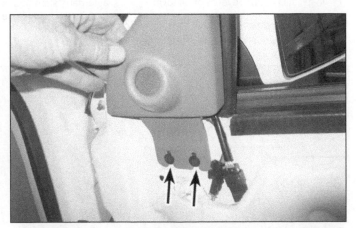

11.7 To detach the upper extension from the door, slide it up and disengage the slots in the two mounting tabs from these two locator pins

11.15 To remove a door speaker, remove these three mounting screws, pull out the speaker and disconnect the electrical connector from the speaker

13.1a On Malibu models, to detach the headlight housing from the vehicle, remove these two bolts

8   To remove the tweeter from the upper extension, rotate it counterclockwise and pull it out.
9   Disconnect the electrical connector from the tweeter.
10   Installation is the reverse of removal.

All Aura and 2008 and later Malibu models

11   Remove the windshield pillar moldings by prying them off with a plastic tool.
12   Disconnect the speaker wiring, then remove the speakers from the trim panels.
13   Installation is the reverse of removal.

Front door speakers

14   Remove the front door trim panel (see Chapter 11).
15   Remove the speaker mounting screws (see illustration).
16   Pull out the speaker and disconnect the electrical connector.
17   Installation is the reverse of removal.

Rear door speakers

18   The rear door speakers, if equipped,

are removed the same way as the front door speakers (see Steps 14 through 16).

Rear speakers (sedans)

19   Remove the rear shelf trim panel.
20   Depress the release tab and pull out the speaker. On 2008 and later models, remove the two speaker bolts to remove the speakers.
21   Disconnect the electrical connector from the speaker.
22   Installation is the reverse of removal.

---

12   Antenna - removal and installation

Fixed mast

MAXX models

1   Using an antenna wrench (available at auto parts stores), unscrew the antenna mast from its mounting base and remove it.
2   Installation is the reverse of removal.

G6 models

3   Open the trunk and remove the right trim panel.
4   Detach the antenna bracket from the panel.
5   Turn the antenna counterclockwise to remove it.
6   Installation is the reverse of removal.

Rear window antenna - grid type

7   The antenna is an integral part of the rear window. To replace the antenna you must replace the rear window.

Roof-mounted antenna (2008 and later AM/FM and all years with XM Satellite Radio, cellular and navigation systems)

8   On these models the antenna is mounted on the roof. To replace it you must remove the headliner and, on vehicles equipped with side-curtain airbags, the side-curtain airbag modules as well. We therefore recommend that you have this type of antenna replaced by a dealer service department or other qualified repair shop.

---

13   Headlight housing - replacement

All G6 and 2007 and earlier Malibu models

1   Remove the headlight housing bolts (see illustrations).
2   Pull out the headlight housing and disconnect the electrical connector (see illustration).
3   Installation is the reverse of removal.
4   Adjust the headlights when you're done (see Section 14).

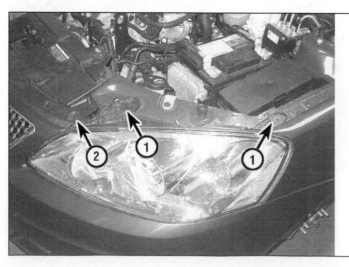

13.1b Headlight housing screws (1) and bumper cover retainer (2) - G6 models

13.2 Pull out the headlight housing and disconnect the electrical connector

14.1a Malibu models, vertical adjustment screw for the left headlight housing (vertical adjustment screw for right headlight housing identical) - Aura models similar

14.1b Headlight vertical adjustment screw (2009 G6 model shown, earlier models may have a different but similar adjustment)

All Aura and 2008 and later Malibu models

5   Remove the front bumper fascia fasteners from the center and side of the headlamp to be removed, then pull the fascia forward to allow clearance enough to remove the headlamp assembly.

6   Pull out the outside edge of the headlight in order to release the up-and-down locator tabs on the rear of the headlight.

7   Pull out the top inside edge of the headlight to release the inboard up/down locator.

8   Pull the headlight out and disconnect the electrical connector, then remove the headlight.

9   Installation is the reverse of removal.

10   Adjust the headlights (see Section 14).

14   Headlights - adjustment

Note: *The headlights must be aimed correctly. If adjusted incorrectly they could blind the driver of an oncoming vehicle and cause a serious accident or seriously reduce your ability to see the road. The headlights should be checked for proper aim every 12 months and any time a new headlight is installed or front end bodywork is performed. It should be emphasized that the following procedure is only an interim step that will provide temporary adjustment until a properly equipped shop can adjust the headlights.*

1   The vertical adjustment screws are located behind each headlight housing (see illustrations). (There are no horizontal adjustment screws.)

2   There are several methods for adjusting the headlights. The simplest method requires masking tape, a blank wall and a level floor.

3   Position masking tape vertically on the wall in relation to the vehicle centerline and the centerlines of both headlights (see illustration).

4   Position a horizontal tapeline in relation to the centerline of all the headlights.

Note: *It might be easier to position the tape on the wall with the vehicle parked only a few inches away.*

5   Adjustment should be made with the vehicle parked 25 feet from the wall, sitting level, the gas tank half-full and no heavy load in the vehicle.

6   Starting with the low beam adjustment, position the high intensity zone so it is two inches below the horizontal line. Adjustment is made by turning the adjusting screw clockwise to raise the beam and counterclockwise to lower the beam.

7   With the high beams on, the high intensity zone should be vertically centered with the exact center just below the horizontal line.

Note: *It might not be possible to position the headlight aim exactly for both high and low beams. If a compromise must be made, keep in mind that the low beams are the most used and have the greatest effect on safety.*

8   Have the headlights adjusted by a dealer service department or service station at the earliest opportunity.

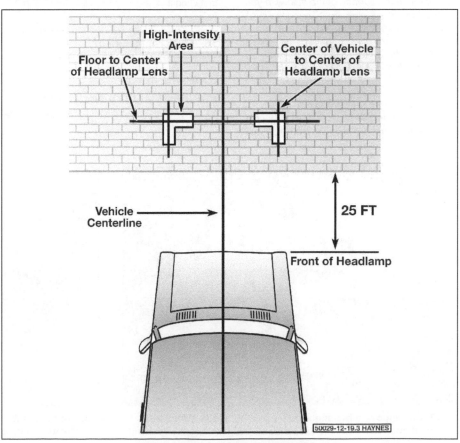

14.3 Headlight adjustment details

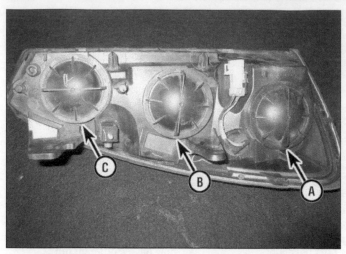

15.2 Each headlight housing contains three bulbs:

A   High beam bulb
B   Low beam bulb
C   Turn signal/parking
     light bulb

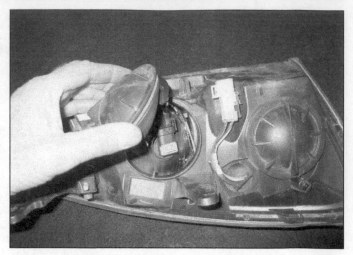

15.3 Remove the dust cover from the headlight bulb that you're going to replace

15   Headlight bulb - replacement

Warning: *Halogen gas-filled bulbs are under pressure and can shatter if the surface is scratched or the bulb is dropped. Wear eye protection and handle the bulbs carefully, grasping only the base whenever possible. Do not touch the surface of the bulb with your fingers because the oil from your skin could cause it to overheat and fail prematurely. If you do touch the bulb surface, clean it with rubbing alcohol.*

1   Remove the headlight housing (see Section 13).
2   Place the headlight housing face down on a clean surface. Lay down a towel or a shop rag to protect the lens (see illustration).
3   Remove the dust cover from the bulb that you're going to replace (see illustration).
4   Disconnect the electrical connector, then rotate the bulb holder counterclockwise and remove it from the housing (see illustration).
5   Installation is the reverse of removal. When installing the new bulb, make sure that you don't touch the glass. If you do, clean it with alcohol and a soft clean cloth before installing it.

16   Bulb replacement

Exterior lights

Front parking/turn signal bulbs
1   Remove the headlight housing (see Section 13).

2   Remove the dust cover from the front parking/turn signal bulb (see illustration 15.3).
3   Disconnect the electrical connector, rotate the bulb holder counterclockwise and pull it out (see illustration 15.4). Remove the bulb from the bulb holder.
4   Installation is the reverse of removal.

Center high-mounted brake light

Sedan models (2007 and earlier models)
5   Open the trunk and locate the center high-mount brake light assembly in the underside of the trunk lid.
6   Remove the high-mounted brake light housing mounting bolts (see illustration).
Note: *On G6 models, the housing, lens and bulbs are a one-piece assembly. No further disassembly is possible.*

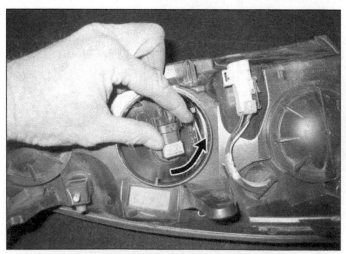

15.4 To remove a headlight bulb from the headlight housing, rotate it counterclockwise 1/4-turn and pull it out of the housing

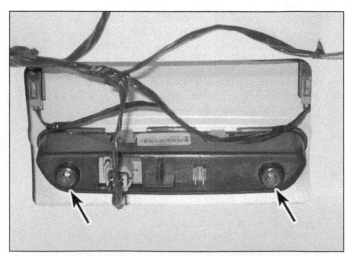

16.6 To detach the center high-mounted brake light housing from the trunk lid, remove these two bolts/nuts (sedan models)

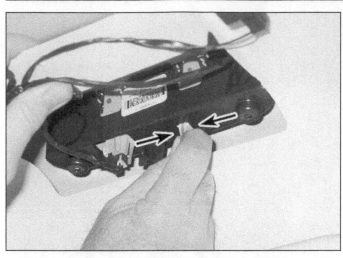

16.7 To remove the black plastic cover from the high-mounted brake light housing, pinch these two tabs together and pull off the cover (sedan models)

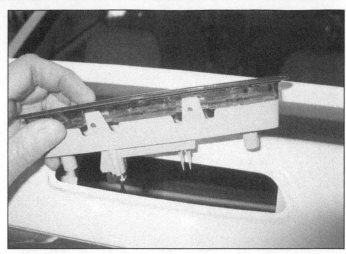

16.8a Working on the outside of the trunk lid, pull the high-mounted brake light housing out of the trunk lid . . .

7   Pinch together the two tabs that protrude through the black plastic cover for the high-mounted brake light housing (see illustration) and remove the cover.
8   Working on the outside of the trunk lid, remove the high-mounted brake light housing and disconnect the electrical connector from the housing (see illustrations).
9   The center high-mounted brake light housing, lens and bulbs are a one-piece assembly. No further disassembly is possible.
10   Installation is the reverse of removal.

MAXX models (2007 and earlier models)
11   Remove the liftgate upper trim panel (see Chapter 11).
12   Remove the high-mounted brake light housing mounting nuts.

13   Working on the outside of the liftgate, remove the high-mounted brake light housing and disconnect the electrical connector from the housing.
14   The center high-mounted brake light housing, lens and bulbs are a one-piece assembly. No further disassembly is possible.
15   Installation is the reverse of removal.

Aura models
16   Remove the rear parcel shelf trim panel (see Chapter 11).
17   Disconnect the electrical connector from the housing.
18   Unclip the housing from the package tray. No further disassembly is possible.
19   Installation is the reverse of removal.

All 2008 and later models except Aura models
20   Remove the rear window shelf panel by folding the seatback down and removing the four bolts and two retainers.
21   Pull the shelf insulator away from the window, disconnect the electrical connector, then remove the stop lamp screws.
22   Installation is the reverse of removal.

License plate light bulbs

Sedan models
23   Open the trunk lid and remove the black plastic cover for the trunk lid latch (see Chapter 11).
24   Remove the license plate light housing from the trunk lid (see illustration).

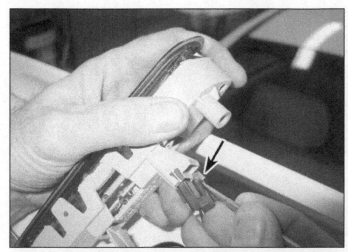

16.8b . . . depress the release tab on the electrical connector and disconnect the connector (sedan models)

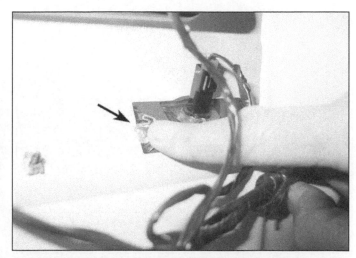

16.24 To remove the license plate light housing from the trunk lid, squeeze the locking tangs and pull out the housing (sedan models)

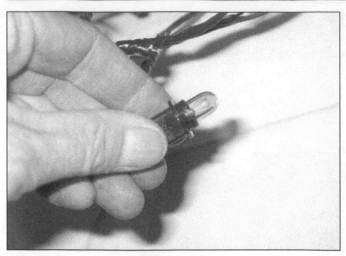

16.25 Remove the bulb socket from the license plate light housing (sedan models)

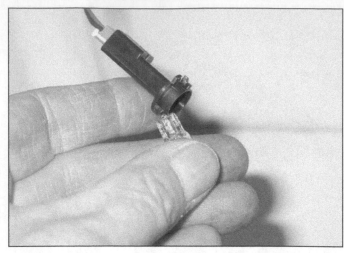

16.26 To remove a license plate light bulb from its socket, pull it straight out

16.33 Remove this retainer, then peel back the carpet from the left or right rear corner of the trunk (sedan models)

25   Remove the bulb socket from the license plate light housing (see illustration).
26   To remove the bulb from the socket, simply pull it straight out of the socket (see illustration).
27   Installation is the reverse of removal.

MAXX models, Aura models and 2008 and later models

28   Remove the two license plate light retaining screws.
29   Rotate the license plate light housing forward, then down, and pull it out through the opening.
30   To remove the light bulb socket from the license plate light housing, rotate it counterclockwise and pull it out of the housing.
31   To remove the bulb from the socket, simply pull it straight out.
32   Installation is the reverse of removal.

Taillight, back-up light, brake light and turn-signal bulbs

Note: *The accompanying photos depict the taillight housing on a 2007 and earlier sedan model, but the taillight housing on other models is quite similar to what you see here.*
33   Open the trunk. Remove the carpeting retainer (see illustration) and peel back the carpeting.

Aura and Malibu models

34   Working inside the trunk, remove the two plastic wing nuts from the left or right rear corner of the trunk (see illustration). Working outside the trunk, remove the two screws from the taillight housing trim next to the trunk opening (see illustration).
Note: *On 2008 and later models, the procedure is similar, however, the taillight housing is attached to the trunk lid instead of the rear corners of the trunk.*

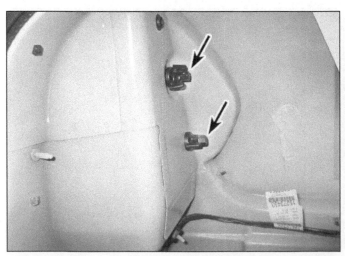

16.34a To detach the taillight/turn signal/brake light/back-up light housing from a sedan model, remove these two plastic wingnuts from the left or right rear corner of the trunk . . .

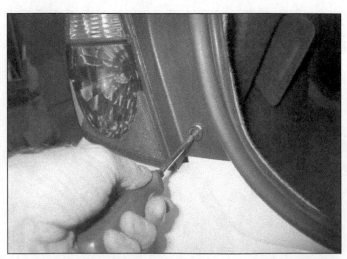

16.34b . . . then, working on the outside, remove the two screws (other screw not shown) from the taillight housing trim next to the trunk

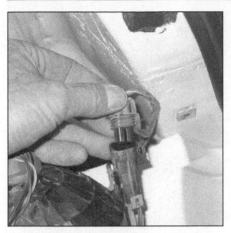

16.35 Pull out the taillight/turn signal/ brake light/back-up light housing and disconnect the electrical connector

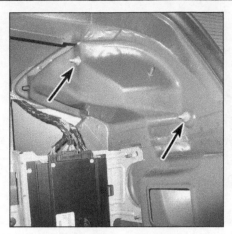

16.36a Taillight housing nuts

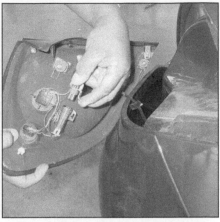

16.36b Slide the taillight assembly out, then disconnect the electrical connector(s)

35   Pull out the taillight housing and disconnect the electrical connector (see illustration).

G6 models

36   Remove the taillight housing nuts, then pull the taillight to the rear (on sedans) or to the side (on coupes) and disconnect the electrical connector (see illustrations).

37   Installation is the reverse of removal. On sedans, make sure to install the locating pin in its retainer as you insert the taillight assembly.

All models

38   Lay the taillight housing on a clean surface and refer to the accompanying taillight bulb guide (see illustration).

39   Remove the socket for the taillight bulb that you want to replace (see illustration).

16.36c Make sure that this locating pin seats properly when you install the taillight

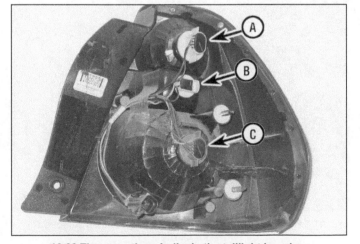

16.38 There are three bulbs in the taillight housing:

A   *Taillight bulb*
B   *Back-up light bulb*
C   *Taillight/brake light/turn-signal light bulb*

16.39 To remove a taillight bulb socket from the taillight housing, rotate the socket counterclockwise and pull it out of the housing

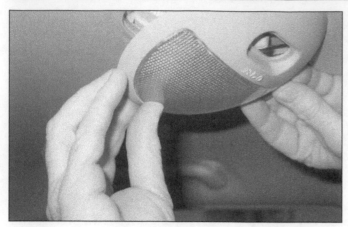

16.42 Using a trim removal tool or a thin, flat piece of wood or plastic, carefully pry loose the dome light lens or the map reading light lens (dome light shown, map reading light similar)

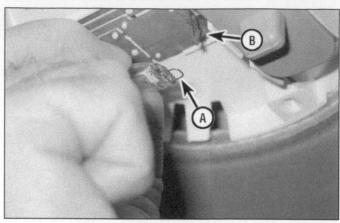

16.43a To remove the bulb from the dome light assembly, disengage the small retaining wire/conductor (A) on each end of the bulb from their corresponding mounting tangs (B)

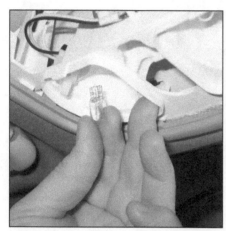

16.43b To replace either of the bulbs from the map reading light assembly, pull the bulb straight out

40   Remove the bulb from its socket by pulling it straight out.
41   Installation is the reverse of removal.

Interior lights
Dome light and map reading light bulbs

42   Pry off the dome light or map reading light lens (see illustration).
43   Remove the light bulb (see illustrations). First, push up on the end of the bulb to disengage the retaining wire from the hook on the mounting tab, then push the tab in slightly and swing the retainer down and past the hook. Once one end is free, simply disengage the other retainer from the hook on the other

mounting tang.
44   Installation is the reverse of removal.

17   Horn - replacement

All G6 and 2007 and earlier Malibu models

Note: *The horns are located at the left front corner of the vehicle, in a void below the left headlight housing and behind the left end of the bumper cover.*
1   Remove the left headlight housing (see Section 13).
2   Locate the horns at the left front corner of the vehicle (see illustration).

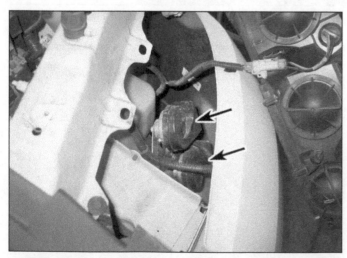

17.2 The horns are located at the left front corner of the vehicle, below the left headlight housing, which you'll have to remove to access the horns

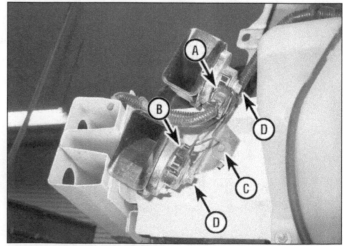

17.3 Horn assembly details (front bumper cover removed for clarity):
A   *Main electrical connector*
B   *Electrical connector for lower horn*
C   *Horn assembly mounting bracket bolt*
D   *Horn mounting bolts*

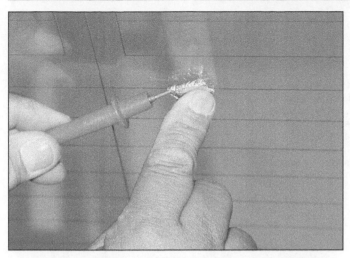

18.4 When measuring the voltage at the rear window defogger grid, wrap a piece of aluminum foil around the positive probe of the voltmeter and press the foil against the wire with your finger

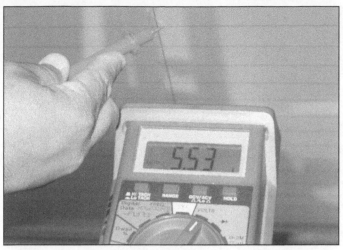

18.5 To determine if a heating element has broken, check the voltage at the center of each element; if the voltage is 5 or 6-volts, the element is unbroken, but if the voltage is 10 or 12-volts, the element is broken between the center and the ground side. If there is no voltage, the element is broken between the center and the positive side

3   Disconnect the electrical connector from the horn (see illustration).
4   Remove the horn mounting bolt and remove the two horns and the mounting bracket as a single assembly.
5   Once the horns and mounting bracket are removed, you can detach either or both horns as necessary from the mounting bracket by unbolting them.
6   Installation is the reverse of removal.

All Aura and 2008 and later Malibu models

Note: *The horn on these models is located inside the left front wheelwell, behind the splash shield.*
7   Loosen the left front wheel lug nuts, raise left front of the vehicle and support the vehicle securely on jack stands, block the other wheels to prevent the vehicle from rolling, remove the left front wheel.
8   Remove the plastic inner wheel well.
9   Disconnect the horn electrical connector, then remove the mounting bolt and horn.
10   Installation is the reverse of removal.

18   Rear window defogger - check and repair

1   The rear window defogger consists of a number of horizontal elements baked onto the glass surface.
2   Small breaks in the element can be repaired without removing the rear window.

Check

3   Turn the ignition switch and defogger system switches to the ON position. Using a voltmeter, place the positive probe against the defogger grid positive terminal and the nega-

tive probe against the ground terminal. If battery voltage is not indicated, check the fuse, defogger switch and related wiring. If voltage is indicated, but all or part of the defogger doesn't heat, proceed with the following tests.
4   When measuring voltage during the next two tests, wrap a piece of aluminum foil around the tip of the voltmeter positive probe and press the foil against the heating element with your finger (see illustration). Place the negative probe on the defogger grid ground terminal.
5   Check the voltage at the center of each heating element (see illustration). If the voltage is 5 or 6-volts, the element is okay (there is no break). If the voltage is zero, the element is broken between the center of the element and the positive end. If the voltage is 10 to 12-volts the element is broken between the center of the element and ground. Check

each heating element.
6   Connect the negative lead to a good body ground. The reading should stay the same. If it doesn't, the ground connection is bad.
7   To find the break, place the voltmeter negative probe against the defogger ground terminal. Place the voltmeter positive probe with the foil strip against the heating element at the positive terminal end and slide it toward the negative terminal end. The point at which the voltmeter deflects from several volts to zero is the point at which the heating element is broken (see illustration).

Repair

8   Repair the break in the element using a repair kit specifically recommended for this purpose, available at most auto parts stores. Included in this kit is plastic conductive epoxy.

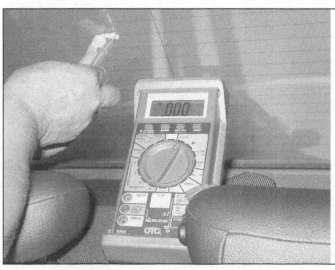

18.7 To find the break, place the voltmeter negative lead against the defogger ground terminal, place the voltmeter positive lead with the foil strip against the heating element at the positive terminal end and slide it toward the negative terminal end. The point at which the voltmeter reading changes abruptly is the point at which the element is broken

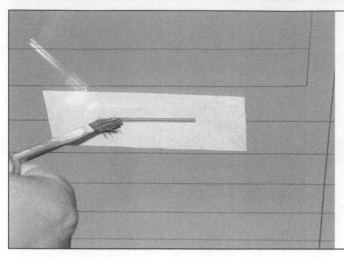

18.13 To use a defogger repair kit, apply masking tape to the inside of the window at the damaged area, then brush on the special conductive coating

9   Prior to repairing a break, turn off the system and allow it to cool off for a few minutes.

10   Lightly buff the element area with fine steel wool, then clean it thoroughly with rubbing alcohol.

11   Use masking tape to mask off the area being repaired.

12   Thoroughly mix the epoxy, following the instructions provided with the repair kit.

13   Apply the epoxy material to the slit in the masking tape, overlapping the undamaged area about 3/4-inch on either end (see illustration).

14   Allow the repair to cure for 24 hours before removing the tape and using the system.

19   Electric side view mirrors - general information

1   Most electric rear view mirrors use two motors to move the glass; one for up and down adjustments and one for left-right adjustments.

2   The control switch has a selector portion that sends voltage to the left or right side mirror. With the ignition ON but the engine OFF, roll down the windows and operate the mirror control switch through all functions (LEFT-RIGHT and UP-DOWN) for both the left and right side mirrors.

3   Listen carefully for the sound of the electric motors running in the mirrors.

4   If the motors can be heard but the mirror glass doesn't move, there's a problem with the drive mechanism inside the mirror.

5   If the mirrors do not operate and no sound comes from the mirrors, check the fuse (see Section 3).

6   If the fuse is OK, remove the mirror control switch. Have the switch continuity checked by a dealership service department or other qualified automobile repair facility.

7   Make sure the mirror is properly grounded.

8   If the mirror still doesn't work, remove the mirror and check the wires at the mirror

for voltage.

9   If there's not voltage in each switch position, check the circuit between the mirror and control switch for opens and shorts.

10   If there's voltage, remove the mirror and test it off the vehicle with jumper wires. Replace the mirror if it fails this test.

20   Cruise control system - general information

1   There are no conventional cruise control system components on these vehicles. The cruise control system is an integral subsystem of the electronic throttle body, which is controlled by the Powertrain Control Module (PCM). If the cruise control system isn't functioning correctly, take the vehicle to a dealer service department or other qualified repair shop for diagnosis.

21   Power window system - general information

1   The power window system operates electric motors, mounted in the doors, which lower and raise the windows. The system consists of the control switches, the motors, regulators, glass mechanisms and associated wiring.

2   The power windows can be lowered and raised from the master control switch by the driver or by remote switches located at the individual windows. Each window has a separate motor, which is reversible. The position of the control switch determines the polarity and therefore the direction of operation.

3   The circuit is protected by a fuse and a circuit breaker. Each motor is also equipped with an internal circuit breaker; this prevents one stuck window from disabling the whole system.

4   The power window system will only operate when the ignition switch is ON. In addition, many models have a window lock-out switch at the master control switch which, when activated, disables the switches at the

rear windows and, sometimes, the switch at the passenger's window also. Always check these items before troubleshooting a window problem.

5   These procedures are general in nature, so if you can't find the problem using them, take the vehicle to a dealer service department or other properly equipped repair facility.

6   If the power windows won't operate, always check the fuse and circuit breaker first.

7   If only the rear windows are inoperative, or if the windows only operate from the master control switch, check the rear window lockout switch for continuity in the unlocked position. Replace it if it doesn't have continuity.

8   Check the wiring between the switches and fuse panel for continuity. Repair the wiring, if necessary.

9   If only one window is inoperative from the master control switch, try the other control switch at the window.

Note: *This doesn't apply to the driver's door window.*

10   If the same window works from one switch, but not the other, check the switch for continuity.

11   If the switch tests OK, check for a short or open in the circuit between the affected switch and the window motor.

12   If one window is inoperative from both switches, remove the trim panel from the affected door and check for voltage at the switch and at the motor while the switch is operated.

13   If voltage is reaching the motor, disconnect the glass from the regulator (see Chapter 11). Move the window up and down by hand while checking for binding and damage. Also check for binding and damage to the regulator. If the regulator is not damaged and the window moves up and down smoothly, replace the motor. If there's binding or damage, lubricate, repair or replace parts, as necessary.

14   If voltage isn't reaching the motor, check the wiring in the circuit for continuity between the switches and motors. You'll need to consult the wiring diagram for the vehicle. If the circuit is equipped with a relay, check that the relay is grounded properly and receiving voltage.

22   Power door lock system - general information

1   A power door lock system operates the door lock actuators mounted in each door. The system consists of the switches, actuators, a control unit and associated wiring. Diagnosis can usually be limited to simple checks of the wiring connections and actuators for minor faults that can be easily repaired.

2   Power door lock systems are operated by bi-directional solenoids located in the doors. The lock switches have two operating positions: Lock and Unlock. When activated, the switch sends a ground signal to the door

lock control unit to lock or unlock the doors. Depending on which way the switch is activated, the control unit reverses polarity to the solenoids, allowing the two sides of the circuit to be used alternately as the feed (positive) and ground side.

3   Some vehicles may have an anti-theft system incorporated into the power locks. If you are unable to locate the trouble using the following general Steps, consult a dealer service department or other qualified repair shop.

4   Always check the circuit protection first. Some vehicles use a combination of circuit breakers and fuses.

5   Operate the door lock switches in both directions (Lock and Unlock) with the engine off. Listen for the click of the solenoids operating.

6   Test the switches for continuity. Remove the switches and have them checked by a dealer service department or other qualified automobile repair facility.

7   Check the wiring between the switches, control unit and solenoids for continuity. Repair the wiring if there's no continuity.

8   Check for a bad ground at the switches or at the control unit.

9   If all but one of the lock solenoids operate, remove the trim panel from the door with the problem (see Chapter 11) and check for voltage at the solenoid while the lock switch is operated. One of the wires should have voltage in the Lock position; the other should have voltage in the Unlock position.

10   If the inoperative solenoid is receiving voltage, replace the solenoid.

11   If the inoperative solenoid isn't receiving voltage, check the relay for an open or short in the wire between the lock solenoid and the control unit.

---

23   Daytime Running Lights (DRL) - general information

---

1   The Daytime Running Lights (DRL) system illuminates the headlights whenever the engine is running. The only exception is with the engine running and the parking brake engaged. Once the parking brake is released, the lights will remain on as long as the ignition switch is on, even if the parking brake is later applied.

2   The DRL system supplies reduced power to the headlights so they won't be too bright for daytime use, while prolonging headlight life.

---

24   Airbag system - general information and precautions

General information

1   All models are equipped with two front airbags, formally known as the Supplemental Inflatable Restraint (SIR) system. This system is designed to protect the driver and the front seat passenger from serious injury in the event of a frontal collision. It consists of an array of external and internal (inside the SDM) information sensors (decelerometers), the Inflatable Restraint Sensing and Diagnostic Module (SDM), the inflator modules (a driver's airbag in the steering wheel and a passenger airbag in the dash) and the wiring and connectors tying all these components together. An optional pair of side-impact airbags, also known as "roof rail" or "side curtain" airbags, is available for protection against side impacts. The side-impact airbags, if equipped, are located along the left and right edges of the headliner, above the doors.

Airbag/inflator modules

Driver's airbag/inflator module

2   The airbag inflator module in the steering wheel consists of a housing, the cushion (airbag), an initiating device and a canister of gas-generating material. The initiator is part of the inflator module deployment loop. When a collision occurs, the SDM sends current through the deployment loop to the initiator. Current passing through the initiator ignites the material in the canister, producing a rapidly expanding gas, which inflates the airbag almost instantaneously. Seconds after the airbag inflates, it deflates almost as quickly through airbag vent holes and/or the airbag fabric.

3   When the SDM sends current to the initiator, it travels through the airbag circuit to the steering column. From there, a "clockspring" on the steering wheel delivers the current to the module initiator. This clockspring assembly, which is the final segment of the airbag ignition circuit, functions as the bridge between the end of the airbag circuit on the (fixed) steering column and the beginning of the circuit on the (rotating) steering wheel. It's designed to maintain a closed circuit between the steering column and the steering wheel regardless of the position of the steering wheel. For this reason, removing and installing the clockspring is critical to the performance of the driver's side airbag. For information on how to remove and install the driver's side airbag, refer to *Steering wheel - removal and installation* in Chapter 10.

Passenger's airbag/inflator module

4   The passenger's airbag/inflator module is mounted above the glove compartment. It's similar in design to the driver's airbag except that it doesn't use a clockspring. When deployed by the SDM, the passenger's airbag bursts through the dashboard above the glovebox. Although this area looks like it's simply part of the dashboard, it's actually a trim cover with a perforated seam that allows the cover to separate from the dash when the passenger's airbag inflates.

Side impact airbag/inflator ("roof rail") modules

5   The (optional) side-impact airbag/inflator ("roof rail") modules are mounted along the outer edges of the headliner, right above the door openings. They extend from the "A-pillar" (front windshield pillar) to the "C-pillar" (rear window pillar). Each module consists of a housing, an inflatable airbag, an initiator and a canister of gas-generating material. Each roof rail module employs its own side impact sensor (SIS), which contains a sensing device that monitors changes in vehicle acceleration and velocity. This data is sent to the SDM, which compares it with its program. When the data exceeds a certain threshold, the SDM determines that the vehicle has been hit hard enough on one side or the other to warrant deployment of the roof rail on that side. The SDM doesn't deploy the roof rail airbags on both sides, just on the side being hit. Then the SDM sends current to the roof rail initiator to inflate the airbag, ripping open the headliner trim as it deploys to protect the occupant(s) on the left or right side of the vehicle. Side impact airbag/inflator modules are long enough to protect the driver and a left-side rear-seat passenger, or a front seat passenger and right-side rear-seat passenger.

Inflatable Restraint Sensing and Diagnostic Module (SDM)

6   The SDM is the computer module that controls the airbag system. Besides a microprocessor, the SDM also includes an array of sensors. Some of them are inside the SDM itself. Other external sensors are located throughout the vehicle. All of the sensors, internal and external, send a continuous voltage signal to the SDM, which compares this data to values stored in its memory. When these signals exceed a threshold value, i.e. when the SDM determines that the vehicle is decelerating more quickly than the threshold value, the SDM allows current to flow through the circuit to the appropriate airbag module(s), which initiates deployment of the airbag(s).

7   For more information about the airbag system in your vehicle, refer to your owner's manual.

Disarming the system and other precautions

Warning: *Failure to follow these precautions could result in accidental deployment of the airbag and personal injury.*

8   Whenever working in the vicinity of the steering wheel, instrument panel or any of the other SIR system components, the system must be disarmed. To disarm the system:

a) *Point the wheels straight ahead and turn the key to the Lock position.*

b) *Disconnect the cable from the negative battery terminal. Refer to Chapter 5, Section 1 for the disconnecting procedure.*

c) *Wait at least two minutes for the back-up power supply to be depleted.*

9   Whenever handling an airbag module, always keep the airbag opening (the trim side) pointed away from your body. Never place the airbag module on a bench or other surface with the airbag opening facing the surface. Always place the airbag module in

a safe location with the airbag opening (the upholstered side) facing up.

10   Never measure the resistance of any SIR component or use any electrical test equipment on any of the wiring or components. An ohmmeter has a built-in battery supply that could accidentally deploy the airbag.

11   Never dispose of a live airbag/inflator module. Return it to a dealer service department or other qualified repair shop for safe deployment and disposal.

12   Never use electrical welding equipment in the vicinity of any airbag components. The connectors for the system are easy to spot because they're bright yellow. Do NOT disconnect or tamper with these connectors, or you run the risk of setting a Diagnostic Trouble Code (DTC) in the SDM. Like the PCM, the SDM has a malfunction indicator light, known as the AIR BAG indicator light, on the instrument cluster. When you turn the ignition key to ON, the SDM checks out all of the SIS components and circuits. If everything is okay,

the AIR BAG indicator light goes off, just like the PCM's Malfunction Indicator Light (MIL). But if there's a problem somewhere, the light stays on, and will remain on until the problem is repaired and the DTC(s) cleared from the SDM's memory.

Impact seat belt retractors

13   All models are equipped with pyrotechnic (explosive) units in the front seat belt retracting mechanisms for both the lap and shoulder belts. During an impact that would trigger the airbag system, the airbag control unit also triggers the seat belt retractors. When the pyrotechnic charges go off, they accelerate the retractors to instantly take up any slack in the seat belt system to more fully prepare the driver and front seat passenger for impact.

14   The airbag system should be disabled any time work is done to or around the seats. **Warning:** *Never strike the pillars or floorpan with a hammer or use an impact-driver tool in these areas unless the system is disabled.*

25   Wiring diagrams - general information

1   Since it isn't possible to include all wiring diagrams for every year covered by this manual, the following diagrams are those that are typical and most commonly needed.

2   Prior to troubleshooting any circuits, check the fuse and circuit breakers (if equipped) to make sure they're in good condition. Make sure the battery is properly charged and check the cable connections (see Chapter 1).

3   When checking a circuit, make sure that all connectors are clean, with no broken or loose terminals. When unplugging a connector, do not pull on the wires. Pull only on the connector housings themselves.

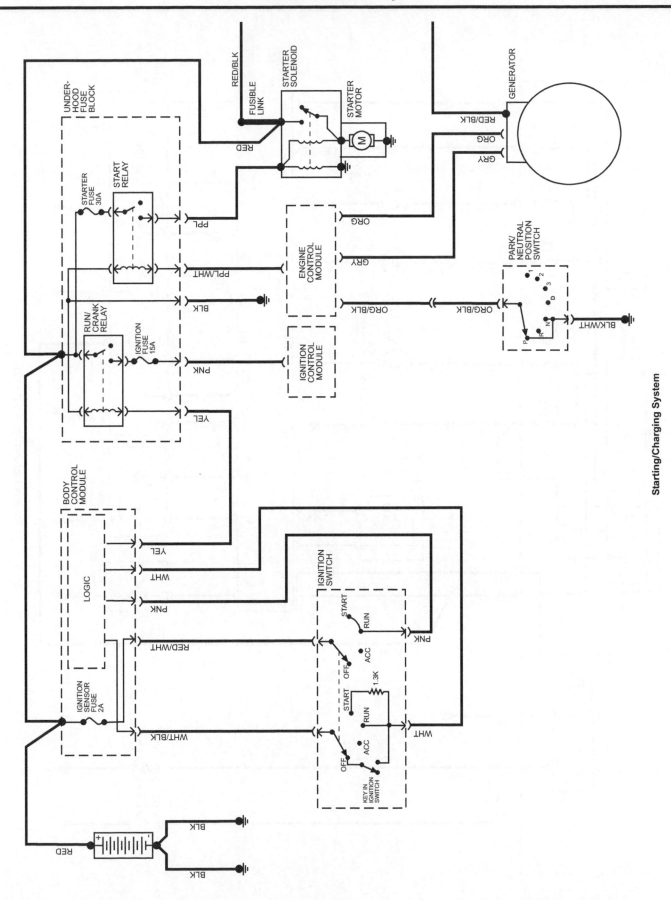

Starting/Charging System

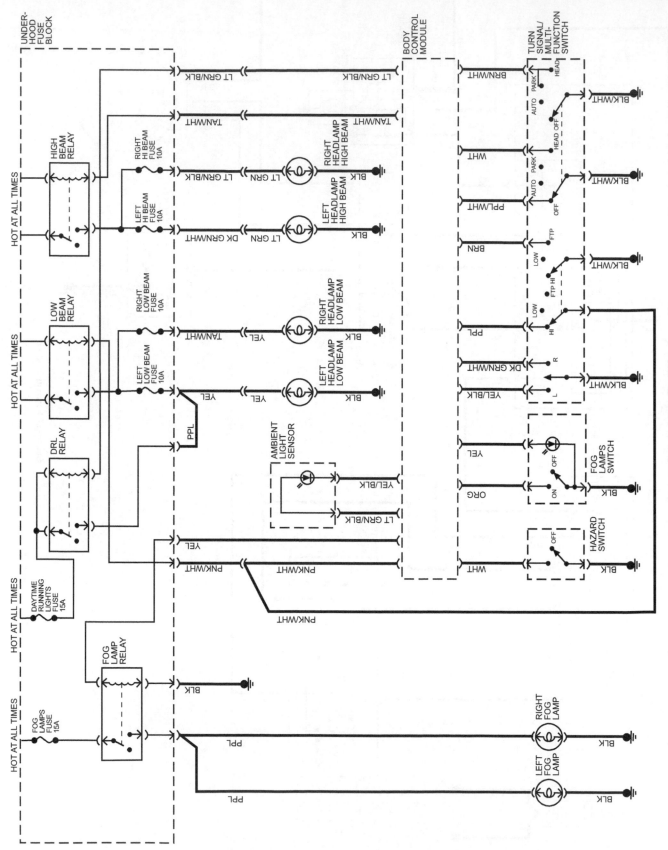

Exterior Lighting System (1 of 2)

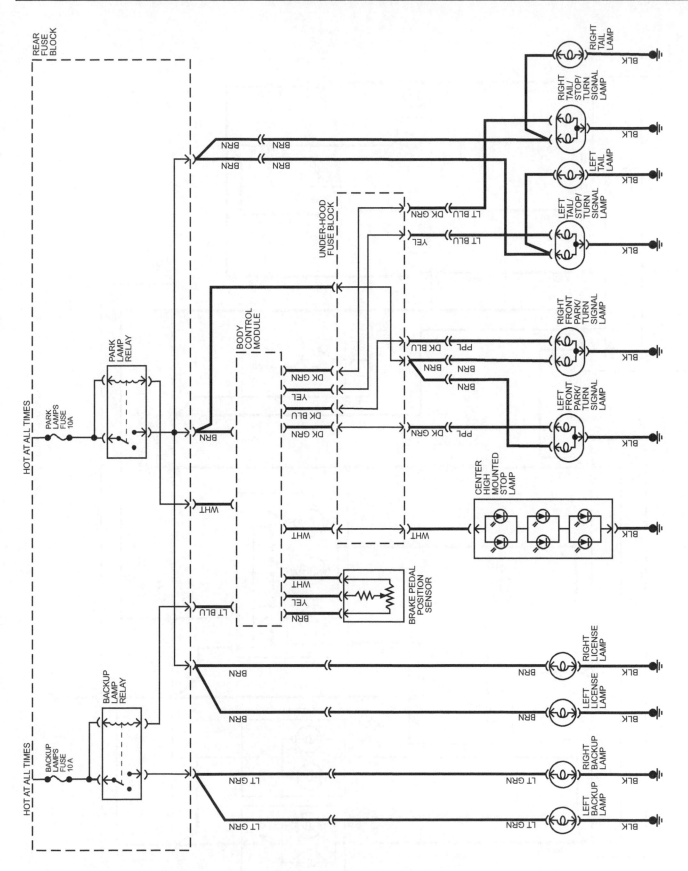

Exterior Lighting System (2 of 2)

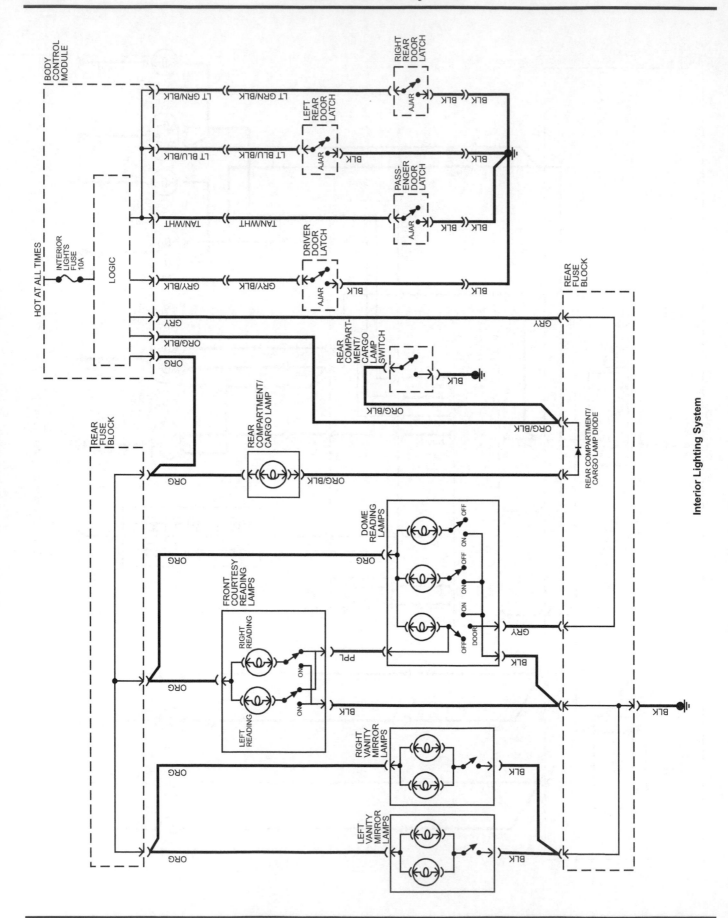

Interior Lighting System

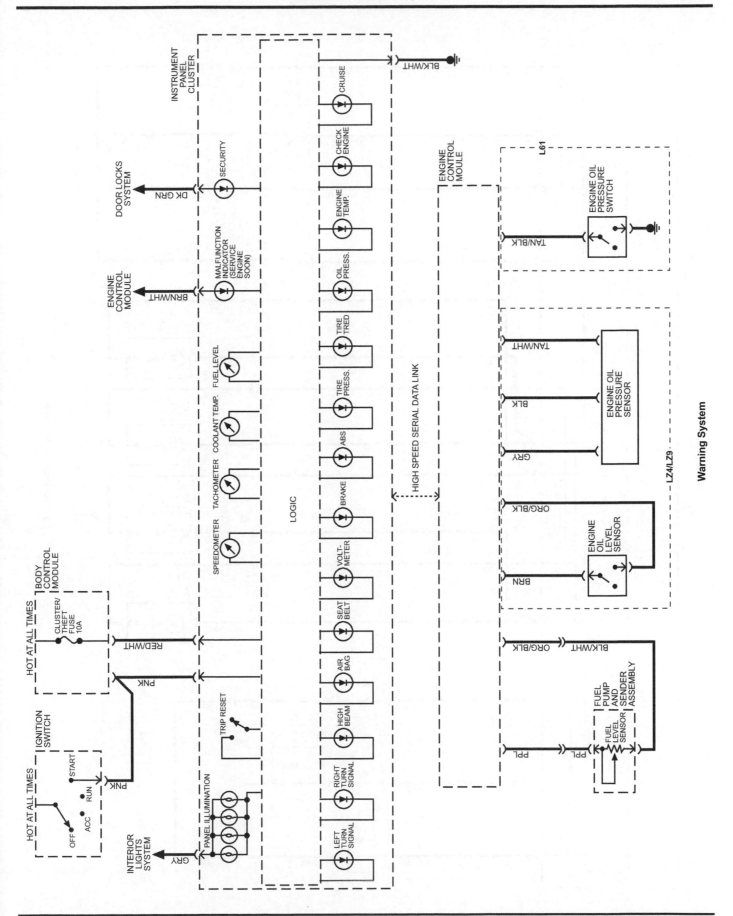

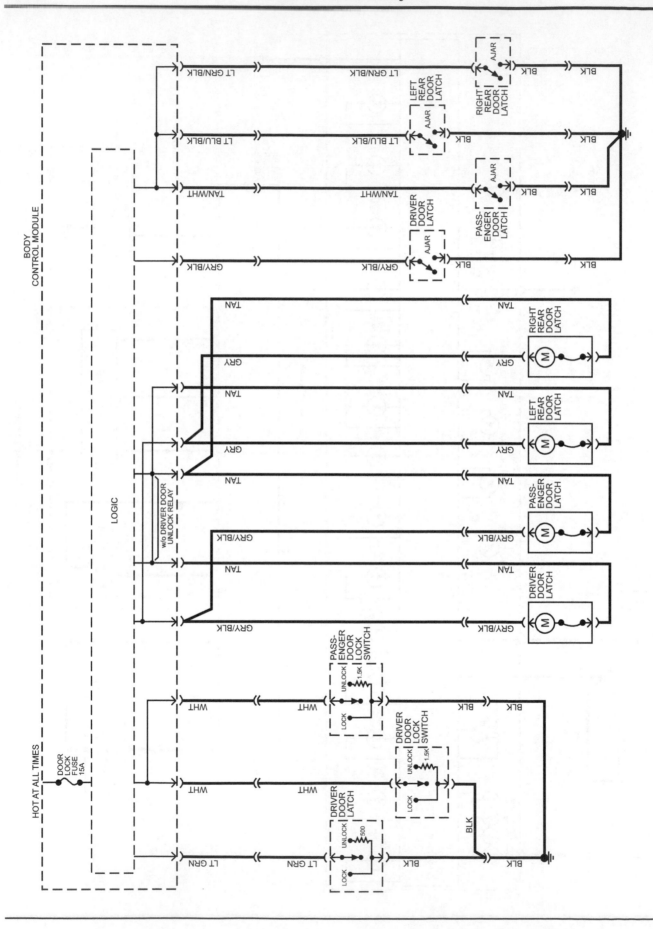

Power Door Locks System

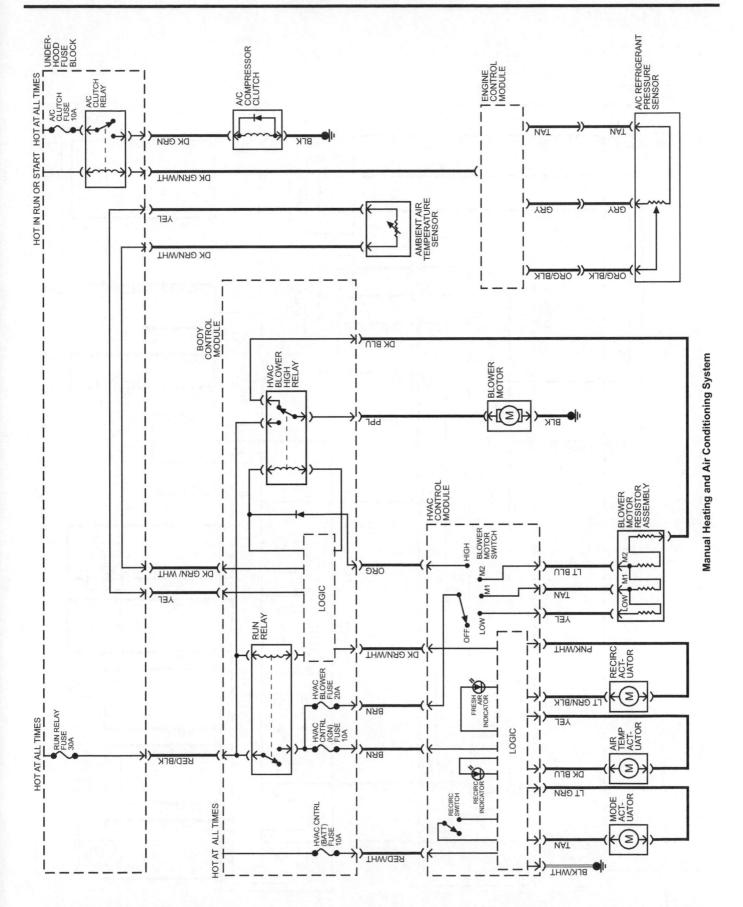

Manual Heating and Air Conditioning System

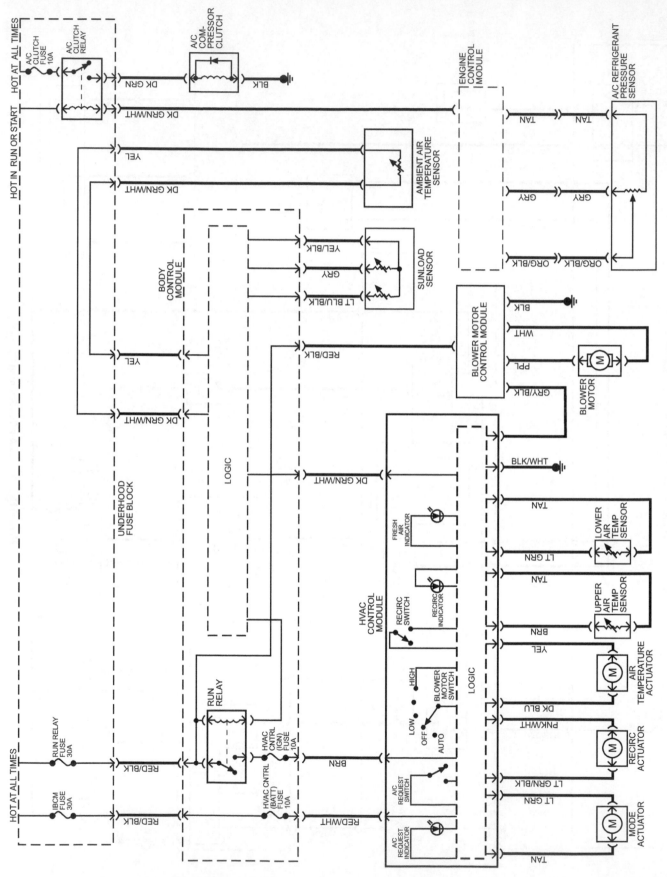

Automatic Heating and Air Conditioning System

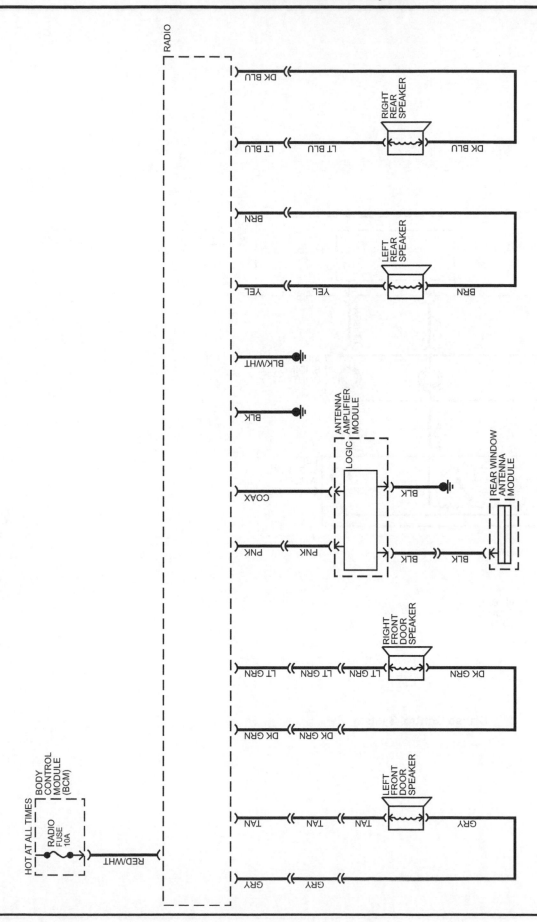

Radio System (base)

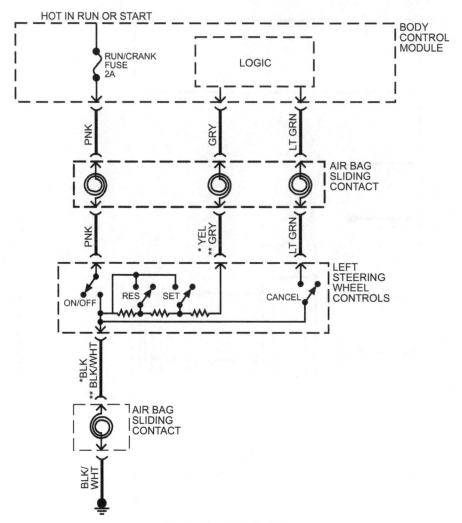

Cruise Control System

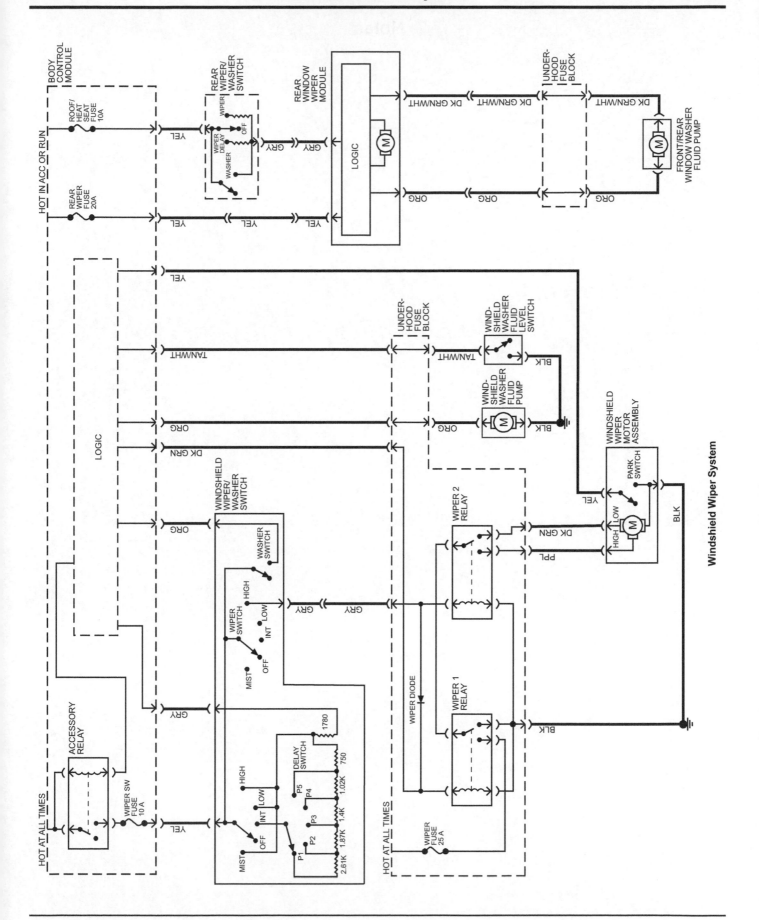

Windshield Wiper System

Notes

Index

Haynes Automotive Manuals

ACURA
12020 Integra '86 thru '89 & Legend '86 thru '90
12021 Integra '90 thru '93 & Legend '91 thru '95
Integra '94 thru '00 - see HONDA Civic (42025)
MDX '01 thru '07 - see HONDA Pilot (42037)
12050 Acura TL all models '99 thru '08

AMC
14020 Mid-size models '70 thru '83
14025 (Renault) Alliance & Encore '83 thru '87

AUDI
15020 4000 all models '80 thru '87
15025 5000 all models '77 thru '83
15026 5000 all models '84 thru '88
Audi A4 '96 thru '01 - see VW Passat (96023)
15030 Audi A4 '02 thru '08

AUSTIN-HEALEY
Sprite - see MG Midget (66015)

BMW
18020 3/5 Series '82 thru '92
18021 3-Series incl. Z3 models '92 thru '98
18022 3-Series incl. Z4 models '99 thru '05
18023 3-Series '06 thru '14
18025 320i all 4-cylinder models '75 thru '83
18050 1500 thru 2002 except Turbo '59 thru '77

BUICK
19010 Buick Century '97 thru '05
Century (front-wheel drive) - see GM (38005)
19020 Buick, Oldsmobile & Pontiac Full-size
(Front-wheel drive) '85 thru '05
Buick Electra, LeSabre and Park Avenue;
Oldsmobile Delta 88 Royale, Ninety Eight
and Regency; Pontiac Bonneville
19025 Buick, Oldsmobile & Pontiac Full-size
(Rear wheel drive) '70 thru '90
Buick Estate, Electra, LeSabre, Limited,
Oldsmobile Custom Cruiser, Delta 88,
Ninety-eight, Pontiac Bonneville,
Catalina, Grandville, Parisienne
19027 Buick LaCrosse '05 thru '13
Enclave - see GENERAL MOTORS (38001)
Rainier - see CHEVROLET (24072)
Regal - see GENERAL MOTORS (38010)
Riviera - see GENERAL MOTORS (38030, 38031)
Roadmaster - see CHEVROLET (24046)
Skyhawk - see GENERAL MOTORS (38015)
Skylark - see GENERAL MOTORS (38020, 38025)
Somerset - see GENERAL MOTORS (38025)

CADILLAC
21015 CTS & CTS-V '03 thru '14
21030 Cadillac Rear Wheel Drive '70 thru '93
Cimarron - see GENERAL MOTORS (38015)
DeVille - see GENERAL MOTORS (38031 & 38032)
Eldorado - see GENERAL MOTORS (38030)
Fleetwood - see GENERAL MOTORS (38031)
Seville - see GM (38030, 38031 & 38032)

CHEVROLET
10305 Chevrolet Engine Overhaul Manual
24010 Astro & GMC Safari Mini-vans '85 thru '05
24013 Aveo '04 thru '11
24015 Camaro V8 all models '70 thru '81
24016 Camaro all models '82 thru '92
24017 Camaro & Firebird '93 thru '02
Cavalier - see GENERAL MOTORS (38016)
Celebrity - see GENERAL MOTORS (38005)
24018 Camaro '10 thru '15
24020 Chevelle, Malibu & El Camino '69 thru '87
Cobalt - see GENERAL MOTORS (38017)
24024 Chevette & Pontiac T1000 '76 thru '87
Citation - see GENERAL MOTORS (38020)
24027 Colorado & GMC Canyon '04 thru '12
24032 Corsica & Beretta all models '87 thru '96
24040 Corvette all V8 models '68 thru '82
24041 Corvette all models '84 thru '96
24042 Corvette all models '97 thru '13
24044 Cruze '11 thru '19
24045 Full-size Sedans Caprice, Impala, Biscayne,
Bel Air & Wagons '69 thru '90
24046 Impala SS & Caprice and Buick Roadmaster
'91 thru '96
Impala '00 thru '05 - see LUMINA (24048)
24047 Impala & Monte Carlo all models '06 thru '11
Lumina '90 thru '94 - see GM (38010)
24048 Lumina & Monte Carlo '95 thru '05
Lumina APV - see GM (38035)
24050 Luv Pick-up all 2WD & 4WD '72 thru '82
24051 Malibu '13 thru '19
24055 Monte Carlo all models '70 thru '88
Monte Carlo '95 thru '01 - see LUMINA (24048)
24059 Nova all V8 models '69 thru '79

24060 Nova and Geo Prizm '85 thru '92
24064 Pick-ups '67 thru '87 - Chevrolet & GMC
24065 Pick-ups '88 thru '98 - Chevrolet & GMC
24066 Pick-ups '99 thru '06 - Chevrolet & GMC
24067 Chevrolet Silverado & GMC Sierra '07 thru '13
24068 Chevrolet Silverado & GMC Sierra '14 thru '19
24070 S-10 & S-15 Pick-ups '82 thru '93,
Blazer & Jimmy '83 thru '94,
24071 S-10 & Sonoma Pick-ups '94 thru '04,
including Blazer, Jimmy & Hombre
24072 Chevrolet TrailBlazer, GMC Envoy &
Oldsmobile Bravada '02 thru '09
24075 Sprint '85 thru '88 & Geo Metro '89 thru '01
24080 Vans - Chevrolet & GMC '68 thru '96
24081 Chevrolet Express & GMC Savana
Full-size Vans '96 thru '19

CHRYSLER
10310 Chrysler Engine Overhaul Manual
25015 Chrysler Cirrus, Dodge Stratus,
Plymouth Breeze '95 thru '00
25020 Full-size Front-Wheel Drive '88 thru '93
K-Cars - see DODGE Aries (30008)
Laser - see DODGE Daytona (30030)
25025 Chrysler LHS, Concorde, New Yorker,
Dodge Intrepid, Eagle Vision, '93 thru '97
25026 Chrysler LHS, Concorde, 300M,
Dodge Intrepid, '98 thru '04
25027 Chrysler 300 '05 thru '18, Dodge Charger
'06 thru '18, Magnum '05 thru '08 &
Challenger '08 thru '18
25030 Chrysler & Plymouth Mid-size
front wheel drive '82 thru '95
Rear-wheel Drive - see Dodge (30050)
25035 PT Cruiser all models '01 thru '10
25040 Chrysler Sebring '95 thru '06, Dodge Stratus
'01 thru '06 & Dodge Avenger '95 thru '00
25041 Chrysler Sebring '07 thru '10, 200 '11 thru '17
Dodge Avenger '08 thru '14

DATSUN
28005 200SX all models '80 thru '83
28012 240Z, 260Z & 280Z Coupe '70 thru '78
28014 280ZX Coupe & 2+2 '79 thru '83
300ZX - see NISSAN (72010)
28018 510 & PL521 Pick-up '68 thru '73
28020 510 all models '78 thru '81
28022 620 Series Pick-up all models '73 thru '79
720 Series Pick-up - see NISSAN (72030)

DODGE
400 & 600 - see CHRYSLER (25030)
30008 Aries & Plymouth Reliant '81 thru '89
30010 Caravan & Plymouth Voyager '84 thru '95
30011 Caravan & Plymouth Voyager '96 thru '02
30012 Challenger & Plymouth Sapporo '78 thru '83
30013 Caravan, Chrysler Voyager &
Town & Country '03 thru '07
30014 Grand Caravan &
Chrysler Town & Country '08 thru '18
30016 Colt & Plymouth Champ '78 thru '87
30020 Dakota Pick-ups all models '87 thru '96
30021 Durango '98 & '99 & Dakota '97 thru '99
30022 Durango '00 thru '03 & Dakota '00 thru '04
30023 Durango '04 thru '09 & Dakota '05 thru '11
30025 Dart, Demon, Plymouth Barracuda,
Duster & Valiant 6-cylinder models '67 thru '76
30030 Daytona & Chrysler Laser '84 thru '89
Intrepid - see CHRYSLER (25025, 25026)
30034 Neon all models '95 thru '99
30035 Omni & Plymouth Horizon '78 thru '90
30036 Dodge & Plymouth Neon '00 thru '05
30040 Pick-ups full-size models '74 thru '93
30042 Pick-ups full-size models '94 thru '08
30043 Pick-ups full-size models '09 thru '18
30045 Ram 50/D50 Pick-ups & Raider and
Plymouth Arrow Pick-ups '79 thru '93
30050 Dodge/Plymouth/Chrysler RWD '71 thru '89
30055 Shadow & Plymouth Sundance '87 thru '94
30060 Spirit & Plymouth Acclaim '89 thru '95
30065 Vans - Dodge & Plymouth '71 thru '03

EAGLE
Talon - see MITSUBISHI (68030, 68031)
Vision - see CHRYSLER (25025)

FIAT
34010 124 Sport Coupe & Spider '68 thru '78
34025 X1/9 all models '74 thru '80

FORD
10320 Ford Engine Overhaul Manual
10355 Ford Automatic Transmission Overhaul
11500 Mustang '64-1/2 thru '70 Restoration Guide
36004 Aerostar Mini-vans all models '86 thru '97
36006 Contour & Mercury Mystique '95 thru '00
36008 Courier Pick-up all models '72 thru '82

36012 Crown Victoria &
Mercury Grand Marquis '88 thru '11
36014 Edge '07 thru '19 & Lincoln MKX '07 thru '18
36016 Escort & Mercury Lynx all models '81 thru '90
36020 Escort & Mercury Tracer '91 thru '02
36022 Escape '01 thru '17, Mazda Tribute '01 thru '11,
& Mercury Mariner '05 thru '11
36024 Explorer & Mazda Navajo '91 thru '01
36025 Explorer & Mercury Mountaineer '02 thru '10
36026 Explorer '11 thru '17
36028 Fairmont & Mercury Zephyr '78 thru '83
36030 Festiva & Aspire '88 thru '97
36032 Fiesta all models '77 thru '80
36034 Focus all models '00 thru '11
36035 Focus '12 thru '14
36045 Fusion '06 thru '14 & Mercury Milan '06 thru '11
36048 Mustang V8 all models '64-1/2 thru '73
36049 Mustang II 4-cylinder, V6 & V8 models '74 thru '78
36050 Mustang & Mercury Capri '79 thru '93
36051 Mustang all models '94 thru '04
36052 Mustang '05 thru '14
36054 Pick-ups & Bronco '73 thru '79
36058 Pick-ups & Bronco '80 thru '96
36059 F-150 '97 thru '03, Expedition '97 thru '17,
F-250 '97 thru '99, F-150 Heritage '04
& Lincoln Navigator '98 thru '17
36060 Super Duty Pick-ups & Excursion '99 thru '10
36061 F-150 full-size '04 thru '14
36062 Pinto & Mercury Bobcat '75 thru '80
36063 F-150 full-size '15 thru '17
36064 Super Duty Pick-ups '11 thru '16
36066 Probe all models '89 thru '92
Probe '93 thru '97 - see MAZDA 626 (61042)
36070 Ranger & Bronco II gas models '83 thru '92
36071 Ranger '93 thru '11 & Mazda Pick-ups '94 thru '09
36074 Taurus & Mercury Sable '86 thru '95
36075 Taurus & Mercury Sable '96 thru '07
36076 Taurus '08 thru '14, Five Hundred '05 thru '07,
Mercury Montego '05 thru '07 & Sable '08 thru '09
36078 Tempo & Mercury Topaz '84 thru '94
36082 Thunderbird & Mercury Cougar '83 thru '88
36086 Thunderbird & Mercury Cougar '89 thru '97
36090 Vans all V8 Econoline models '69 thru '91
36094 Vans full size '92 thru '14
36097 Windstar '95 thru '03, Freestar & Mercury
Monterey Mini-van '04 thru '07

GENERAL MOTORS
10360 GM Automatic Transmission Overhaul
38001 GMC Acadia '07 thru '16, Buick Enclave
'08 thru '17, Saturn Outlook '07 thru '10
& Chevrolet Traverse '09 thru '17
38005 Buick Century, Chevrolet Celebrity,
Oldsmobile Cutlass Ciera & Pontiac 6000
all models '82 thru '96
38010 Buick Regal '88 thru '04, Chevrolet Lumina
'88 thru '04, Oldsmobile Cutlass Supreme
'88 thru '97 & Pontiac Grand Prix '88 thru '07
38015 Buick Skyhawk, Cadillac Cimarron,
Chevrolet Cavalier, Oldsmobile Firenza,
Pontiac J-2000 & Sunbird '82 thru '94
38016 Chevrolet Cavalier & Pontiac Sunfire '95 thru '05
38017 Chevrolet Cobalt '05 thru '10, HHR '06 thru '11,
Pontiac G5 '07 thru '09, Pursuit '06
& Saturn ION '03 thru '07
38020 Buick Skylark, Chevrolet Citation,
Oldsmobile Omega, Pontiac Phoenix '80 thru '85
38025 Buick Skylark '86 thru '98, Somerset '85 thru '87,
Oldsmobile Achieva '92 thru '98, Calais '85 thru '91,
& Pontiac Grand Am all models '85 thru '98
38026 Chevrolet Malibu '97 thru '03, Classic '04 thru '05,
Oldsmobile Alero '99 thru '03, Cutlass '97 thru '00,
& Pontiac Grand Am '99 thru '03
38027 Chevrolet Malibu '04 thru '12, Pontiac G6
'05 thru '10 & Saturn Aura '07 thru '10
38030 Cadillac Eldorado, Seville, Oldsmobile
Toronado & Buick Riviera '71 thru '85
38031 Cadillac Eldorado, Seville, DeVille, Fleetwood,
Oldsmobile Toronado & Buick Riviera '86 thru '93
38032 Cadillac DeVille '94 thru '05, Seville '92 thru '04
& Cadillac DTS '06 thru '10
38035 Chevrolet Lumina APV, Oldsmobile Silhouette
& Pontiac Trans Sport all models '90 thru '96
38036 Chevrolet Venture '97 thru '05, Oldsmobile
Silhouette '97 thru '04, Pontiac Trans Sport
'97 thru '98 & Montana '99 thru '05
38040 Chevrolet Equinox '05 thru '17, GMC Terrain
'10 thru '17 & Pontiac Torrent '06 thru '09

GEO
Metro - see CHEVROLET Sprint (24075)
Prizm - '85 thru '92 see CHEVY (24060),
'93 thru '02 see TOYOTA Corolla (92036)
40030 Storm all models '90 thru '93
Tracker - see SUZUKI Samurai (90010)

(Continued on other side)

Haynes Automotive Manuals (continued)

NOTE: If you do not see a listing for your vehicle, please visit haynes.com for the latest product information and check out our Online Manuals!

GMC
Acadia - see *GENERAL MOTORS (38001)*
Pick-ups - see *CHEVROLET (24027, 24068)*
Vans - see *CHEVROLET (24081)*

HONDA
42010 **Accord CVCC** all models '76 thru '83
42011 **Accord** all models '84 thru '89
42012 **Accord** all models '90 thru '93
42013 **Accord** all models '94 thru '97
42014 **Accord** all models '98 thru '02
42015 **Accord** '03 thru '12 & **Crosstour** '10 thru '14
42016 **Accord** '13 thru '17
42020 **Civic 1200** all models '73 thru '79
42021 **Civic 1300 & 1500 CVCC** '80 thru '83
42022 **Civic 1500 CVCC** all models '75 thru '79
42023 **Civic** all models '84 thru '91
42024 **Civic & del Sol** '92 thru '95
42025 **Civic** '96 thru '00, **CR-V** '97 thru '01 & **Acura Integra** '94 thru '00
42026 **Civic** '01 thru '11 & **CR-V** '02 thru '11
42027 **Civic** '12 thru '15 & **CR-V** '12 thru '16
42030 **Fit** '07 thru '13
42035 **Odyssey** all models '99 thru '10
Passport - see *ISUZU Rodeo (47017)*
42037 **Honda Pilot** '03 thru '08, **Ridgeline** '06 thru '14 & **Acura MDX** '01 thru '07
42040 **Prelude CVCC** all models '79 thru '89

HYUNDAI
43010 **Elantra** all models '96 thru '19
43015 **Excel & Accent** all models '86 thru '13
43050 **Santa Fe** all models '01 thru '12
43055 **Sonata** all models '99 thru '14

INFINITI
G35 '03 thru '08 - see *NISSAN 350Z (72011)*

ISUZU
Hombre - see *CHEVROLET S-10 (24071)*
47017 **Rodeo** '91 thru '02, **Amigo** '89 thru '94 & '98 thru '02 & **Honda Passport** '95 thru '02
47020 **Trooper** '84 thru '91 & **Pick-up** '81 thru '93

JAGUAR
49010 **XJ6** all 6-cylinder models '68 thru '86
49011 **XJ6** all models '88 thru '94
49015 **XJ12 & XJS** all 12-cylinder models '72 thru '85

JEEP
50010 **Cherokee, Comanche & Wagoneer Limited** all models '84 thru '01
50011 **Cherokee** '14 thru '19
50020 **CJ** all models '49 thru '86
50025 **Grand Cherokee** all models '93 thru '04
50026 **Grand Cherokee** '05 thru '19 & **Dodge Durango** '11 thru '19
50029 **Grand Wagoneer & Pick-up** '72 thru '91 Grand Wagoneer '84 thru '91, Cherokee & Wagoneer '72 thru '83, Pick-up '72 thru '88
50030 **Wrangler** all models '87 thru '17
50035 **Liberty** '02 thru '12 & **Dodge Nitro** '07 thru '11
50050 **Patriot & Compass** '07 thru '17

KIA
54050 **Optima** '01 thru '10
54060 **Sedona** '02 thru '14
54070 **Sephia** '94 thru '01, **Spectra** '00 thru '09, **Sportage** '05 thru '20
54077 **Sorento** '03 thru '13

LEXUS
ES 300/330 - see *TOYOTA Camry (92007, 92008)*
ES 350 - see *TOYOTA Camry (92009)*
RX 300/330/350 - see *TOYOTA Highlander (92095)*

LINCOLN
MKX - see *FORD (36014)*
Navigator - see *FORD Pick-up (36059)*
59010 **Rear-Wheel Drive Continental** '70 thru '87, **Mark Series** '70 thru '92 & **Town Car** '81 thru '10

MAZDA
61010 **GLC (rear-wheel drive)** '77 thru '83
61011 **GLC (front-wheel drive)** '81 thru '85
61012 **Mazda3** '04 thru '11
61015 **323 & Protogé** '90 thru '03
61016 **MX-5 Miata** '90 thru '14
61020 **MPV** all models '89 thru '98
Navajo - see *Ford Explorer (36024)*
61030 **Pick-ups** '72 thru '93
Pick-ups '94 thru '09 - see *Ford Ranger (36071)*
61035 **RX-7** all models '79 thru '85
61036 **RX-7** all models '86 thru '91
61040 **626 (rear-wheel drive)** all models '79 thru '82
61041 **626 & MX-6 (front-wheel drive)** '83 thru '92
61042 **626** '93 thru '01 & **MX-6/Ford Probe** '93 thru '02
61043 **Mazda6** '03 thru '13

MERCEDES-BENZ
63012 **123 Series Diesel** '76 thru '85
63015 **190 Series** 4-cylinder gas models '84 thru '88
63020 **230/250/280** 6-cylinder SOHC models '68 thru '72
63025 **280 123 Series** gas models '77 thru '81
63030 **350 & 450** all models '71 thru '80
63040 **C-Class:** C230/C240/C280/C320/C350 '01 thru '07

MERCURY
64200 **Villager & Nissan Quest** '93 thru '01
All other titles, see FORD Listing.

MG
66010 **MGB** Roadster & GT Coupe '62 thru '80
66015 **MG Midget, Austin Healey Sprite** '58 thru '80

MINI
67020 **Mini** '02 thru '13

MITSUBISHI
68020 **Cordia, Tredia, Galant, Precis & Mirage** '83 thru '93
68030 **Eclipse, Eagle Talon & Plymouth Laser** '90 thru '94
68031 **Eclipse** '95 thru '05 & **Eagle Talon** '95 thru '98
68035 **Galant** '94 thru '12
68040 **Pick-up** '83 thru '96 & **Montero** '83 thru '93

NISSAN
72010 **300ZX** all models including Turbo '84 thru '89
72011 **350Z & Infiniti G35** all models '03 thru '08
72015 **Altima** all models '93 thru '06
72016 **Altima** '07 thru '12
72020 **Maxima** all models '85 thru '92
72021 **Maxima** all models '93 thru '08
72025 **Murano** '03 thru '14
72030 **Pick-ups** '80 thru '97 & **Pathfinder** '87 thru '95
72031 **Frontier** '98 thru '04, **Xterra** '00 thru '04, & **Pathfinder** '96 thru '04
72032 **Frontier & Xterra** '05 thru '14
72037 **Pathfinder** '05 thru '14
72040 **Pulsar** all models '83 thru '86
72042 **Roque** all models '08 thru '20
72050 **Sentra** all models '82 thru '94
72051 **Sentra & 200SX** all models '95 thru '06
72060 **Stanza** all models '82 thru '90
72070 **Titan pick-ups** '04 thru '10, **Armada** '05 thru '10 & **Pathfinder Armada** '04
72080 **Versa** all models '07 thru '19

OLDSMOBILE
73015 **Cutlass** V6 & V8 gas models '74 thru '88
For other OLDSMOBILE titles, see BUICK, CHEVROLET or GENERAL MOTORS listings.

PLYMOUTH
For PLYMOUTH titles, see DODGE listing.

PONTIAC
79008 **Fiero** all models '84 thru '88
79018 **Firebird** V8 models except Turbo '70 thru '81
79019 **Firebird** all models '82 thru '92
79025 **G6** all models '05 thru '09
79040 **Mid-size Rear-wheel Drive** '70 thru '87
Vibe '03 thru '10 - see *TOYOTA Corolla (92037)*
For other PONTIAC titles, see BUICK, CHEVROLET or GENERAL MOTORS listings.

PORSCHE
80020 **911 Coupe & Targa** models '65 thru '89
80025 **914** all 4-cylinder models '69 thru '76
80030 **924** all models including Turbo '76 thru '82
80035 **944** all models including Turbo '83 thru '89

RENAULT
Alliance & Encore - see *AMC (14025)*

SAAB
84010 **900** all models including Turbo '79 thru '88

SATURN
87010 **Saturn** all S-series models '91 thru '02
Saturn Ion '03 thru '07- see *GM (38017)*
Saturn Outlook - see *GM (38001)*
87020 **Saturn L-series** all models '00 thru '04
87040 **Saturn VUE** '02 thru '09

SUBARU
89002 **1100, 1300, 1400 & 1600** '71 thru '79
89003 **1600 & 1800** 2WD & 4WD '80 thru '94
89080 **Impreza** '02 thru '11, **WRX** '02 thru '14, & **WRX STI** '04 thru '14
89100 **Legacy** all models '90 thru '99
89101 **Legacy & Forester** '00 thru '09
89102 **Legacy** '10 thru '16 & **Forester** '12 thru '16

SUZUKI
90010 **Samurai/Sidekick & Geo Tracker** '86 thru '01

TOYOTA
92005 **Camry** all models '83 thru '91
92006 **Camry** '92 thru '96 & **Avalon** '95 thru '96
92007 **Camry, Avalon, Solara, Lexus ES 300** '97 thru '01
92008 **Camry, Avalon, Lexus ES 300/330** '02 thru '06 & **Solara** '02 thru '08
92009 **Camry, Avalon & Lexus ES 350** '07 thru '17
92015 **Celica Rear-wheel Drive** '71 thru '85
92020 **Celica Front-wheel Drive** '86 thru '99
92025 **Celica Supra** all models '79 thru '92
92030 **Corolla** all models '75 thru '79
92032 **Corolla** all rear-wheel drive models '80 thru '87
92035 **Corolla** all front-wheel drive models '84 thru '92
92036 **Corolla & Geo/Chevrolet Prizm** '93 thru '02
92037 **Corolla** '03 thru '19, **Matrix** '03 thru '14, & **Pontiac Vibe** '03 thru '10
92040 **Corolla Tercel** all models '80 thru '82
92045 **Corona** all models '74 thru '82
92050 **Cressida** all models '78 thru '82
92055 **Land Cruiser** FJ40, 43, 45, 55 '68 thru '82
92056 **Land Cruiser** FJ60, 62, 80, FZJ80 '80 thru '96
92060 **Matrix** '03 thru '11 & **Pontiac Vibe** '03 thru '10
92065 **MR2** all models '85 thru '87
92070 **Pick-up** all models '69 thru '78
92075 **Pick-up** all models '79 thru '95
92076 **Tacoma** '95 thru '04, **4Runner** '96 thru '02 & **T100** '93 thru '08
92077 **Tacoma** '05 thru '18
92078 **Tundra** '00 thru '06 & **Sequoia** '01 thru '07
92079 **4Runner** all models '03 thru '09
92080 **Previa** all models '91 thru '95
92081 **Prius** all models '01 thru '12
92082 **RAV4** all models '96 thru '12
92085 **Tercel** all models '87 thru '94
92090 **Sienna** all models '98 thru '10
92095 **Highlander** '01 thru '19 & **Lexus RX330/330/350** '99 thru '19
92179 **Tundra** '07 thru '19 & **Sequoia** '08 thru '19

TRIUMPH
94007 **Spitfire** all models '62 thru '81
94010 **TR7** all models '75 thru '81

VW
96008 **Beetle & Karmann Ghia** '54 thru '79
96009 **New Beetle** '98 thru '10
96016 **Rabbit, Jetta, Scirocco & Pick-up** gas models '75 thru '92 & Convertible '80 thru '92
96017 **Golf, GTI & Jetta** '93 thru '98, **Cabrio** '95 thru '02
96018 **Golf, GTI, Jetta** '99 thru '05
96019 **Jetta, Rabbit, GLI, GTI & Golf** '05 thru '11
96020 **Rabbit, Jetta & Pick-up** diesel '77 thru '84
96021 **Jetta** '11 thru '18 & **Golf** '15 thru '19
96023 **Passat** '98 thru '05 & **Audi A4** '96 thru '01
96030 **Transporter 1600** all models '68 thru '79
96035 **Transporter 1700, 1800 & 2000** '72 thru '79
96040 **Type 3 1500 & 1600** all models '63 thru '73
96045 **Vanagon Air-Cooled** all models '80 thru '83

VOLVO
97010 **120, 130 Series & 1800 Sports** '61 thru '73
97015 **140 Series** all models '66 thru '74
97020 **240 Series** all models '76 thru '93
97040 **740 & 760 Series** all models '82 thru '88
97050 **850 Series** all models '93 thru '97

TECHBOOK MANUALS
10205 **Automotive Computer Codes**
10206 **OBD-II & Electronic Engine Management**
10210 **Automotive Emissions Control Manual**
10215 **Fuel Injection Manual** '78 thru '85
10225 **Holley Carburetor Manual**
10230 **Rochester Carburetor Manual**
10305 **Chevrolet Engine Overhaul Manual**
10320 **Ford Engine Overhaul Manual**
10330 **GM and Ford Diesel Engine Repair Manual**
10331 **Duramax Diesel Engines** '01 thru '19
10332 **Cummins Diesel Engine Performance Manual**
10333 **GM, Ford & Chrysler Engine Performance Manual**
10334 **GM Engine Performance Manual**
10340 **Small Engine Repair Manual**, 5 HP & Less
10341 **Small Engine Repair Manual**, 5.5 thru 20 HP
10345 **Suspension, Steering & Driveline Manual**
10355 **Ford Automatic Transmission Overhaul**
10360 **GM Automatic Transmission Overhaul**
10405 **Automotive Body Repair & Painting**
10410 **Automotive Brake Manual**
10411 **Automotive Anti-lock Brake (ABS) Systems**
10420 **Automotive Electrical Manual**
10425 **Automotive Heating & Air Conditioning**
10435 **Automotive Tools Manual**
10445 **Welding Manual**
10450 **ATV Basics**

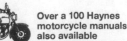

Over a 100 Haynes
motorcycle manuals
also available

10/22

Haynes North America, Inc. • (805) 498-6703 • www.haynes.com